ASTROPHYSICS

ASTROPHYSICS

Decoding the Stars

JUDITH IRWIN
Queen's University
Kingston, Canada

WILEY

Registered Offices
John Wiley & Sons, Inc., 111 River Street, Hoboken, NJ 07030, USA
John Wiley & Sons Ltd, The Atrium, Southern Gate, Chichester, West Sussex, PO19 8SQ, UK

For details of our global editorial offices, customer services, and more information about Wiley products visit us at www.wiley.com.

Wiley also publishes its books in a variety of electronic formats and by print-on-demand. Some content that appears in standard print versions of this book may not be available in other formats.

Library of Congress Cataloging-in-Publication Data

Names: Judith Irwin, 1954- author.
Title: Astrophysics : decoding the stars / Judith Irwin.
Description: Hoboken, New Jersey : Wiley, [2024] | Includes index.
Identifiers: LCCN 2023003634 (print) | LCCN 2023003635 (ebook) | ISBN 9781119623557 (hardback) | ISBN 9781119623540 (adobe pdf) | ISBN 9781119623571 (epub)
Subjects: LCSH: Stars. | Astrophysics.
Classification: LCC QB801 .I79 2024 (print) | LCC QB801 (ebook) | DDC 523.8–dc23/eng20230415
LC record available at https://lccn.loc.gov/2023003634
LC ebook record available at https://lccn.loc.gov/2023003635

Hardback ISBN: 9781119623557

Cover image: © Alan Dyer/Getty Images
Cover design: Wiley

Set in 9/13pt Ubuntu by Straive, Chennai, India
Printed and bound by CPI Group (UK) Ltd, Croydon CR0 4YY

C9781119623557_260723

To my children, Alex and Irene – two bright stars

Contents

Preface

This text is intended for upper-level undergraduate students. At my university, the course is also double-numbered as a graduate course. This means that most students have one astronomy course behind them, as well as at least one physics course at an introductory level. The text *Astrophysics, Decoding the Cosmos, Second Edition*, by myself and also published by Wiley, is a useful companion to this one, but it is not an absolute prerequisite to understanding the material presented here. For example, I include some basic standard astrophysics in Chapter 2 for the sake of completion so that students can find most of what they need with this text alone. At the same time, I present a number of *Online Resources* at the back of each chapter that permit further exploration of the topics covered and that are sometimes used in problems.

This is a 'transition text'. By that, I mean it provides enough detail and physics to satisfy a student who wants to feel competent in stellar basics but who will not be pursuing astrophysics at a more advanced level. At the same time, it can be a launching point for students who do plan on graduate work. I also ask graduate students to delve into a topic and present it to the class. This is a rich addition to the course, and I also learn from their presentations.

The problem with stars is that it is impossible not to become engrossed in the subject matter. Almost every section, or even paragraph, could easily morph into an entire book. One 'descends' into the complexity of the subject quite quickly. However, I have attempted to provide clarity and present the important, time-worn facts that a newcomer to the subject might appreciate. Numerous 'side boxes' also allow for brief introductions to interesting, related topics. A brief, final 'round-table' chapter highlights a few more hypothetical or unusual stars.

When all is said and done, go outside and look at the stars.

Judith Irwin
Township of South Frontenac, Ontario
September, 2022

Acknowledgements

I wish there were words besides 'thank you' to express a deeper appreciation to those who have spent their time poring over some of the material in this text. But as I am no poet, a sincere thank you is all that can be mustered here. I hope that the people who read this will add the appropriate embellishment! To begin with, thanks to my husband, Richard Henriksen, who read each chapter and gave his kind input on flow and on physics. And thank you to my students at Queen's, who insisted on accuracy and clarity.

Many others have also made suggestions, offered information or given constructive criticism. Thanks to Mats Carlsson for preparing the probability density distribution plots of the solar atmosphere in Appendix B. Thanks also to Emanuele Tognelli for assistance with the Pisa Stellar Evolution models and providing important input regarding pre-main sequence evolution. The variable star expert, John Percy, made valuable suggestions for Chapter 9. Thank you to Warrick Ball for help with astero-seismology information. Paul Charbonneau, an expert on solar theory, read Chapter 1 and provided important clarifications. Jieun Choi made valuable comments on stellar evolution in Chapter 8 and also provided her delightful H-R diagram animation in Online Resource 8.2. Both Laura Fissell and Sarah Sadavoy graciously provided feedback on the star formation section. Thank you to Adam Burrows and Ashley Ruiters for updates on sections involving supernovae. And many thanks to Aarchi Shah for proofreading.

I wish to thank all of those individuals who generously allowed me to reproduce their figures, especially Alan Dyer (https://amazingsky.net) for his front cover photo. Finally, thank you to those who came before me in writing textbooks on stars. I have greatly benefitted from your work. These authors include Francis LeBlanc, Bradley Carroll and Dale Ostlie, Rudolf Kippenhahn et al., A.C. Phillips, Walter Maciel, Dina Prialnik, Markus Aschwanden, Arvind Bhatnagar and William Livingston, Henny Lamers and Emily Levesque, and of course, Sir Arthur S. Eddington.

Judith Irwin

Introduction

... it is reasonable to hope that in a not too distant future we shall be competent to understand so simple a thing as a star.
The Internal Constitution of the Stars, by Arthur S. Eddington [86]

Stars are the 'integers' of our universe. There are about 10^{11} stars in the Milky Way alone, and countless other star-filled galaxies fill our universe. Apart from *dark matter*, whose nature is currently unknown, stars are the dominant component of the mass of any galaxy. They also emit their light at wavelengths that can be detected by the human eye, so historically they have been accessible to us. Long before radio telescopes, long before space-based infra-red (IR), ultra-violet (UV), X-ray or γ-ray telescopes, and long before Galileo turned his optical telescope on the Milky Way, we could identify these integers as pin-points of light in the sky. Stars are the first point of connection between human beings and the larger universe within which we live.

From the earliest times, we have tried to comprehend and explain the stellar firmament. Mythologies arose, from arctic Inuit tales of a brother and sister whose disagreement sent them apart from each other to become the Sun and Moon, to Greek stories of heroes and gods, to Mayans who saw the Milky Way as a road for the souls of the underworld – each civilization has attempted to make sense of the heavens. The same Sun and the same Moon shine down on all of us. If we are at the same latitude on Earth, we see the same stars as well, providing a commonality of experience and a continuity in our search to understand these twinkling messengers.

In our modern age, this search has evolved and developed into a scientific framework, one of observation, measurement and analysis. But putting it all together still requires a kind of story-telling. The story must be consistent, fit the facts, and be numerically accurate. It must make sense and not contradict what we know from other established knowledge. Ultimately, though, the final scientific step requires a synthesis of this knowledge, leading to a story that makes sense. Stars have provided rich fodder for such a process.

The Danish astronomer Tycho Brahe (1546–1601) pre-dates the telescope, but using as precise instruments as possible at the time, he made careful observations of stellar and planetary positions. One could argue that *astronomy*, as a science that involves accurate positional measurements, dates from that time. Shortly thereafter, Johannes Kepler (1571–1630) used the planetary measurements of Brahe to derive his three laws of planetary motion. Kepler's Laws endure to this day, but it wasn't until Isaac Newton (1642–1727) that Kepler's laws could be derived from the Universal Law of Gravitation, and it wasn't until Albert Einstein (1879–1955) that Newton's laws were put into the context of the General Theory of Relativity.

But what about *astrophysics*? When did astronomy morph into astrophysics?

This transition started with *stars* ('of the stars' is *astron*, in Greek ἀστρον). As the atomic nucleus was being explored about 100 years ago, so was the stellar core. It is worth pausing to think about this. Our modern view of the universe as an expanding space filled with the tiny perturbations of gravitational waves, the existence of black holes, powerful jets emerging from active galactic nuclei, pulsars that rotate with a higher accuracy than man-made clocks, exploding massive stars – all of that knowledge has arisen in only 100 years, just a tiny dot on the human evolutionary roadway. It started with the *physics of stars*, with an attempt to understand what actually powers the stars, and the implications of that power.

There were a number of players in this process of discovery, but one individual does stand out as, arguably, the first astrophysicist: Sir Arthur S. Eddington (1882–1944), shown in Fig. I.1 (*Left*). Famous for his solar eclipse expedition of 1919, which supported Einstein's General Theory of Relativity, Eddington both advanced and popularized the physics of stars. Eddington's engaging and foundational book, *The Internal Constitution of the Stars* (1926) [86], set out the principles of physics as applied to stars, much of which stands the test of time.

Eddington's students included Subrahmanyan Chandrasekhar, Leslie Comrie, Cecilia Payne-Gaposchkin, Hermann Bondi, Georges Lemaître and Alice Vibert Douglas, each of whom nudged, pushed or hurled the new field of astrophysics forward. Dr Alice

Figure I.1 Two early astrophysicists. *Left*: The eminent Sir Arthur Stanley Eddington (b. 1882 Kendal, d. 1944 Cambridge). Credit: The Library of Congress / Wikimedia commons / Public domain. *Right*: Eddington's biographer and student, Dr Alice Vibert Douglas (b. Montreal 1894, d. Kingston 1988). Credit: Notman & Son Photographers. Queen's University Archives, Kingston.

Vibert Douglas (Fig. I.1 *Right*) also became Eddington's biographer [309]. Each has an asteroid named after them: (2761) Eddington and (3269) Vibert Douglas. The early development of stellar astrophysics firmly establishes this field as a *mature* discipline, yet one in which much active, dynamic and engaging research is being carried out today. This text is designed to explore both well-established nuts and bolts as well as modern puzzles of the field.

I.1 THE SIMPLE PHYSICAL STAR

In this text, we consider a 'star' to be a *self-gravitating object that is undergoing sustained nuclear fusion (referred to as 'burning') of non-isotopic hydrogen in the stellar interior*. This may be a mouthful, but there are reasons for being so specific.

Many objects are closely related, but we do not call them stars. For example, gaseous planets like Jupiter may be formed by processes similar to stars. However, we do not consider them to be stars because they cannot sustain interior hydrogen fusion reactions.[1] Objects like white dwarfs, pulsars and black holes are *stellar remnants*. They are not stars for the same reason. A *nova* is the brightening of a white dwarf star because of nuclear reactions at the surface when mass falls onto it, but this does not make a nova a star by our requirement of *interior* nuclear reactions. The burning of deuterium, which is an isotope of hydrogen (one proton and one neutron in its nucleus), occurs at a lower temperature than normal hydrogen burning. Therefore, deuterium reactions can occur in the late stages of stellar formation and can also occur in substellar objects whose masses are higher than 13 times the mass of Jupiter. These are also excluded from our pantheon of bona fide stars because deuterium is an isotope of hydrogen.

On the other hand, a red giant, depending on its evolutionary stage, may be undergoing hydrogen burning in an interior shell and not its core. Such an object, by our definition, is indeed a star. The stars, then, include all objects that are on the *main sequence* (see Fig. I.2 in Sect. I.4) which, by definition, represent those that are burning hydrogen into helium in their cores (Sect. 5.4) as well as all objects that have evolved off of the main squence prior to ejecting most of their mass (Sect. 8.2).

You can see that we have sought a physical definition of a star, but historically, such understanding was not necessarily known. For example, a white dwarf just looks like a faint star in the sky, so it is common to see 'white dwarf star' in the literature. Similarly, 'neutron star' is commonly used. The word 'nova' has been simplified from *stella nova*, meaning 'new star'. Again, such terminology predates the physics.

How could our self-gravitating star be 'simple', as the title to this section suggests? In fact, if we include stars of different masses as well as stellar remnants, just about every subfield of physics is represented: molecular, atomic and nuclear

---

[1] However, according to [190], substellar objects between 60 and 80 times the mass of Jupiter can burn a small amount of hydrogen, though not in a sustained way. See also Sect. 7.4.1.

physics; particle physics; mechanics, magneto-hydrodynamics, electromagnetic theory and thermodynamics; quantum mechanics; radiative processes in physics; and special and general relativity. Indeed, it is difficult to find any area of physics that *is not* represented. Even condensed-matter physics shows up in the study of neutron stars (e.g. see [250]). And the new field of gravitational wave physics had its observational foundations from studies of neutron stars and black holes in binary systems [e.g. 154]. Clearly, not every one of these areas of physics is required in any single object, but the properties of the range of stars and stellar remnants encompass them all. This hardly seems simple.

The simplicity, then, must arise from the *dominant* force that is at work. This force is *gravity*, with its well-known $1/r^2$ behaviour associated with the minimum energy geometry of the sphere. Our conclusion is that *isolated stars are spherical*. This is not to say that other forces can be completely ignored; regions in which nuclear energy generation is occurring, regions in which stars are convective, stellar atmospheres, or active surface regions are examples of where gravity must be tempered with other forces that are important on smaller scales. However, stars are large, and big things feel the force of gravity as the dominant force *globally*.

In the comedy series *The Big Bang Theory*, Leonard Hofstadter tells this joke: 'There's this farmer, and he has these chickens, but they won't lay any eggs. So, he calls a physicist to help. The physicist then does some calculations, and he says, um, I have a solution, but it only works with spherical chickens in a vacuum.' That more or less sums it up. To the zeroth order, stars are beautifully spherical chickens in a vacuum – but the devil is in the details.

I.2 THE DOMINANCE OF GRAVITY FOR STARS

Gravity is not actually a very strong force when masses are low. Let us compare, for example, the Coulomb force on an electron, F_C, due to another electron that is 1 cm away, compared to the gravitational force of attraction between these two charged particles, F_G. We will put in the values of the relevant quantities in cgs units,[2] which is standard in astrophysics. For examples that compare cgs to SI units,[3] see the Introduction of [154]. Here, we write out the numerical values of the constants explicitly, but we will use their symbols in subsequent sections. Values of numerical constants can be found in Appendix A.

$$\frac{F_C}{F_G} = \frac{\frac{e^2}{r^2}}{\frac{G\, m_e^2}{r^2}} = \frac{\frac{(4.8 \times 10^{-10})^2}{(1)^2}}{\frac{(6.67 \times 10^{-8})(9.1 \times 10^{-28})^2}{(1)^2}} = 4.2 \times 10^{42} \tag{I.1}$$

---

[2] Centimetre-gram-second.
[3] Système International d'Unités.

where e is the electrostatic unit (esu), r is the separation between the two charges (cm), G is the universal gravitational constant (cm^3 g^{-1} s^{-2}) and m_e is the mass of the electron (g). First of all, both forces are $1/r^2$ forces, so the distances between the particles doesn't matter for this comparison. Secondly, the Coulomb force clearly 'wins', and these two electrons will rapidly accelerate away from each other.

Suppose we put our two electrons on the surface of the Sun and compare the electrostatic force between them to the gravitational force exerted on either one of them by the Sun. In that case,

$$\frac{F_C}{F_G} = \frac{\dfrac{e^2}{r^2}}{\dfrac{G\, m_e M_\odot}{R_\odot^{\,2}}} = \frac{\dfrac{(4.8 \times 10^{-10})^2}{(1)^2}}{\dfrac{(6.67 \times 10^{-8})(9.1 \times 10^{-28})(1.99 \times 10^{33})}{(6.96 \times 10^{10})^2}} = 9.2 \times 10^3 \qquad \text{(I.2)}$$

where $M_\odot$ is the mass of the Sun and $R_\odot$ is the radius of the Sun.[4] Remarkably, the Coulomb force is still clearly the dominant force by a factor of about 10,000/1. Thus, for any case in which charge is separated, the Coulomb force must not be neglected and can actually dominate, as in this example.

Similarly, we could ask about the Lorentz force associated with a magnetic field, F_B. The magnetic field on the surface of the Sun is $B \approx 1$ Gauss (represented by Roman G), although regions near sunspots can be much higher: for example, 2000 G. Let us consider the rather extreme case of regions near sunspots in which the hot magnetic plasma[5] reaches velocities, v, at least of the order of the solar escape velocity (Eq. 1.27) of 618 km s^{-1} (see also Sect. 1.2.5). We know that such velocities are achieved because of the presence of a *solar wind* of particles that continuously escapes from the Sun. Then let us again consider the magnitude of the forces acting on an electron.

$$\frac{F_B}{F_G} = \frac{e\left(\dfrac{v}{c}\right) B}{\dfrac{G\, m_e M_\odot}{R_\odot^{\,2}}} = \frac{(4.8 \times 10^{-10})\left(\dfrac{618 \times 10^5}{3.0 \times 10^{10}}\right)(2000)}{\dfrac{(6.67 \times 10^{-8})(9.1 \times 10^{-28})(1.99 \times 10^{33})}{(6.96 \times 10^{10})^2}} = 7.9 \times 10^{13} \qquad \text{(I.3)}$$

Again, we see that the Lorentz force can exceed the force due to gravity by many orders of magnitude. One could easily lower v to be much less than the escape velocity and/or lower the magnetic field strength and still conclude the dominance of F_B over F_G.

Our conclusion from these examples is that *local* electromagnetic forces on a star's surface can be far greater than the force of gravity and must be taken into account to explain *local dynamics*. For the Sun, solar prominences, coronal mass ejections and the solar wind are good examples of local forces dominating over gravity (e.g. Sect. 1.2.5).

---

[4] Note that the subscript $\odot$ refers to the Sun for these and any other Solar quantities.
[5] A plasma is an ionized gas.

We could replace our electrons with more massive protons, and the numbers would change, but the conclusions would not.

For the star as a whole, however, *gravity wins globally*. This is due to the fact that stars, like just about everything else in the natural universe, are electrically neutral. For example, we do not see an entire star with a significant net charge.

Then, let us look at the entire star, rather than a pocket of activity on its surface. We will consider u_E, u_B and u_G, the *energy densities* (erg cm^{-3}) of the electric field, E, the magnetic field, B, and the gravitational field, respectively,

$$u_E = \frac{E^2}{8\pi}, \ u_B = \frac{B^2}{8\pi}, \ u_G \approx \left(\frac{9}{20\pi}\right)\frac{GM^2}{R^4} \tag{I.4}$$

The third expression is an approximation because of an assumption that the star has a constant density, which we adopt to obtain an order-of-magnitude result. It has been obtained from the gravitational binding energy (the absolute value of the gravitational potential energy,[6] U_G) divided by the stellar volume. For a star as a whole, we will take $E \approx 0$ and consider only the energy density of the magnetic field.

Then, to an order of magnitude, using the same values for the Sun as were used in Eq. I.3,

$$\frac{u_B}{u_G} \approx 0.3 \ \frac{B^2 R^4}{GM^2} \approx 10^{-10} \tag{I.5}$$

Now it is clear that gravity dominates globally. This example may not be a perfect one-to-one comparison because the gravitational energy density applies to the entire star, whereas the magnetic field is highly variable with location. Taking the magnetic field value at the base of the convection zone, $\approx 10^4$ G [97], would increase this ratio by an order of magnitude. Nevertheless, we would have to modify the magnetic field by many orders of magnitude before it became the dominant global force.

Are there any other forces that could rival gravity? We know that the Earth has a slight bulge at its equator due to its rotational motion. What about stars? For this comparison, we will look at the magnitudes of the rotational kinetic energy, U_{rot} (erg), and the gravitational potential energy:

$$\frac{U_{rot}}{|U_G|} \approx \frac{\frac{1}{2}I\omega^2}{\frac{3}{5}\frac{GM^2}{R}} \approx \frac{\frac{1}{2}\left(\frac{2}{5}MR^2\right)\omega^2}{\frac{3}{5}\frac{GM^2}{R}} \approx \frac{1}{3}\frac{R^3\omega^2}{GM} \approx 10^{-5} \tag{I.6}$$

Again, the approximation is because of an assumed constant density for the sphere, and we have used the corresponding expressions for moment of inertia, I, and gravitational energy (see Eq. I.15). The evaluation is for an angular velocity of the Sun at its

---

[6] The gravitational potential energy (a negative quantity) is minus the work done by gravity in moving a set of particles from infinity to their current configuration. The gravitational binding energy is the inverse, i.e. the energy required to move all particles from their current configuration to infinity. See also Footnote 1 in Chapter 7.

surface, $\omega = 2\pi/P = 2.9 \times 10^{-6}$ s^{-1}, where the adopted rotational period is $P = 25$ days applicable to the equator.[7] The resulting ratio would be the same if we used energy densities, since this simply requires a division by volume in both the numerator and denominator.

The result of Eq. I.6 clearly shows that the rotational energy is very small compared to the gravitational energy for the Sun. In fact, [178] have shown that the angular difference between the polar and equatorial radii is only 7.20 milli-arcsec – that is, 0.02% of the solar diameter – and shows no variation over the 11-year sunspot cycle. Our Sun is very round, indeed!

Our conclusion that the magnitude of the gravitational potential energy for stars is greater than their rotational energies must clearly be the case. Were that not true, stars would regularly be flying apart from centrifugal forces! There are stars, though, with rotational energies that are sufficient to distort them into oblate spheroids (see Box I.1 on page xxi and Prob. I.2). These are exceptions, though, and most likely result from tidal stresses and/or mass transfer between stars in binary systems [e.g. 73, 288]. Also, stars tend to slow down in their rotations with time, a topic to which we shall return in Box 8.5 on page 202. In short, spherical symmetry is a realistic and, fortunately, simplifying approximation for stars.

Box I.1

Starring ... Regulus (α Leonis)

Distance: $d = 24.2$ pc; Spectral Type: B7 V or B8 IV
Effective Temperature: $T_{eff} = 14231 \pm 314$ K [308]
Mass: M = 3.4 M$_\odot$ [118]

Regulus is the brightest star in the constellation of Leo, the Lion. It has an extremely rapid rotation, with a period of 15.9 hr [215] and a spin velocity of ≈ 317 km s^{-1} (86% of the escape velocity). Regulus is in a binary system [118] with a companion star that is likely a low-mass white dwarf. It may be that mass transfer from the companion in its red giant phase prior to it becoming a white dwarf caused mass transfer and angular momentum transfer onto Regulus. The result was the spin-up of Regulus to its 'fast-rotator' status and resulting oblate spheroid geometry. The polar and equatorial radii of Regulus are $R_p = 3.14$ R$_\odot$ and $R_e = 4.16$ R$_\odot$, respectively. An oblate spheroid is called a *Maclaurin Spheroid*.

---

[7] The Sun rotates more quickly at the equator than near the poles.

I.3 THE NUMERICAL AND ANALYTICAL STAR

To what extent can we understand the physics of, in Eddington's words, 'so simple a thing as a star'?

Surely, the modern 'star-builder' must run complicated numerical calculations to predict the internal structures of stars and the even more complicated atmospheric calculations that take into account the variety of spectral lines that we see at the surface. Evolving a star through its evolutionary stages is yet another level of difficulty, requiring much computational effort. Indeed, all modern approaches to stellar structure and evolution require the use of such numerical codes.

However, to obtain a physical intuition about stars, it is often best to make some reasonable simplifying assumptions. This allows us to probe the behaviour of the star, and the mathematics can be handled in a tractable fashion. Such an analytical approach was illustrated in Eq. I.6, in which we assumed that the star was of uniform density. This simple assumption allowed us to reach the correct conclusion that a star's rotational energy is smaller than its gravitational energy. Was that process justified?

Stars do not have constant densities, but rather the density varies (declines) with the radius out to the surface at R. In general, the equation for potential energy of a uniform density star given in Eq. I.15 can be generalized for the non-uniform case,

$$U_G = -k\frac{GM^2}{R} \qquad (I.7)$$

where k is a constant and equal to $3/5 = 0.6$ for a uniform sphere. How much does k vary if the density is not constant?

Prob. I.3 suggests how one might calculate U_G and the value of k for a specific declining density distribution. To answer this question simply, though, let us consider two extreme, and completely unphysical, stars that have the same total mass, M, and same radius, R. Star A has virtually all of its mass concentrated right at the center in a region of size $r = 0.2R$, with negligible mass out to $r = R$. Such a star approximates a constant-density sphere but over a small central region only. Then R in Eq. I.7 just becomes $0.2R$ and k = 3. Star B has all of its mass in a shell at a radius $r = R$, so it is essentially a hollow sphere. For a very thin shell, it can be shown that k = 1/2. Therefore, for these extreme density distributions, k departs from the constant-density case only by factors of a few. A realistic stellar density distribution is nowhere near these extremes. Therefore, simplifying our star by assuming constant density was quite reasonably justified for order of magnitude results.

There will be many examples in this text that serve to illustrate the important physics in a process without resorting to numerical calculations. Indeed, in many cases, an analytical approach is quite sufficient for understanding (e.g. Sect. 6.3). Should further numerical work be required, the simple case can often be used as a starting point for iterative numerical work.

I.4 THE THEORETICAL AND OBSERVATIONAL STAR

Our previous sections have already illustrated how theory is wedded to observation. I know of no other area of astrophysics in which theory and observation are so beautifully intermeshed. Moreover, I know of no other area of astrophysics in which theory works *so well*. Those who are actively engaged in stellar research could very well challenge such a statement since they know the complications, pitfalls and subtleties inherent in their subdiscipline. Nevertheless, I think it is safe to say that we have a pretty good understanding of stellar structure. We have a fairly good understanding of stellar evolution. And we have a not-so-good understanding of star formation. Theory is essential. After all, we could not possibly wait a billion years to watch how a star evolves. On the other hand, sometimes even simple theoretical arguments can reveal that future.

Eddington was one of the first to realize and emphasize this. In [86], he says,

> We can imagine a physicist on a cloud-bound planet who has never heard tell of the stars, calculating the ratio of radiation pressure to gas pressure for a series of globes of gas of various sizes, starting, say, with a globe of mass 10 gm., then 100 gm., 1000 gm., and so on, so that his nth globe contains 10^n gm. ... Just for the particular range of mass about the 33rd to 35th globes the [results] become interesting ... Regarded as a tussle between matter and aether (gas pressure and radiation pressure) the contest is overwhelmingly one-sided except between Nos. 33 - 35, where we may expect something interesting to happen.
>
> What 'happens' is the stars.
>
> We draw aside the veil of cloud beneath which our physicist has been working and let him look up at the sky. There he will find a thousand million globes of gas nearly all of mass between his 33rd and 35th globes – that is to say, between 1/2 and 50 times the sun's mass.

Effectively, Eddington is saying that a theorist could have predicted the existence of stars without ever having seen one!

Box I.2

Magnitudes

The *apparent magnitude*, m_ν, of a star is the magnitude that would be measured at the Earth's surface, corrected for the atmosphere. It is related to a star's flux density, f_ν (cgs units of erg s^{-1} cm^{-2} Hz^{-1}), measured in some frequency band centered at ν, by

$$m_\nu - m_{\nu_0} = -2.5 \, log\left(\frac{f_\nu}{f_{\nu_0}}\right) \tag{I.8}$$

(*continued*)

(continued)

where m_{v_0} and f_{v_0} are the apparent magnitude and flux density, respectively, of a reference star whose values are known. This and subsequent relations can also be expressed in terms of wavelength, i.e. m_λ and f_λ (cgs units of erg s^{-1} cm^{-2} cm^{-1}), where f_λ is measured in some waveband centered at λ.

Apparent magnitude is usually specified by the observing band explicitly. For example, in the optical V-band (λ 545 nm),

$$V - V_0 = -2.5 \, log\left(\frac{f_V}{f_{V_0}}\right) \tag{I.9}$$

The *colour index* is the difference in magnitude for a given star between two different observing bands: for example, between the B-band (λ 438 nm) and V-band,

$$B - V = -2.5 \, log\left(\frac{f_B}{f_V}\right) \tag{I.10}$$

The *absolute magnitude*, M_λ, is the magnitude of a star if it were placed at a distance of 10 pc. Plotting an absolute magnitude removes any differences in brightness due to distance. For the V-band,

$$V - M_V = -5 + 5 \, log(d) \tag{I.11}$$

where the distance, d, is in pc. The quantity $V - M_V$ is called the *distance modulus*.

A *bolometric* magnitude is any magnitude that has been integrated over all wavelengths so that it is no longer waveband-specific. For an absolute bolometric magnitude, M_{bol}, and luminosity, L(erg s^{-1}), of a star,

$$M_{bol} - M_{bol\odot} = -2.5 \, log\left(\frac{L}{L_\odot}\right) \tag{I.12}$$

where $M_{bol\odot}$ and $L_\odot$ are the absolute bolometric magnitude and luminosity, respectively, of the Sun. These are given in Eqs. 1.17 and 1.1, respectively.

But lest the theorist, Eddington, get the last word, let's look at a rather important *observational* tool, namely the *H-R diagram* (Hertzsprung-Russell diagram) shown in Fig. I.2. An H-R diagram plots a measure of the star's luminosity, usually absolute magnitude in some spectral band (see Box I.2 on page xxiii) on the y-axis, against a measure of the star's temperature, usually colour index or *spectral type* (e.g. Appendix F), on the x-axis. An H-R diagram that plots observational quantities is also called a *colour-magnitude* or *CM* diagram and is in the 'observational plane'. Note that stellar distances must be known in order to plot absolute magnitude. The x-axis is oriented such that higher temperatures are towards the *left* on the diagram,

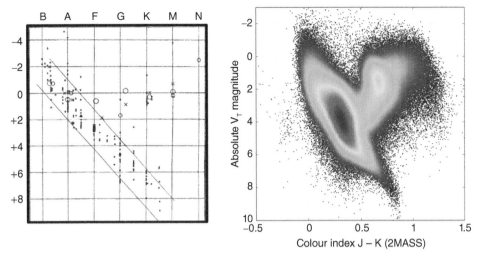

Figure I.2 One hundred years separates these two H-R diagrams. *Left*: H-R diagram from the year 1914 [269] / Springer Nature / Public Domain, showing absolute magnitude (y-axis) against spectral classes B to N (x-axis labelled at the top). Different symbols represent different groups of stars. Parallel lines enclose the *main sequence*. *Right*: The first H-R diagram released in the year 2015 from Gaia satellite data containing ≈ 1 million stars for which distances have been measured. The absolute magnitude in the V_T passband, which is centered at λ 530 nm [e.g. 319], has been plotted on the y-axis. The x-axis shows the *colour index*, found from the difference between magnitudes in the J and K bands (λ 1.25 μm and λ 2.20 μm, respectively [e.g. 144]). The colours give the density of stars in the area of the plot, with dark red representing the highest density and blue the lowest; there are so many points that the colours appear continuous. Copyright: ESA/Gaia/DPAC/IDT/FL/DPCE/AGIS.

and the magnitude scale is set such that numerically *lower* values correspond to more luminous stars. Therefore, cooler dimmer stars are at the bottom right, and hotter brighter stars are at the top left.

Figure I.2 illustrates impressive progress in the quality of the H-R diagram, mainly because of improvements in distance measurements that are required to obtain absolute magnitudes, as well as telescope sensitivity. The *Left* H-R diagram is a plot from a 1914 paper by Henry Norris Russell [269]. This figure shows a few dozen stars for which distances were known at the time. The *main sequence*, enclosed by two parallel lines, is just barely detectable, stretching from upper left to lower right. Stars that are on the main sequence are those in which hydrogen is being converted into helium in their cores. The *red giant branch* consists of a few stragglers that stretch towards the upper right. This region is populated by stars that have evolved off of the main sequence once their core hydrogen has been depleted.

Now compare this plot to the *Right* figure, which was released in 2015, approximately 100 years later. The main features of the H-R diagram, the main sequence and the red giant branch, still jump out. However, this plot is based on distances obtained by the Gaia satellite (see Box 3.1 on page 60), supplemented with data from an earlier satellite named Hipparcos as well as colour information from ground-based observatories. One million stars with distance measurements from Gaia are represented in this image. Subsequent data releases have increased this number to 1.8 *billion* stars. It is fair to say that observational progress has been impressive!

The H-R diagram shows that *stars do not have arbitrary properties*. If they did, then this diagram would just be a scatter-gram. Rather, the observational properties of stars are governed by physical properties that we will explore in much more detail later in this text. For now, suffice it to say that the organization of stars on this plot represents the most fundamental of all observational and theoretical information about stars that is available to us.

Online Resources

I.1 *Stellarium Planetarium software*: stellarium.org
I.2 *Other free planetarium packages*:
     https://www.deepsky2000.com/top-10-free-astronomy-software

PROBLEMS

I.1. (a) Reproduce the calculations of Eq. I.2 for a type B0V star (see the data in Table F.1).
      (b) Repeat part (a) using SI units and confirm that the same answer is obtained.
      (c) Is F_G for the more massive B0V star larger or smaller than that of the Sun? Explain.

I.2. (a) The potential energy of a shell within a star is

$$dU_G = -\frac{GM_r \, dm}{r} \tag{I.13}$$

where M_r is the mass interior to the radius, r, and dm is the mass of the shell (see Fig. 6.1). The mass of any shell within a star at radius r is

$$dm = 4 \, \pi \, r^2 \, \rho(r) \, dr \tag{I.14}$$

where $\rho(r)$ is the density of the shell.[8] Show that the total gravitational potential energy of a uniform density sphere is given by

$$U_G = -\left(\frac{3}{5}\right) \frac{GM^2}{R} \tag{I.15}$$

where M, R, are the total mass and radius of the star, respectively.

---

[8] Note that Eq. I.14 is one of the equations of stellar structure, which we will see again in Sect. 6.1.1.

I.3. Suppose that the density of a fictitious star with a total mass M and radius R has a density profile described by

$$\rho(r) = \rho_c(1 - r/R) \qquad (1.16)$$

where ρ_c is the density at the center of the star.

(a) Show that the mass M_r interior to r is given by

$$M_r = 4\pi\rho_c \left(\frac{r^3}{3} - \frac{r^4}{4R} \right) \qquad (1.17)$$

Set $r = R$ to find a simple expression for the total mass of this star. How much does this result differ from a uniform-density sphere whose density is ρ_c?

(b) Find an expression for the total gravitational potential energy of this star, U_G. Express the result in terms of M and R only (i.e. eliminate ρ_c).

(c) What is the value of k in Eq. I.7 for this star? Compare your result to the value of k for a uniform-density star of the same total mass and radius, and comment on the difference.

I.4. To an order of magnitude, how high would the magnetic field have to be in a neutron star for the magnetic energy density to rival its gravitational energy density? (See Sect. 10.4 for information on neutron stars.) Compare the results to typical magnetic fields of *magnetars*, which are objects that have the strongest magnetic fields known.

I.5. (a) Find $U_{rot}/|U_G|$ for the star Regulus (Box I.1 on page xxi), assuming uniform density. Do you expect this ratio to increase, decrease or stay the same if a declining density distribution is used instead of a uniform density? [Optional Challenge: Research the properties of MacLaurin spheroids and repeat this question.]

I.6. Access or download a planetarium software program (examples are provided in the Online Resources on page xxvi), and become familiar with its operation.

(a) What is your latitude and longitude? Set up the virtual sky for your location and for midnight tonight.

(b) Find the brightest star that is visible, and list its altitude and azimuth.[9] List the star's V magnitude, B - V colour index (see Box I.2 on page xxiii) and spectral type. Use the calibration tables of Appendix F to find its mass and effective temperature.

---

[9] Altitude is the angle from the horizon (altitude = 0 degrees) upwards towards the zenith (altitude = 90 degrees). Azimuth is the angle measured around the horizon starting from the north (azimuth = 0) and moving eastwards (e.g. due east has azimuth = 90 degrees, and due west has azimuth = 270 degrees).

(c) What planets, if any, are visible? Assume that the limiting magnitude is $V = 6$. List the altitudes and azimuths of the planets and the constellations they are in. If there are no planets up, adjust the time or date until at least one is visible.

I.7. Go outside and *look at a star!* Find a bright star in the sky that you previously did not know by name, and find its common name and host constellation.

(a) For your observation,

(i) Specify the date, time, and weather conditions.
(ii) Specify your latitude and longitude.
(iii) Estimate the altitude of the star (the width of a closed fist held out at arm's length subtends approximately 10 degrees).

(b) Do an internet search to find some information about this star, and write a *brief* paragraph about it. For example, is it a red giant or a main sequence star? Is there anything else of interest that you would like to convey?

Chapter 1
The Closest Star

The point of living is to study the Sun.

Anaxagoras (499–428 BC) [29]

It takes quite a leap of imagination to connect the Sun to the myriad pinpoints of starlight in the night sky. Nevertheless, this connection appears to have been made by Anaxagoras, a Greek citizen from Clazomenae (in present-day Turkey) who took up residence in Athens. The claim of Anaxagoras that the Sun was not a god, but rather a 'fiery rock', led to his imprisonment and eventual banishment from Athens sometime around 438 BC [116].

We have come a long way since Anaxagoras, amassing a wealth of information from both theoretical and observational advances in stellar astrophysics. Such studies have not only revealed that the stars are 'suns' like ours (e.g. see Box 1.4 on page 26) but also piece together a compelling picture of star formation and evolution that will be addressed later in this text. In this chapter, it is important to take a close look at the Sun as a

Astrophysics: Decoding the Stars, First Edition. Judith Irwin.
© 2023 John Wiley & Sons Ltd. Published 2023 by John Wiley & Sons Ltd.

sort of 'reference star' for future considerations. We will not focus too heavily on how the information is obtained, but rather on the results. The *how* will come later.

1.1 THE SUN – FIRST AMONG EQUALS

Figure 1.1 nicely illustrates our previous contention (Sect. I.1) that stars are simple – but with fundamental underlying complexities. These images were taken on the *same day* but in different wavebands. The *Left* image shows what we see in visible

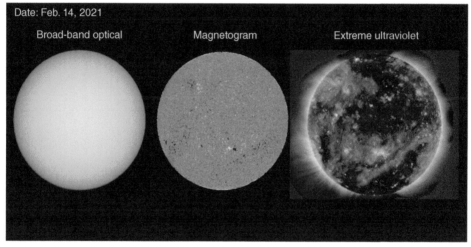

Figure 1.1 Images of the Sun taken on Feb. 14, 2021. **Left**: Visible (broad-band continuum) light showing the solar photosphere. To see where we are today with sunspot cycles, see Online Resource 1.2. **Center**: Magnetogram, showing the magnetic field in both the photosphere and the lower chromosphere. Yellow and dark navy blue represent higher fields that have positive polarity (fields pointing towards the observer) and negative polarity (fields pointing away from the observer), respectively. **Right**: Composite image taken at three different extreme ultraviolet (EUV) wavelengths. Red (λ 211 Å) represents emission whose primary ion is FeXIV (ionized iron with 13 electrons gone); the characteristic temperature is 2×10^6 K and the primary target is the active region corona. Green (λ 193 Å) represents emission whose primary ions are FeXII and FeXXIV (iron with 11 and 23 electrons gone, respectively); the characteristic temperatures are 1.6×10^6 K and 2×10^7 K, respectively, and the primary targets are the corona and hot flare plasma. Blue (λ 171 Å) shows emission from FeIX (8 electrons gone); for this colour, the characteristic temperature is 0.6×10^6 K, and the targets are the quiet corona and upper transition region [192]. Credit for left and right images: NASA/Goddard/SDO AIA Team. Credit for center image: National Solar Observatory/CC BY 4.0.

light, roughly at wavelengths to which our eyes are most sensitive. The Sun is an opaque ball of hot gas. Although visible light could easily show sunspots, the solar face is remarkably clear of such features on the date when the observations were made. Indeed, this picture of our star is as simple as it gets! The *Right* image has been formed from three different extreme ultraviolet (EUV) images. Now we see the dynamic complexity of our closest star. Active regions are clearly seen as well as gas that streams outwards from the surface in the beginning of the *solar wind*. Clearly, adopting appropriate observing wavelengths ('tuning' the observing frequency) gives us very different information about this or any other star. The *center* image shows a *magnetogram* that indicates the magnetic field structure. Bright yellow and dark navy blue regions indicate where the magnetic field is very strong. When yellow, the fields are pointing towards the observer, and when dark blue, fields are pointing away from the observer. We will see later (Sect. 1.4) that the magnetic field is a driver of solar activity. It is worth spending a few moments comparing these images.

We know more about the Sun than any other star. This knowledge includes some rather precise values for luminosity, mass, radius, effective temperature and age, which are, respectively [76],

$$L_\odot = (3.8275 \pm 0.0014) \times 10^{33} \text{ erg s}^{-1} \tag{1.1}$$

$$M_\odot = (1.98892 \pm 0.00013) \times 10^{33} \text{ g} \tag{1.2}$$

$$R_\odot = (6.9599 \pm 0.0001) \times 10^{10} \text{ cm} \tag{1.3}$$

$$T_{\text{eff}\odot} = 5772.0 \pm 0.8 \text{ K} \tag{1.4}$$

$$t_\odot = (4.571 \pm 0.0044) \text{ Gyr} \tag{1.5}$$

The Sun's absolute magnitude (see Box I.2 on page xxiii for magnitude information) in a variety of filters is [324]

$$M_{U\odot} = 5.61 \ (\textit{Johnson}, \ \lambda = 0.3611 \ \mu m) \tag{1.6}$$

$$M_{B\odot} = 5.44 \ (\textit{Johnson}, \ \lambda = 0.4396 \ \mu m) \tag{1.7}$$

$$M_{V\odot} = 4.81 \ (\textit{Johnson}, \ \lambda = 0.5511 \ \mu m) \tag{1.8}$$

$$M_{R\odot} = 4.43 \ (\textit{Johnson}, \ \lambda = 0.6582 \ \mu m) \tag{1.9}$$

$$M_{I\odot} = 4.10 \ (\textit{Johnson}, \ \lambda = 0.8034 \ \mu m) \tag{1.10}$$

$$M_{J\odot} = 3.67 \ (2\textit{MASS}), \ \lambda = 1.2393 \ \mu m) \tag{1.11}$$

$$M_{H\odot} = 3.32 \ (2\textit{MASS}), \ \lambda = 1.6495 \ \mu m) \tag{1.12}$$

$$M_{K_s}\odot = 3.27 \ (2\textit{MASS}), \ \lambda = 2.1638 \ \mu m) \tag{1.13}$$

where the designation in parentheses indicates the magnitude system along with the effective wavelength[1] of the band center. The star Vega has been used as a calibrator

---

[1] This is actually the 'pivot' wavelength, which gives an exact relation between F_ν and F_λ. For a mathematical description, see [299].

(the 'vegamag' system). Then the colour index, J - K = M_J - M_K = 0.40, so the Sun would sit near coordinates (0.40, 4.81) in the colour-magnitude diagram shown in Fig. I.2.

In the filters used by the Gaia satellite (Box 3.1 on page 60, and Table 3.1), the absolute magnitude and colour index are, respectively [45],

$$M_{G\odot} = 4.67 \quad (\textit{Gaia}, \ \lambda = 639.74 \ \text{nm}) \tag{1.14}$$

$$\left(G_{BP} - G_{RP}\right)_{\odot} = 0.82 \tag{1.15}$$

The bolometric apparent and absolute magnitude of the Sun are [204]

$$m_{bol\,\odot} = -26.832 \tag{1.16}$$

$$M_{bol\,\odot} = 4.74 \tag{1.17}$$

The bolometric correction, BC, is the difference between the magnitude of a star (either apparent or absolute) in the V-band and the star's magnitude when integrated over all wavebands. It can be expressed for any star (e.g. Appendix F), but for the Sun, the value is

$$BC_{\odot} = m_{bol\,\odot} - V_{\odot} = M_{bol\,\odot} - M_{V\odot} = -0.07 \tag{1.18}$$

Images and videos of the Sun are some of the most spectacular astronomical pictures in existence. For impressive videos of the Sun 'in action', see Online Resource 1.3. Yet, in spite of this in-depth knowledge, there is still much that we don't know about our nearest stellar neighbour. Its immediate and life-giving effects on the Earth are reasons to explore it exhaustively, both from space (Online Resource 1.8) and from the ground (Online Resource 1.9). A recent example is the Parker Solar Probe (Online Resource 1.10), launched on Aug. 12, 2018. Its main goals are to understand the physical mechanisms for accelerating solar wind particles and heating the solar corona. At its closest perihelion[2] orbit, it will come to a distance of 6.2 million km (8.86 $R_{\odot}$) from the Sun's surface in a region of intense heat. This is the closest any human-made object has been to a stellar surface, and it represents a bold attempt to 'touch a star'.

What does it mean, though, to talk about the 'surface' of a ball of gas? To understand this, let us look in some detail at the *solar atmosphere*.

1.2 THE SOLAR ATMOSPHERE

1.2.1 Physical Overview

Figure 1.2 illustrates the three main regions that make up the solar atmosphere: the *photosphere* (yellow region), the *chromosphere* (green region) and the *corona* (orange region). There is an additional *transition region* that marks the strong

---

[2] *Perihelion* is the distance of closest approach to the Sun. For a star, this would be called *periastron*.

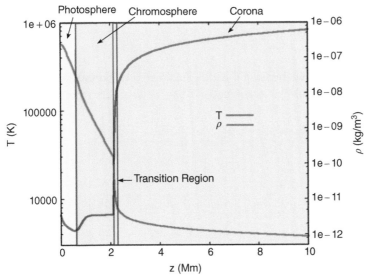

Figure 1.2 Temperature (red curve, **left** labelling) and density (green curve, **right** labelling) profiles in the solar atmosphere, from [122]. The photosphere (yellow), chromosphere (green), transition region (blue) and corona (orange) are marked. Notice that non-cgs units are used, e.g. 1 Mega-meter (Mm) = 1000 km. On the x-axis, zero corresponds to the solar surface at $\tau_{500\ nm} = 1$. Credit: Adapted from José Juan González Avilés.

gradients in temperature and density between the chromosphere and corona. Plots of atmospheric pressure, density and temperature with height can be found in Appendix B.

The magnetic field plays an important role in the solar atmosphere. Wherever the magnetic pressure dominates over gas pressure, the magnetic forces will dictate structure. It is straightforward to compare the two. The magnetic pressure (dyn cm^{-2})[3] is

$$P_B = \frac{B^2}{8\,\pi} \tag{1.19}$$

for B in Gauss. Notice that this is the same as the central equation of Eq. I.4 (energy density and pressure have the same dimensions). The gas pressure, which we specify as P without a subscript, is given by the *ideal gas law* (see also Sect. 2.1.1),

$$P = n\,k\,T \tag{1.20}$$

where n is the particle density (cm^{-3}), which includes all particles (ions, neutrals and free electrons), k is Boltzmann's constant and T is the temperature.

---

[3] This equation refers to the magnetic energy density, which is equivalent to the magnetic pressure when the fields are 'tangled' or isotropic over some scale. For non-isotropic conditions, see e.g. [19].

Whether or not magnetic forces dominate, then, depends on the ratio of these two quantities, called β:

$$\beta \equiv \frac{P}{P_B} \tag{1.21}$$

If $\beta > 1$, then gas pressure dominates; and if $\beta < 1$, then magnetic pressure dominates. For example, the average value of β in the darkest part of sunspots (the *umbra*),

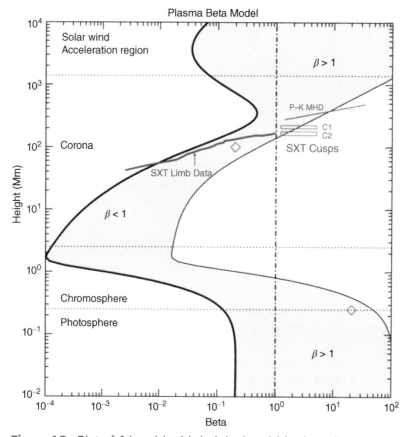

Figure 1.3 Plot of β (x-axis) with height (y-axis) in the solar atmosphere. Here, β assumes that the gas is completely ionized. The vertical dash-dotted line shows $\beta = 1$, where gas and magnetic pressures are in balance. To the right of the line, gas pressure dominates and to the left of the line, magnetic pressure dominates. The shaded region corresponds to field lines that originate from the solar surface between a sunspot with $B = 2500$ G (**left** heavier curve) and a plage region with $B = 150$ G (**right** curve). By an altitude of 10 Mm, the corresponding magnetic field strengths are $B = 300$ G (heavier curve) and $B = 40$ G (lighter curve), respectively. For other labelling, see [112]. With permission of Springer Nature. Credit: G.A. Gary.

where $B \sim 1500$ Gauss, is $\beta \approx 0.85$, indicating magnetic pressure dominance, although some umbral regions have $\beta > 1$ as well [55]. A plot showing β for certain conditions in the solar atmosphere is given in Fig. 1.3.

As Fig. 1.3 shows, β varies considerably with height and depends on the surface magnetic field from which the higher fields originate. Generally, though, magnetic pressures dominate in the chromosphere and corona. This is because gas pressure declines rapidly with height, as shown in Fig. B.4 of Appendix B, whereas the magnetic field strength does not decline so strongly.

Similar to β is a comparison between the *sound speed* and the *Alfvén speed* (Prob. 1.1). These speeds indicate how quickly a signal can propagate when a medium is perturbed. The sound speed, c_s, is the speed of a *sound wave*, which is a propagating pressure perturbation of the gaseous medium. Sound waves are longitudinal, i.e. the pressure perturbations are in the direction of the wave propagation. The Alfvén speed, c_A, is the speed of an *Alfvén wave*, which is a propagating perturbation along magnetic field lines. Alfvén waves are transverse, like the waves that propagate when a guitar string is plucked. The sound speed is given by

$$c_s = \sqrt{\gamma_a \frac{P}{\rho}} \qquad (1.22)$$

where ρ is the mass density of the gas (g cm^{-3}) and γ_a (unitless) is called the *adiabatic index*. The adiabatic index takes on values between 1 and 5/3 for a gas (see Sect. 2.1.7). The Alfvén speed is

$$c_A = \frac{B}{\sqrt{4\pi\rho}} \qquad (1.23)$$

where B is the magnetic field strength (Gauss) and ρ is the mass density. As an example, in the umbrae of sunspots, typical values are $c_s = 7.3$ km s^{-1} and $c_A = 8.7$ km s^{-1} [55]. So a magnetic perturbation propagates faster than a gas perturbation in such regions.

Box 1.1

Mean Free Path and Optical Depth

The *mean free path*, $\bar{l}$ (cm), of a photon is the average distance it could travel before interacting with a particle,

$$\bar{l} = \frac{1}{n\,\sigma_y} = \frac{1}{\kappa_y\,\rho} = \frac{1}{\alpha_y} \qquad (1.24)$$

where σ_y (cm^2) is the *effective cross-section* of the particle for the conditions of interest, n (cm^{-3}) is the density, κ_y (cm^2 g^{-1}) is the *mass absorption coefficient*, ρ (g cm^{-3}) is the mass density and α_y (cm^{-1}) is the *absorption coefficient*. Thus, $\bar{l}$ depends on the type of material, its density and the frequency of the photon.

(continued)

The *optical depth*, τ_ν (unitless), is the number of mean free paths that a photon could take along a line of sight (ignoring sideways motions) through the cloud. The change in optical depth in differential form, $d\tau_\nu$, is

$$d\tau_\nu = -\sigma_\nu \, n \, dr = -\kappa_\nu \, \rho \, dr = -\alpha_\nu \, dr \tag{1.25}$$

The negative sign is because one measures τ_ν *into* the cloud as seen by the observer, whereas the coordinate system, r, is normally measured along the direction travelled by the photon.

Thus, if $\tau_\nu = 1$, the cloud is 'just' *optically thick*, and a photon, on average, can travel one mean free path through the material. To the external observer, the cloud would appear opaque if $\tau_\nu \geq 1$, in which case the observer could not see beyond a distance, $\bar{l}$, into the cloud. If $0 < \tau_\nu < 1$, then the cloud is *optically thin* and would appear semi-transparent, like a partially transparent cloud in the daytime sky. An observer could see into and through such a cloud and could also see a background source, if present.

1.2.2 The Photosphere

The photosphere is the region that starts where the optical depth (Box 1.1 on Page 7) of the Sun becomes unity, i.e. the Sun becomes opaque. This is an effective definition of the Sun's 'surface' and is the surface that we see optically. The models that produced Fig. 1.2 and the figures shown in Appendix B have set their zero point in height at $\tau_{\lambda 500\ nm} = 1$. Notice that the wavelength of λ 500 nm has been specified since optical depth is frequency dependent.[4]

The photosphere then extends to a height of about 600 km [254], covering the region within which the temperature declines with height (Fig. 1.2) until it reaches a minimum of 4850 K [14]. This thickness is only 0.09% of the solar radius $R_\odot$. At a distance of 1 AU, 600 km subtends an angle of 0.8 arcsec. The human eye can resolve approximately 1 arcmin,[5] so the depth of the photosphere is quite negligible from our vantage point and the Sun's limb appears sharp to the human eye – easily verified by looking at the Sun through a thin, but safe, layer of clouds.

Because the temperature declines with height in the photosphere, the Sun appears darker towards the limb in comparison to the center of its (projected) disk. This is called *limb darkening* and is readily apparent in Figs. 1.1 and 1.9. When we look along a line of sight that is near the solar limb, the observed photons are coming from the outer part of the photosphere where it is cooler, whereas photons that emerge from

---

[4] This value of the optical depth is essentially equivalent to $\tau_{Ross} = 2/3$ [132], where *Ross* stands for the Rosseland mean opacity (Sect. 4.2.1).
[5] This varies with individual and age.

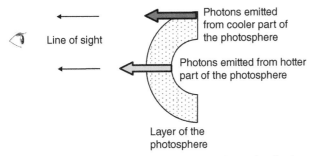

Figure 1.4 Geometry showing how the solar limb appears redder and fainter in comparison to the center. Photons arriving from a region near the limb come from a higher layer of the photosphere, corresponding to cooler temperatures, as plotted in Fig. 1.2.

the center of the Sun's disk originate from a hotter, deeper layer. Figure 1.4 illustrates this effect. Lower temperatures result in fainter emission. They also correspond to emission whose spectrum peaks at longer (redder, rather than yellower) wavelengths, consistent with *black body* emission (Sect. 2.2).

The fact that the temperature of the photosphere declines with height is also the reason *spectral lines* are seen in *absorption*, rather than emission. Figure 1.5 provides an illustration. A spectrum of the Sun shows numerous dark lines, called *Fraunhofer lines*, that result from quantum transitions within various constituents in the photosphere. Background photons from the black body spectrum are present over a wide range of wavelengths. Those photons that are at wavelengths corresponding to these lines excite upwards electronic transitions in these foreground elements. As a result, background photons at those wavelengths are removed from the line of sight, resulting in dark lines in the spectrum.

The Earth's atmosphere scatters sunlight around us, so Fraunhofer lines can easily been seen by using a simple hand-held spectrograph and looking towards the blue sky in a direction away from the Sun on a cloudless day. Analysis of these lines provides information on the temperature structure of the Sun and also on its chemical composition and elemental abundances. Similar studies of other stars provide powerful tools for understanding the composition and physical properties of the different kinds of stars that populate the H-R diagram (Fig. 1.2). More will be said about such studies in Sect. 3.3.

In visible light, the most distinguishing features of the photosphere are sunspots (Fig. 1.9). These are cooler regions, approximately 2000 K cooler than the rest of the surface, that are a consequence of magnetic activity (Sect. 1.4). Sunspots consist of a darker *umbra* in the center with typical magnetic field strengths of 2 to 4 kG, surrounded by a weaker *penumbra* with field strengths of ~ 1 kG [323]. Sunspots act as the 'footprint' for prominences and flare activity, and they increase and decrease

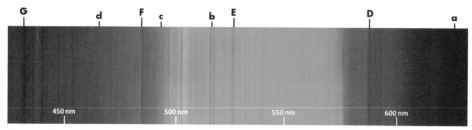

Figure 1.5 The many dark vertical lines are absorption lines in the Sun's atmosphere, called *Fraunhofer lines*. They are due to spectral line transitions within various elements that absorb light from the background as they transition from a lower to a higher energy state. The background is the black body spectrum, commonly observed as the rainbow. Several lines are labelled as follows.

G (λ 430.8 → 434.0 nm): a blend of lines from Ca, Fe, and H (Hγ);
d (λ 466.8 nm): Fe;
F (λ 486.1 nm): H (Hβ);
c (λ 495.8 nm): Fe;
b (λ 516.7 → 518.4 nm): a blend of lines from Mg and Fe;
E (λ 527.0 nm): Fe;
D (λ 587.6 → 589.6 nm): a blend of lines from He and Na;
a (λ 627.7 nm): O_2.
Credit: © Mátyás Molnár.

in number over an 11-year cycle. Another visible photospheric feature is a *granulation* pattern that is a consequence of *convection* (Fig. 4.4). Convection will be discussed in Sect. 4.3.

1.2.3 The Chromosphere

The chromosphere starts at a height of 600 km, where the temperature is lowest, and extends to a height of about 2000 km (Fig. 1.2). Originally, the chromosphere ('colour ball') obtained its named from the reddish colour of Hα emission[6] that was observed just above the lunar limb during a total solar eclipse. A more physical definition [43] is that this is a region in which *local thermodynamic equilibrium* (LTE, see Box 1.2 on page 12) no longer holds, but hydrogen is still predominantly neutral (with some partial ionization).

The chromosphere is a dynamic region whose structure is dominated by magnetic activity ($\beta < 1$, Fig. 1.3). Notable chromospheric features, most readily seen at the solar limb in Hα light, are *prominences* and *spicules* (Fig. 1.6).

Prominences are filamentary, loop-like structures that are anchored to underlying magnetic polarity reversals in sunspots. When viewed against the solar disk, rather

---

[6] See Box 1.3 on Page 14 for spectral line nomenclature.

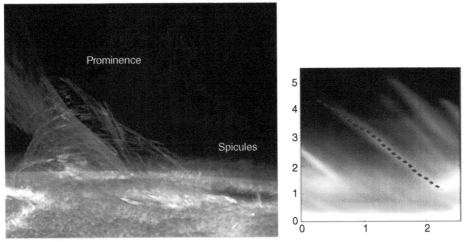

Figure 1.6 Images near the limb of the Sun taken in Hα light, showing the solar chromosphere region. **Left**: The large filamentary loop is a solar prominence. The filaments follow the direction of the magnetic field. Smaller, hair-like features are spicules. This image was taken on Jan. 12, 2007 by the Hinode spacecraft, a collaboration between the space agencies of Japan, the United States, the United Kingdom and Europe. Credit: NASA/JAXA. **Right**: Close-up of *spicules*. Labels are in units of Mm (1 Mm = 1000 km). The blue dots delineate a single spicule. Credit: [278] Sharma et al. (2017), IOP Publishing.

than at the limb, they appear as filaments. Prominences appear to be examples of *magnetic flux ropes* [318], which are tube-like regions within which magnetic field lines twist or wrap around a central axis [117]. The review by [117] likens the structure of prominences to a 'magnetic skeleton' upon which 'flesh and blood' can be added via the energetics revealed by magneto-dynamic simulations. These dynamic structures often become the origin for much larger solar eruptions that extend into the solar corona (Sect. 1.2.5).

Spicules are essentially jets that transport energy from the photosphere into the chromosphere and corona. They cover the surface of the sun, making the Sun's limb appear 'hairy'. Typical spicule velocities (transverse or radial) are tens of km s^{-1} [278] but sometimes are up to 100 km s^{-1} [271]. They constantly change, with life-times of order of a few minutes, as a video at Online Resource 1.4 illustrates. Some results support the hypothesis that fast spicules originate from *magnetic reconnection* (Fig. 1.8).

Bright regions called *plages* also surround sunspots with magnetic fields that are typically between 100 and 500 G [323], somewhat weaker than sunspot penumbrae but still much higher than the net background surface magnetic field of the quiet Sun of $B \approx 0.1$ to 0.5 G.

Box 1.2

Thermodynamic Equilibrium (TE) and Local Thermodynamic Equilibrium (LTE)

A gaseous object that has uniform temperature, T, is said to be in *thermal equilibrium*. T is a description of the random motions of particles within the gas (i.e. kinetic temperature, Sect. 2.1.3). In addition, there can be a radiation field throughout the object that has a characteristic temperature called the *radiation temperature*, T_R (Sect. 2.2). If the radiation and particles become 'trapped together', reaching equilibrium with each other, then $T = T_R$ and the object is said to be in *thermodynamic equilibrium* (TE). For the radiation to be trapped, the object must be opaque ($\tau_\nu > 1$). At any point within such an object, the radiation field can be described by the Planck curve (Eq. 2.48 or 2.49). Since the temperature is constant everywhere, there is no net *flux* of radiation through the object.

A stellar interior is not in TE because the center of a star is hotter than its surface. However, it is in *local thermodynamic equilibrium* (LTE). The mean free path of a photon within the Sun is of order $\bar{l} \approx 1$ cm ($\tau_\nu \gg 1$). Photons therefore 'leak through' the star diffusively, taking a random walk from the core to the surface and having many interactions en route. We can make a crude estimate of the solar temperature gradient from $\nabla T = (T_c - T_s)/R_\odot \approx T_c/R_\odot = 2.3 \times 10^{-4}$ K cm^{-1}, where $T_c = 1.6 \times 10^7$ K is the central temperature, $T_s = 5781$ K is the surface temperature and $R_\odot$ is the solar radius. Then over a mean free path of 1 cm, the change in temperature is only $\approx 10^{-4}$! This difference in temperature is negligible, approximating the condition of constant T. That is, LTE is TE over a 'local' mean free path. On the other hand, this small temperature difference is also what *drives* the outwards radiative flux, a point to which we will return in Sects. 4.2.3 and 4.3.2.

1.2.4 The Transition Region

The transition region (TR, Fig. 1.2) is a narrow layer (approximately 100 km thick) within which the temperature rises dramatically from about 10,000 K to 800,000 K [296]. The region is essentially a 'discontinuity' in the temperature and density of the solar atmosphere and acts like a boundary between the partially ionized chromosphere ($T \approx 10^4$ K) and the fully ionized corona ($T \approx 10^6$ K). Any transport of material from the surface to the corona must clearly pass through this region. As with the other parts of the solar atmosphere, the structure in this region is governed by magnetic phenomena, and the geometry is likely highly corrugated [162].

Like the adjacent regions below and above, the TR is non-uniform and highly dynamic, so it is often simply referred to as the high-temperature gradient region, rather than a 'layer' as if it were a passive sheet. Motions through this region can be verified from the redshifts and blueshifts of spectral lines as material falls down (recedes with respect to the observer) and upwells (advances with respect to the observer), respectively. Downflows dominate in the lower (cooler) TR, and upflows dominate in the higher (hotter) TR. The upflows generally correspond to the legs of magnetic loops that stretch from the chromosphere to the corona and may represent the beginnings of the solar wind. The reason for the redshift/blueshift split with height is not entirely understood. However, it may be an observational consequence of mass circulation in which hot gas is injected rapidly towards the corona (blueshifts in higher temperature gas), after which the heated plasma slowly cools and descends (redshifts at lower temperatures) [297].

Hundreds of spectral lines are seen in emission in the TR, especially in the ultraviolet (UV) and extreme ultraviolet (EUV). As the temperature increases, the ionization of different species can occur (see Box 1.3 on Page 14 for nomenclature), so the temperature gradient is also an ionization gradient. Moving from low to high temperature in the TR is like climbing a steep staircase, with each step (of irregular height) corresponding to the formation temperature of a newly ionized species from which new spectral lines become visible. At the lower end, for example, we move from partially ionized hydrogen (observable via the Lyβ line at λ 1026 Å) to OII (λ 718 Å), CIII (λ 977 Å), OIII (λ 703 Å), OIV (λ 790 Å), OV (λ 760 Å), OVI (λ 1032 Å), Mg VI/Ne VI (λ 402 Å), Ne VII (λ 465 Å), Ne VIII (λ 770 and 780 Å) and Mg X (λ 625 Å) near the top [296]. The wavelengths listed above in parentheses correspond to lines that are commonly observed for the ionized species listed. As the observer tunes an observation to the wavelength of a different line, the various steps of the ladder are illuminated, allowing the TR to be finely 'sliced up' from an observational perspective.

Why does such a strong temperature gradient occur in such a narrow region?

The temperature at any point is determined by a balance between heating rates and cooling rates. Solar radiative heating is present in the photosphere. However, other sources of heating must be present to increase the temperature in the chromosphere and corona. This extra heating is not completely understood (see Sect. 1.2.6) but may include acoustic shocks, Alfvén waves, a combination of Alfvén and sound waves (*magnetosonic waves*), magnetic reconnection (Fig. 1.8) [14] or other non-radiative sources. The heating is non-uniform; but as an exercise, it is helpful to imagine a steady heating source that starts in the photosphere and continues all the way into the corona, and which is roughly constant with height. If the cooling rate matched the heating rate, then the temperature would not change. But this is clearly not the case.

What, then, dominates the cooling? A number of cooling channels exist, but in the chromosphere below the transition region, the dominant ones are radiative transitions from Mg II, Ca II and Fe II. Partially ionized *metals*[7] in this region provide a

---

[7] Any element heavier than helium is referred to as a 'metal' in astronomy (Sect. 2.1.2).

steady source of free electrons ($n_e \approx 10^{-4}\, n_H$). When they recombine with the ions, downwards bound-bound transitions result in spectral lines whose radiation leaves the region. The result is cooling. A higher electron density generally results in more cooling because more collisions can occur (and therefore more recombinations), resulting in more lines (cf. the effect of n_e in Eq. 2.26).

The particle density, meanwhile, decreases precipitously with height in the chromosphere (Fig. 1.2), and so too might n_e except that hydrogen, the most abundant element by far, begins to ionize at temperatures of about 5000 K. By the time the temperature reaches 8000 K, hydrogen is fully ionized and $n_e \approx n_H$. Therefore, free electrons with their corresponding cooling continue to provide some offset to the heating in the mid to upper chromosphere.

The heating continues with height, but now there are essentially no new electrons available. With hydrogen completely ionized, n_e cannot increase any more, and adequate radiative cooling channels are no longer available. Instead, there is a catastrophic thermal instability, and the temperature rises abruptly through the transition region to coronal values of 10^6 K.

It is interesting that the main cooling channel available to the corona is *electron conduction* (cf. Sect. 4.4) from the corona back down to the TR (recall from thermodynamics that heat travels from high to low temperatures). Therefore, the TR acts as an energy source for upflowing energy and waves, as well as an energy sink for the upper corona [11]. Conductive transport, however, is insufficient to cool the million-degree gas in the corona.

Box 1.3

Ionization and Hydrogen Line Nomenclatures

An ionization state is referred to with Roman numerals. For example, neutral hydrogen is HI and ionized hydrogen is HII. Neutral carbon is CI, singly ionized carbon is CII, doubly ionized carbon is CIII and so on. Extremely high ionization states are possible, given the right conditions and element. For example, lines from Fe XXV, which has had 24 electrons removed, have been observed in solar flares [e.g. 252], and lines from Fe XXVI, with 25 electrons stripped and only one remaining in a bound state, have been observed in X-ray binaries [e.g. 152].

Hydrogen is the most common element in the universe and can show a variety of spectral lines depending on the physical conditions in the gas. These spectral lines correspond to *bound-bound transitions* within this atom, i.e. an electron is involved in a quantum jump between two bound states. Any spectral line transitions whose lower energy level is the ground state (principal quantum number n = 1) is in the *Lyman series*. For example, a transition between n = 2 and n = 1 is called Ly α, a transition between n = 3 and n = 1 is Ly β, n = 3 and n = 1 is Ly γ and so on. The Lyman lines are in the UV part of the spectrum.

Lines from hydrogen whose lower level is state, n = 2, are called Balmer lines. They are designated H α for transitions between n = 3 and n = 2, H β for n = 4 and n = 2, H γ for n = 5 and n = 2, etc. The Balmer lines are in the optical part of the spectrum.

Other named series for hydrogen are Paschen and Bracket (lower level of n = 3 and n = 4, respectively). For an illustration of these series and other details of the hydrogen atom, see the freely available supplementary material, Appendix C, of the Student Companion Website to [154] at https://bcs.wiley.com/he-bcs/Books?action=index&itemid=1119623685&bcsid=12073.

1.2.5 The Corona

The strikingly beautiful, but faint, solar corona is the outermost part of the solar atmosphere and can be seen by eye during a total solar eclipse when the moon blocks out the blindingly bright disk of the Sun (Fig. 1.7). Both large closed loops as well as open streamers can be seen in the corona. Numerous named features have also been identified, including *helmet streamers*, *loop arcades*, *flares* and *microflares*, and *sigmoid features*, among others, mostly named according to their behaviour and appearance (see [11] for a more complete list). As with other parts of the atmosphere, the magnetic field plays a critical role in the structure of the observed features, and $\beta < 1$ throughout most of the coronal region (Fig. 1.3).

The solar corona is very large in comparison to rest of the atmosphere (Fig. 1.2) and transitions to the solar wind, which consists of highly energetic particles that travel far out into interplanetary space. This transition from the solar corona to the solar wind can be considered to occur at the *Alfvén surface*, which is the 'surface' at which the radial motion of the accelerating solar material exceeds the Alfvén speed, c_A (Eq. 1.23). Particles that exceed c_A are *super-Alfvénic*, and it is no longer possible for any material, or indeed information, to return downwards, because outwards motions exceed the rate at which downwards waves can propagate. The material has essentially become disconnected from its initial magnetic conditions. As long as magnetic forces dominate, the Alfvén surface will not be at the same location as a surface that is defined by the *escape velocity*, the latter dictated by gravity alone (Prob. 1.2). By measuring outbound and inbound motions of material in the solar corona, [75] have measured the Alfvén surface to be at extremely high altitudes of 12 to 15 solar radii! This marks a boundary between the solar corona and interplanetary space.

The most breathtaking coronal features are *coronal mass ejections* (CMEs), as Online Resource 1.6 illustrates. These can be seen optically because optical photons

Figure 1.7 The solar corona, seen in exquisite detail from Mongolia during the August 2008 solar eclipse. Large, delicate loops and streamers can be seen in whitish colouration. The comparatively small pinkish features near the limb are prominences. Even features on the foreground moon are visible in this digitally enhanced optical image. Credit: Miloslav Druckmüller, Peter Aniol, Martin Dietzel and Vojtech Rušin; image processing by Miloslav Druckmüller.

from the photosphere are Thomson scattered[8] in the corona. CMEs expel huge masses of plasma into the interplanetary medium, of order $M_{CME} \approx 10^{15} \rightarrow 10^{19}$ g [323] (cf. the mass of the most massive asteroid, Ceres, at $M = 10^{19}$ g), and outflow speeds are from many hundreds to several thousand km s^{-1}. The total energy release (gravitational plus kinetic plus magnetic) is of order $10^{29} \rightarrow 10^{32}$ erg.

Although CMEs are the most energetic events in the solar atmosphere, the mass loss rate from CMEs is still less than about 1% of the more consistent solar wind associated with quiet regions [11]. Together, the outflowing steady and episodic solar wind particles contribute to a kind of 'space weather' in the solar system. Magnetic flux also leaves the Sun. As magnetic fields spiral outwards with the rotating Sun, these fields also affect the dynamics of charged solar wind particles. Particles that are accelerated in shocks or magnetic reconnection events are simply called *solar energetic particles* (SEPs), with energies reaching hundreds of MeV per nucleon and

---

[8] Thomson scattering is the scattering of photons from free electrons when the photon energy is less than the rest mass energy of an electron. The scattering is wavelength-independent, meaning the scattered photon has the same wavelength as the incoming photon.

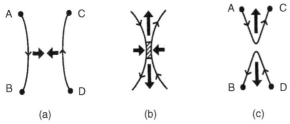

Figure 1.8 Simplified view of magnetic reconnection. **(a)** Two magnetic field lines are approaching each other in the directions given by the arrows. **(b)** As they come very close, a localized diffusion region (shaded) is formed, and vertical forces occur. **(c)** The original field lines are broken and reconnect in a different magnetic field topology. Source: Adapted from [256].

contributing the bulk of the flux of *cosmic rays* (CRs) at the lower-energy end of the CR spectrum (see Fig. 1.8 of [154]). CME events can lead to strong auroral activity on Earth, geomagnetic storms, and satellite malfunctions. They are also a potential danger for astronauts [307]. As a result, considerable interest has been expended on this Sun-Earth connection.

1.2.6 Energy Source for Heating the Solar Atmosphere

It is clear that the solar atmosphere requires a source of energy beyond radiative heat. A critical energy source is related to magnetic fields, but finding the exact mechanism remains elusive in spite of much theoretical and observational effort. An important unsolved problem is how magnetic energy is converted into heat – essentially, how particle motions become randomized. Reference [67] has tabulated a list of possible *magneto-hydrodynamic* (MHD) models, which fall into four categories:

a. *Wave dissipation models*: In this class, Alfvén waves are generated at the photosphere and then transported to the corona, where they are finally dissipated as heat. The method of wave damping differs between variants of these models.

b. *Turbulence models*: Turbulence exists in many astrophysical plasmas, and energies tend to cascade from large to small MHD scales, dissipating heat in the process. Details of the dissipation are not certain and could occur episodically, rather than steadily, with small nanoflare-like bursts.

c. *Footprint stressing models*: Here, the magnetic field in the corona becomes twisted and braided by slow motions at the footprint of the field loop. As magnetic field lines of different polarity come close to each other, they reconnect suddenly, releasing energy. This *magnetic reconnection* rearranges the magnetic field topology into a lower energy state and is capable of releasing energy as a result (Fig. 1.8). In this class, there are numerous small-scale reconnection events in a kind of 'steady-state', and numerous nano-flares can occur. As before, assumptions need to be made as to how the energy is dissipated.

d. *Taylor relaxation models*: These models also rely on magnetic reconnection. However, large, highly twisted field lines are indeed observed in the Sun, and these models take these twists to be a reservoir of energy. The magnetic *helicity*, which describes the topology of the field lines, such as how much they are twisted or knotted, is conserved in these models on large scales. Although magnetic reconnection may alter the helicity of individual flux tubes, the global helicity of the entire system is constant. In these models, the heating rate increases with increasing twist.

1.3 THE SOLAR INTERIOR

The solar interior can be divided into three sections, each of which is marked in Fig. 1.9:

- *The core*: Located in the inner 25% of the Sun. By a radius of 0.25 $R_\odot$, 99% of the Sun's energy has been generated by nuclear reactions. We will consider this region in detail in Chapter 5.
- *The radiative zone*: From 25% to about 70% of the radius of the Sun, energy is transported by radiative transport. We will consider these details in Section 4.2.3.
- *The convective zone*: From about 70% of the radius to the surface, the Sun is convective. This means the gas motion resembles 'boiling' action and energy is transported by the collective motions of convective cells. Convection will be discussed in Section 4.3.

The interior is optically thick (Box 1.1 on page 7) and so cannot be observed directly via radiation. There are other ways of probing the interior, however, and surface values can also be measured, providing boundary conditions and constraints on physical models. Fortunately, this problem has had much attention, leading to standard solar models, which will be considered next.

1.3.1 The Standard Solar Model (SSM)

Ongoing efforts for many decades have led to the concept of the *Standard Solar Model* (SSM). An SSM results from evolving a solar mass object that is initially chemically homogeneous, up to the age of the Sun. A choice of initial chemical composition

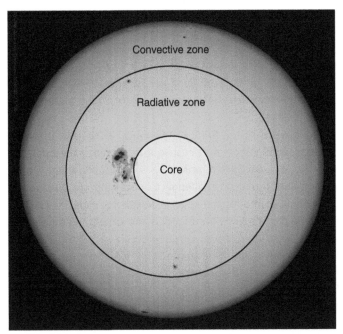

Figure 1.9 The solar surface showing sunspots and limb darkening. The interior regions, core (radius $r < 25\%\ R_\odot$), radiative zone (r from 25% to 70% $R_\odot$) and convective zone (r from 70% to $R_\odot$) are also shown. The sunspots are on the surface, not the interior. Credit: Reproduced from NASA, https://images.nasa.gov/details-GSFC 20171208, last accessed Oct 03, 2022.

must be adopted as well as a *mixing length parameter* (see Fig. 4.5), α_{ml}, which relates to the convection zone and will be discussed further in Sect. 4.3.2. A good fit of SSM models to the observables suggests that the mixing length parameter is $\alpha_{ml} = 2.18 \pm 0.05$ [313].The initial chemical composition is not arbitrary because one can observe the current metallicity of the photosphere, which depends on the initial values. Specifically, the ratio of metals to hydrogen at the surface, $(Z/X)_\odot$ (see Sect. 2.1.2 for abundance nomenclature), must match observations.

One might think that theory, constrained by observations of values at the solar surface, is the only tool available to us. However, we now have two relatively new ways of 'seeing' into the Sun. The first is neutrino fluxes from nuclear reactions in the core. Neutrinos emerge unhindered from the core, and a decade of neutrino measurements have now been brought to bear on the problem. Recent values for neutrino fluxes can be found in [26]. The second is *helioseismology*, a recent and productive way of probing the solar interior by studying the various oscillations on the Sun. This is analogous to seismologically probing of the interior of the Earth by studying the propagation of waves made from earthquakes. The more general term for stars is *asteroseismology*, which will be discussed further in Sect. 9.2. For the Sun, helioseismology gives us a variety of values. One of these is the depth of the convective envelope of the Sun, whose lower radius is $R_c = 0.713\ R_\odot$ and is known to an accuracy of ~0.2%. The *sound speed profile* (c_s as a function of radius) also results and is now known to an accuracy of ~0.1% [310]. Helioseismological values like this strongly constrain structural models of the solar interior.

The metal abundance at the surface of the Sun $(Z/X)_\odot$ (Sect. 2.1.2), and therefore the initial metallicity, is extremely important because it determines the opacity (Sect. 2.3); and as we will see in Sect. 6.3, the opacity as a function of radius is a key parameter in deriving any stellar model. A challenge to obtaining good values of $(Z/X)_\odot$, though, is knowing how to model the solar atmosphere, especially the oxygen abundance, because some oxygen lines are blended and others need non-LTE corrections [23].

Two standard solar models have emerged. One model, the GS98 SSM (updated[9] to 2016), uses a 1D stellar atmosphere model with $(Z/X)_\odot = 0.0229$. A 1D model provides average values for the atmosphere and is typical of the kind of models that are used for other stars. A competing model is the AGSS09met SSM, in which $(Z/X)_\odot = 0.0178$. The AGSS09met surface abundances were determined from simulations that applied 3D hydrodynamic models, including turbulence. A detailed comparison between these two SSMs can be found in [23], important differences being how the spectral lines of C, O, N and Ne are treated. For example, C and O appear to give more consistent results for AGSS09met, whereas N is more consistent in GS98. However, the 1D GS98 model gives a better match to the helioseismology results than the 3D AGSS09met. Although either model is believed to be accurate to about the 1% level [76], the 25% difference in surface abundances has come to be known as the *solar abundance problem.*

In Appendix C, we provide SSM model data for the updated GS98 SSM in Table C.1. For this model, we also show plots of various solar properties as a function of radius in the appendix figures. The largest region of uncertainty comes from the sub-convection region between 0.65 and 0.7 $R_\odot$ [310]. Called the *tachocline* (Fig. 1.11), it can be seen as a problematic jump in the hydrogen and helium abundances in Fig. C.4. The remaining plots show smooth increases or decreases. Notice the plot of luminosity, shown in Fig. C.3. The stellar *core* is, by definition, the central region within which nuclear reactions are taking place. From the center of a star to the 'edge' of the core, the luminosity of the star increases. Once the radius has reached 25% of $R_\odot$, 99% of the Sun's luminosity has been achieved. This 25% solar core is drawn in Fig. 1.9. Once the radius exceeds the core radius, the luminosity is then constant. The remaining 75% of the Sun is 'just' envelope through which the luminous emission must travel. We will consider how this happens in Chapter 4.

---

[9] Examples of updates include improved opacities, nuclear reaction rates and others; see [313].

1.3.2 Solar Rotation

Telescopic observations of sunspots were first carried out in the early 17th century by Galileo Galilei, Thomas Harriot, Christoph Scheiner and others. It was not long before the motion of sunspots across the surface of the Sun was carefully documented, as shown by the sketch by Scheiner in the year 1625 (Fig. 1.10). Even the Sun's tilt with respect to the *ecliptic* (the line that designates the planetary orbital plane) can be seen in the figure. It was soon realized that sunspots at higher latitudes[10] drift more

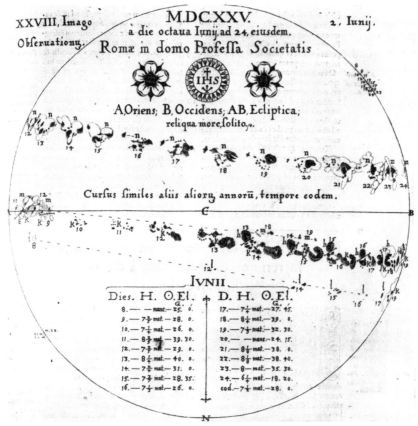

Figure 1.10 Solar rotation can be seen in this 1625 drawing by Christoph Scheiner [10]. Several sunspot complexes have been sketched over a sequence of 17 days. Their motion across the face of the Sun as well as their changing appearance can be seen. The table lists the day (Dies. or D.), the hour (H.) and the solar elevation (⊙.El.) of the observations. The horizontal line is the ecliptic, and the angle that the line of sunspots makes with the ecliptic reveals the Sun's ~7° tilt. Courtesy of Rainer Arlt, https://www.e-rara.ch/zut/doi/10.3931/e-rara-556, last accessed Oct 03, 2022.

---

[10] Just like the Earth, the solar equator has a latitude of 0° and the two poles are at ±90°.

slowly across the Sun's face than spots near the equator. This was a clear indication that solar rotation is not perfectly rigid at the surface.

While surface rotation has been mapped for over 400 years, it has only been in the last few decades that *internal rotation* has been measurable using helioseismological data [151]. Figure 1.11 shows the internal rotation of the Sun and the surface rotation at different latitudes. Rotation in the core has been difficult to measure, but newer results indicate that the core is rapidly rotating with a seven-day period, a factor of 3.8 faster than the radiative envelope [104]. From the outer core to the surface, a good map of rotation can be seen. For example, up to the base of the convection zone, the rotational period, P, is *approximately* constant, with a period of about 27 days (plus or minus one day). Thus, the solar interior is undergoing rigid or *solid body rotation*. If an object rotates like a solid body, the rotational velocity, v, within the region increases steadily with radius,

$$v = \frac{2\pi r}{P} \cos\theta = \omega r \cos\theta \tag{1.26}$$

where r is the radius, ω is the angular velocity (radians s^{-1}) and θ is the latitude. For example, at the equator, $\theta = 0°$ and at the pole, $\theta = 90°$ where there is no rotation. Rotation within the convection zone, on the other hand, is *differential* and

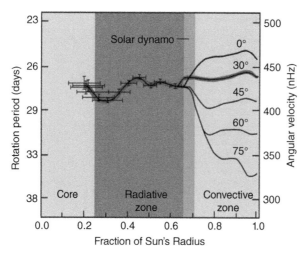

Figure 1.11 Internal rotation of the Sun. The core, radiative zone and convective zones are marked. The subconvection region, called the *tachocline*, is where the solar dynamo (marked) is thought to be active. The rotation period (**left**) and $\omega/2\pi$ (**right**) are marked, along with the different curves that correspond to different latitudes on the surface. The equator corresponds to 0°. Credit: Courtesy of Kenneth R. Lang, *The Life and Death of Stars*, Fig. 5.8, Cambridge University Press, 2013.

varies significantly with solar latitude. At the equator, the rotation at the surface is somewhat faster than the interior: it rotates once every 25 days, corresponding to a rotation speed of 2 km s^{-1}. Close to the poles, a complete rotation at the surface occurs every 34 days.

1.4 THE MAGNETIC SUN

Threaded through this narrative about the Sun – and through the Sun itself – is the presence of the magnetic field. Sunspots, spicules, prominences, flares and coronal heating are all related to magnetic activity. The best-known variation in this activity is revealed by the sunspot cycle in which, every 11 years, sunspots go from a maximum number to a minimum number to maximum again. Related magnetic activity follows the same cycle. Sunspots occur within a latitude range of $\pm 35°$, and as the sunspot cycle progresses towards maximum, sunspots appear at lower latitudes closer to the equator.[11] Sunspots occur when magnetic flux breaks through the surface of the Sun (see Fig. 1.12 *Right*), and each pair maintains the same sense of polarity in a given hemisphere and cycle. If the leading spots in the northern hemisphere are north poles and trailing spots are south poles, then leading spots in the southern hemisphere are south poles and trailing spots are north poles.[12] Eleven years later, this sense reverses for both hemispheres, so it takes 22 years for the polarity to return to its previous orientation. Magnetic reversal at the surface occurs near sunspot maximum.

Magnetic fields are ubiquitous in astronomy, from the interstellar medium (ISM) to stars, galaxies, and the intergalactic medium, and just how these fields originate is a question that has garnered much attention in the literature [e.g. 322]. However, if a very weak *seed field* is present, say a remnant of the formation of the Sun, a well-known mechanism that can amplify the field strength to observable values is the *magnetic dynamo*. There are different kinds of dynamos, but the basic idea is that the energy of motion (kinetic energy) is converted into magnetic energy via bulk motions of a conducting fluid. The field is embedded in the fluid and follows its motion, i.e. the field is *frozen in*.

Figure 1.12 *Left* shows an example of an Ω dynamo in which a seed field that is *poloidal* (a vertical north-south, or pole-to-pole, field) can be converted into a *toroidal field* (a field whose orientation is around the solar equator like a torus). This can happen because of the Sun's differential rotation. The Sun's faster rotation near the equator drags the field forward compared to higher latitudes, resulting in a conversion to an east-west field. There is also an increase in the density of field lines in the toroid, so the magnetic field strength increases (it amplifies). A toroidal field is necessary to account for the roughly east-west orientation of sunspot pairs as magnetic flux tubes emerge through the surface.

---

[11] A plot of the latitude of sunspots as a function of time over several cycles is called the *butterfly diagram*.
[12] This is called *Hale's polarity law*.

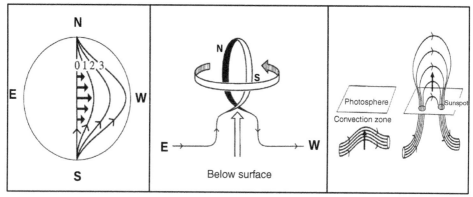

Figure 1.12 **Left**: Illustration showing how poloidal field lines can convert into toroidal field lines as a result of differential rotation. An originally poloidal (N/S) field line on the solar surface (#0) is eventually stretched out laterally because of the faster rotation at the equator, until it resembles a toroidal (E/W) field (#3). This is an example of the Ω dynamo. **Center**: Illustration showing how toroidal fields can be converted into poloidal fields. Rising plasma due to convection twists because of the coriolis force, creating poloidal fields again. Here the direction N/S is into/out-of the plane of the paper, and the semi-circular arrow shows the direction of the twist. **Right**: Illustration showing how rising magnetic flux due to convection can poke through the solar surface, producing sunspots. Credit: Adapted from [256].

Figure 1.12 *Center* shows an example of an α dynamo in which rising fields in convective cells will rotate because of the *coriolis force* (similar to hurricane rotation on the Earth),[13] converting toroidal fields into poloidal fields again. With many convective cells, amplification can occur when many of the resulting poloidal fields combine. Downwards motions are also occurring, but there is an asymmetry because solar density is declining outwards. This means rising material expands, whereas falling material contracts. There is also an asymmetry in the geometry because rising material occurs more towards the center of convective cells and then falls at the boundaries (Fig. 4.4).

An $\alpha \Omega$ dynamo combines the two effects and forms the basis of the original dynamo mechanism proposed by Eugene Parker in 1955 [240], after whom the Parker Probe has been named. There have been many approaches towards the solar dynamo, including drivers that involve turbulence and dynamos that act mainly in the tachocline rather than throughout the convection zone (for a more complete list, see [256] or [50]). Recently, however, the *flux-transport dynamo* has gained traction.

---

[13] The Coriolis force is $\vec{F} = -2\,m\,\vec{\omega} \times \vec{v}$, where m is the mass of the moving material, $\vec{\omega}$ is the angular rotation of the Sun (E-W in Fig. 1.12 *Center*) and $\vec{v}$ is the velocity of the moving material (up or down in the figure). By the right-hand rule, the force will be into or out of the page, twisting a loop.

There are variants of this model, but again, poloidal magnetic fields can be converted into toroidal fields (Fig. 1.12 *Left*) and then poloidal (Fig. 1.12 *Center*), as described earlier. However, a key component is the presence of a slow, steady *meridional flow*.

Meridional flow acts like a conveyor belt, taking plasma and embedded magnetic flux from the equator to the poles. Material cannot simply accumulate at the poles, however, so a global circulation pattern must exist, with the return flow back towards the equator at the bottom of the convection zone [58]. These flows were originally measured at the surface to be only ≈11 m s^{-1} [138], so they are difficult to measure, being only a small fraction of solar rotational motion (2 km s^{-1} at the equator) and a small fraction of horizontal convective motion (≈7 km s^{-1} [230]).

Recent studies using helioseismological data have now advanced this field considerably with flow measurements throughout the subsurface layers, including detection of the return flow towards the equator. Moreover, the *variation* in meridional flow speed patterns has been shown to be related to the 11-year solar cycle [172]. Figure 1.13 shows the global pattern of the flow and reveals a *single* meridional flow 'cell' in each hemisphere. Remarkably, the time that it takes for plasma to make one complete revolution around the convection zone is approximately 22 years. This matches the complete solar sunspot cycle time in which magnetic polarity also returns to its original orientation, supporting flux transport dynamo action in the Sun.

The magnetic field of the Sun is the underlying driver of energetic phenomena that is visible at the surface. Even a bald-faced, sunspot-free Sun, such as shown in visible light in Fig. 1.1 *Left*, reveals complex magnetic fields (*Center*) and related high-energy chromospheric activity (*Right*). The magnetic cycle strongly modulates the frequency of eruptive phenomena such as flares and coronal mass ejections, and even changes the solar luminosity at the ~0.1% level. Cosmic ray flux, auroral activity and potential threats to satellites (Sect. 1.2.5) are all ultimately connected to the solar magnetic field and its rhythms. See Online Resource 1.11 for *space weather* predictions.

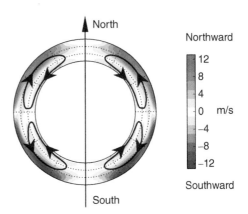

Figure 1.13 Diagram showing a vertical slice through the Sun. The meridional flow circulation is depicted with arrows in the convection region. The colour legend indicates the speeds of the flow [119]. Credit: ©MPS/Z.-C. Liang.

Box 1.4

Starring … 18 Scorpii – a Solar Twin

Distance: $d = 14.1$ pc; Spectral Type: G2 Va
Effective Temperature: $T_{eff} = 5{,}817 \pm 4$ K
Radius: $R = 1.010 \pm 0.009\ R_\odot$;
Mass: $M = 1.03 \pm 0.03\ M_\odot$
Luminosity: $L = 1.05 \pm 0.02\ L_\odot$ [253]

The star 18 Scorpii (see red arrow on map) has the same stellar classification as the Sun, which is a G2 V star, where 'V' is the Roman numeral for 5 and refers to *dwarf* stars (those that are on the main sequence in the H-R diagram in Fig. I.2 *Right*). The additional designation, 'a', means 18 Scorpii is an 'extremely luminous' dwarf. At an apparent magnitude of $V = 5.503$, this star is visible to the naked eye in a clear dark sky.

Credit: IAU and *Sky & Telescope* magazine (Roger Sinnott and Rick Fienberg). Licensed under the Creative Commons Attribution 3.0 License.

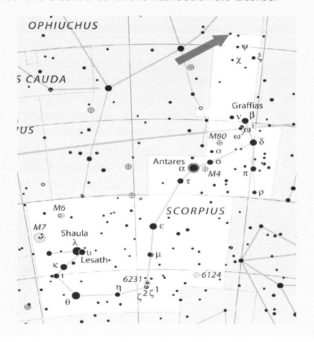

Online Resources

1.1   *The Sun now in UV light*: https://sdo.gsfc.nasa.gov/data
1.2   *Plot of sunspot cycles (current)*: https://www.swpc.noaa.gov/products/solar-cycle-progression

1.3 *NASA videos of the Sun*: https://www.nasa.gov/mission_pages/sunearth/videos/index.html

1.4 *NASA video of spicules*: https://www.youtube.com/watch?v=AiK-Ial3zh0

1.5 *NASA video of solar prominences*: https://www.youtube.com/watch?v=X5Zqo9pkvhc

1.6 *NASA/ESA/SOHO video of coronal mass ejection*: https://www.youtube.com/watch?v=gcn24Qz6zbs

1.7 *Highlights from the Solar Dynamics Observatory*: https://www.youtube.com/watch?v=mvPH_gDMarw

1.8 *List of space probes for studying the Sun*: https://en.wikipedia.org/wiki/List_of_Solar_System_probes#Solar_probes

1.9 *Ground-based solar observatories*: https://en.wikipedia.org/wiki/List_of_solar_telescopes

1.10 *The Parker Solar Probe*: https://www.nasa.gov/content/goddard/parker-solar-probe

1.11 *Space Weather Prediction Center*: https://www.swpc.noaa.gov

1.12 *Standard Solar Model (SSM) data (GS98)*: https://www.ice.csic.es/personal/aldos/Solar_Data.html
N.B. This resource is used in a number of other problems in this text. It is useful to download the data into a spreadsheet or other computer algebra aid and save it for future problems.

PROBLEMS

1.1. Show that the square of the ratio of sound speed to Alfvén speed, $(c_s/c_A)^2$, agrees with β to within a factor of two.

1.2. For a spherical gravitationally bound object, the escape velocity is

$$v_{esc} = \sqrt{\frac{2\,G\,M}{r}} \qquad\qquad r \geq R \qquad\qquad (1.27)$$

where r is measured from the center of the object and R is its radius.

(a) Compute the escape velocity and the Alfvén velocity for the magnetic field strengths corresponding to the two curves shown in Fig. 1.3 at an altitude of 10 Mm. Figure 1.2, or Figure B.2 in Appendix B will be useful.

(b) If material at this height were moving outwards at a speed of 1000 km s^{-1}, is it certain that this material will continue into interplanetary space? Consider both magnetic field strengths.

1.3. This problem uses GS98 Standard Solar Model (SSM) data from Online Resource 1.12. Ensure that you know what each data column means.

(a) At a radius of $R = 0.50\,R_\odot$, list the following quantities, specifying units:

(i) the solar mass within this radius, M_R

    (ii)   the temperature, $T(R)$

    (iii)   the mass density, $\rho(R)$

    (iv)   the pressure, $P(R)$

    (v)   the luminosity, $L(R)$

    (vi)   the hydrogen abundance fraction, $X(R)$

    (vii)   the sodium abundance fraction, $f(Na)$

    (viii) the iron abundance fraction, $f(Fe)$

(b)   Add up all of the model abundances and specify (percentage) how close the result is to 1.

1.4.   This problem uses GS98 SSM data from Online Resource 1.12.

  (a)   Plot a graph of the sound speed (km s^{-1}) as a function of radius (km) within the Sun. State any assumptions.

  (b)   Fit a curve to the plot of part (a). Try to get a reasonably good fit between the outer limit of the core and the lower limit of the convection zone. [HINT: try a polynomial.]

  (c)   If a disturbance occurred at the base of the convection zone, how long would it take the resulting wave to reach the top of the core?

1.5.   Consider a fictitious star whose density profile follows the relation

$$\rho = \rho_c \left[1 - \left(\frac{r}{R} \right)^2 \right] \tag{1.28}$$

where ρ_c is the central density of the star and R is its radius.

  (a)   Show that the mass, M_r, interior to a radius, r, is given by

$$M_r = 4\pi \rho_c \left(\frac{r^3}{3} - \frac{r^5}{5R^2} \right) \tag{1.29}$$

and find an expression for the total mass of the sphere.

  (b)   Show that the average density of this star $\bar{\rho} = 0.4\,\rho_c$.

  (c)   Suppose that this star had the same mass, radius and central density as the Sun. Find the radius within which 90% of the stellar mass is contained.

  (d)   Is the Sun's density distribution steeper or shallower than this fictitious star? Explain.

Chapter 2
The Gaseous and Radiative Star – The Basics

Dishevelled atoms tear along at 100 miles a second, their normal array of electrons being torn from them in the scrimmage. The lost electrons are speeding 100 times faster. ... The music of the spheres has almost a suggestion of – jazz.

A.S. Eddington [309]

In this chapter, we focus on some of the nuts and bolts that are needed to understand the physics of stars. First we will consider the **gaseous (material) star**, then the **radiative star** and finally the **interaction between matter and radiation**.

2.1 THE GASEOUS STAR

2.1.1 The Ideal Gas

An ideal gas is, by definition, any gas that obeys the *ideal gas law* (also called the *perfect gas law*) that relates the *particle pressure*, P (dyn cm^{-2}), volume, V (cm^3), total number of free particles, N, and temperature, T (K) of a gas. The number density,

Astrophysics: Decoding the Stars, First Edition. Judith Irwin.
© 2023 John Wiley & Sons Ltd. Published 2023 by John Wiley & Sons Ltd.

n (cm^{-3}), or mass density, ρ (g cm^{-3}), may replace N and V, provided the appropriate units are used. The ideal gas law can be written in several equivalent ways, i.e.

$$PV = \mathcal{N}\mathcal{R}T \qquad (2.1)$$

$$PV = NkT \qquad (2.2)$$

$$P = nkT \qquad (2.3)$$

$$P = \frac{\rho kT}{\mu m_H} \qquad (2.4)$$

where $\mathcal{N}$ is the number of *moles* (mol.),[1] $\mathcal{R}$ is the universal gas constant (8.314 $\times$ 10^7 erg mol.$^{-1}$ K^{-1}), k is Boltzmann's constant (1.381 $\times$ 10^{-16} erg K^{-1}), m_H is the mass of the hydrogen atom (1.67 $\times$ 10^{-24} g $\approx$ m$_p$ $\approx$ m$_n$, where m$_p$ is the mass of a proton and m$_n$ is the mass of a neutron) and μ is the mean molecular weight (unitless, see Sect. 2.1.4).

The ideal gas law is an example of an *equation of state*, which is an equation that describes how the physical properties of a material relate to each other at any location. Such a relation is necessary when building a stellar model (Chapter 6). It can be derived by assuming that interactions between particles are dominated by *elastic collisions*: that is, any exchange of energy during a collision is strongly dominated by an exchange of *kinetic energy*. That should be the case as long as particles do not come so close together that interatomic forces become important (e.g. see Chapter 5 of [273]).

A rule of thumb (though not a perfect condition) involves comparing the size of a particle to the average space between particles, $\bar{r}$,

$$\bar{r} = \frac{1}{n^{1/3}} \qquad (2.5)$$

where n is the number density of the particles. One measure of atomic particle size is the radius, r, of the first 'orbital' (principal quantum number n = 1) as defined classically by the de Broglie wavelength,

$$\lambda_{dB} = \frac{h}{p} = 2\pi r_{(n=1)} \qquad (2.6)$$

where h is Planck's constant and p is the momentum of the particle. If $\bar{r}$ approaches $r_{(n=1)}$, then non-ideal conditions must be considered. This is certainly the case for stellar remnants such as white dwarfs (Sect. 10.2) and neutron stars (Sect. 10.5) and is also true in the high-density cores of some stars. For our own Sun, the ideal gas law is applicable throughout (Prob. 2.6) [61].

Finally, for any function, f, that is a function of two variables x and y, the full derivative can be written with partial derivatives as coefficients,

$$df = \left.\frac{\partial f}{\partial x}\right|_y dx + \left.\frac{\partial f}{\partial y}\right|_x dy \qquad (2.7)$$

---

[1] A mole is equal to the number of atoms in 12 grams of pure carbon-12. That number is N_A = 6.022 $\times$ 10^{23}, where N_A is called *Avogadro's number*.

where $\frac{\partial f}{\partial x}\big|_y$ means the function f is differentiated with respect to x, holding y constant. Since pressure is a function of two variables, i.e. $P = P(\rho, T)$,

$$dP = \frac{\partial P}{\partial \rho}\bigg|_T d\rho + \frac{\partial P}{\partial T}\bigg|_\rho dT \qquad (2.8)$$

which leads to

$$\frac{dP}{P} = \frac{d\rho}{\rho} + \frac{dT}{T} \qquad (2.9)$$

2.1.2 Abundances and Metallicity

Stellar abundance generally refers to the proportion of various elements that are found in a star. In astronomy, the dominant element is, by far, hydrogen, followed by helium and then all other elements. Because of these approximate proportions, astronomers consider these constituents separately, with 'all other elements' referred to collectively as *metals*. Thus the term 'metals' is not used as a chemist would; rather, it simply means every element that is heavier than He. The *mass fractions* of hydrogen, helium and metals are, respectively,

$$X \equiv \frac{M_H}{M} \qquad (2.10)$$

$$Y \equiv \frac{M_{He}}{M} \qquad (2.11)$$

$$Z \equiv \frac{M_m}{M} \qquad (2.12)$$

where M refers to the total mass and the subscript, m, means metals. It is clear that

$$X + Y + Z = 1 \qquad (2.13)$$

An important concept is *solar abundance*, which refers to the abundances of the various elements as seen in the Sun. According to [12], the photospheric solar mass fractions are (uncertainties of a few percent)

$$X = 0.7381, \quad Y = 0.2485, \quad Z = 0.0134, \quad Z/X = 0.0181 \qquad (2.14)$$

and the *bulk* values, i.e. the values applicable to the entire Sun, are

$$X_\odot = 0.7154, \quad Y_\odot = 0.2703, \quad Z_\odot = 0.0142, \quad Z_\odot/X_\odot = 0.0199 \qquad (2.15)$$

Notice that the bulk values have higher fractions of helium and metals than the surface values. In the core of the Sun, there is a higher fraction of He because of nuclear reactions that are converting H into He (Sect. 5.4). In addition, though, there are combined effects of thermal diffusion, gravitational settling and radiative acceleration. In general, the heavier particles will descend in the Sun with time, in comparison to hydrogen, lowering the photospheric metals in comparison to the interior [12]. Values involving Z may be referred to as the *metallicity* with some small

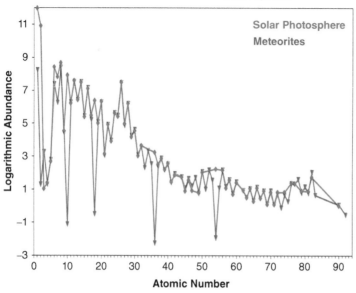

Figure 2.1 Solar abundances as a function of atomic number as observed in the present-day solar photosphere (red) and meteorites (teal blue), using data from [12]. Logarithmic values are compared to the hydrogen abundance, which is defined, by convention, to be $log(H) = 12.0$. Notice the low values of the gases H, He, Ne, Ar, Kr and Xe (n = 1, 2, 10, 18, 36 and 54, respectively) in meteorites compared to the solar photosphere.

variation, underscoring some of the challenges outlined in Sect. 1.3.1. For example, [306] quotes $Z_{\odot} = 0.0196 \pm 0.0014$.

Solar photospheric abundances are plotted as a function of atomic number in Fig. 2.1. This plot gives the *number fractions* of each element compared to hydrogen, whose logarithmic value has, by convention, been set to 12.0. For example, in the photosphere, $log(H) = 12.0$ and $log(He) = 10.93$. Therefore, for every 10^{12} hydrogen particles, there are 8.5×10^{10} helium particles, or a number fraction of 8.5%. More directly, $log(He/H) = 10.93 - 12 = -1.07$, so $He/H = 10^{-1.07} = 8.5\%$. The *mass fractions* of helium (Eq. 2.14 or 2.15) are much higher because He is more massive than H.

As Fig. 2.1 indicates, both the solar photosphere and meteorites show similar abundances, arguing for a similar origin. With the exception of about a 10% variation due to gravitational settling, photospheric abundances are believed to reflect the abundances at the time of the formation of the Solar System.

There are a few exceptions to this good agreement, however. The most obvious ones are volatile (low boiling point) gases such as H, He, Ne, Ar and Kr, which are highly depleted in meteorites because they easily escape during the formation of solid meteoritic material in the early solar nebula.

A different exception is lithium, which is depleted by a factor of about 150 in the solar photosphere compared to meteorites. The study of lithium depletion is an active area of research, but it appears to be related to the fact that this element is relatively easily destroyed in stars, needing a temperature of 'only' 2.5×10^6 K. Over time, therefore, lithium should gradually decrease in the Sun compared to meteorites.

We will return to stellar chemical compositions in general in Sect. 3.3, but it is worth emphasizing the results for the Sun here because (to within errors) we do not see strong evidence for departures from these values anywhere else in the disk of our Galaxy,[2] either in stars or interstellar gas clouds. This means 'solar abundance' is a pretty good estimate of an abundance for any other star or cloud in the *disk* (but not the halo) of the Milky Way.

2.1.3 The Maxwell-Boltzmann Velocity Distribution and Gas Temperature

Gas particles that have undergone elastic collisions, as is the case for an ideal gas in an isotropic density distribution, will relax to a velocity distribution called the *Maxwell-Boltzmann velocity distribution*. The number density of particles with speeds between v and $v + dv$, $n(v)$ in units of $cm^{-3} \left(\frac{cm}{s} \right)^{-1}$, is

$$n(v) \, dv = n \left(\frac{m}{2\pi kT} \right)^{3/2} exp \left(-\frac{mv^2}{2kT} \right) 4\pi v^2 \, dv \tag{2.16}$$

where m is the mass of a gas particle, v is the particle speed and k is Boltzmann's constant. An integration of Eq. 2.16 over all velocities would return the total number density, n (cm^{-3}), of particles of mass m in the gas.

Eq. 2.16 *defines* the gas *kinetic temperature*, T, which is the same T as appears in Eqs. 2.1 through 2.4. Whenever a gas is referred to as 'thermal', the implication is that its particle distribution obeys Eq. 2.16. An example of a 'non-thermal' gas is a distribution of cosmic rays whose velocities are relativistic and whose particles follow a power law rather than a Maxwell-Boltzmann distribution. Usually, when 'temperature' is specified without a qualifier, it is kinetic temperature that is implied.

For particles of mass m, T is related to the average of the squares of the particle speeds, i.e. the *mean-square* particle speed, $< v^2 >$.

$$\frac{1}{2} m <v^2> = \frac{3}{2} kT = u_{p.th} \tag{2.17}$$

where $u_{p.th}$ is the average kinetic energy of a particle in this thermal distribution. If v is replaced by the *relative velocity* between particles, then the mass, m, should be

---

[2] Note that when Galaxy is specified with a capital 'G', it refers to our own Milky Way Galaxy. Other galaxies are specified with a small 'g'.

changed to the *reduced mass*, μ_m (g). For two particles, the reduced mass is

$$\mu_m \equiv \frac{m_1\, m_2}{m_1 + m_2} \tag{2.18}$$

so, for example, two protons of mass m_p would have a reduced mass of $\mu_m = 1/2\ m_p$.[3]

It is straightforward to find the velocity of the peak of the Maxwell-Boltzmann distribution, i.e. the *most probable velocity*, v_{mp}, by differentiating Eq. 2.16 with respect to v and setting the result to zero. Similarly the *mean* velocity can be found from $<v> = \left[\int n(v)v\ dv\right] / \left[\int n(v)dv\right]$. One obtains

$$v_{mp} = \sqrt{\frac{2\,kT}{m}} \qquad <v> = \sqrt{\frac{8\,kT}{\pi m}} \tag{2.19}$$

From Eqs. 2.16 and 2.19, it can be seen that the distribution shifts to higher velocities for lighter particles. Physically, it is easy to understand this. More massive particles are 'sluggish' compared to lighter particles at the same temperature. For example, we can compare free electrons and free protons in an ionized gas at some temperature T. The electrons will be travelling $\sqrt{m_p/m_e} = 43$ times faster. Often this is sufficient to consider the protons to be at rest in comparison to the electrons.

2.1.4 The Mean Molecular Weight

The mean molecular weight (unitless) is just as it sounds, i.e. the average (mean) mass of all *free* particles in units of the hydrogen mass,

$$\mu \equiv \frac{<m>}{m_H} = \frac{1}{m_H\, N} \sum_i m_i \tag{2.20}$$

where m_i is the mass of the *i*th particle and N is the total number of particles.

Suppose a gas consists of pure neutral atomic hydrogen, for example, containing N_H particles in total. Then

$$\mu = \frac{1}{m_H\, N_H} \left[m_{H1} + m_{H2} + \cdots + m_{HN}\right] = \frac{N_H\, m_H}{m_H\, N_H} = 1 \tag{2.21}$$

where the subscripts $1, 2$, etc. refer to particle 1, particle 2, etc.

For a completely ionized pure atomic hydrogen gas, there are equal numbers of protons and electrons. The electrons contribute negligibly to the mass but contribute non-negligibly to the number of free particles. Then the average mass is just the mass of a proton, but the total number of particles is $N = N_H + N_e = 2\ N_H$. Consequently, $\mu = 1/2$. Notice that the pressure of such a gas (Eq. 2.4) will be twice the pressure of a neutral gas since both protons and electrons contribute to the pressure.

A gas of pure neutral helium would have a mean molecular weight of 4, and a gas of pure neutral argon would have a mean molecular weight of 40. It is clear that both the *composition* of the gas as well as its *ionization state* are important in determining the value of μ.

---

[3] This result derives from the *two-body problem* in classical mechanics.

In Prob. 2.1, equations are given for the mean molecular weight for two extremes: when all gas particles are neutral and when all gas particles are completely ionized. The latter means that *every* atom has lost *all* of its electrons. In fact, in the deep interiors of stars where the temperature is very high, complete ionization is not a bad representation of reality. However, in the outer regions of stars, and especially in stellar atmospheres, only a fraction of the electrons of a given element may have left the atom, while some remain. We will consider this in the next subsection.

2.1.5 Fractional Ionization

To understand stellar properties, it is important to take account of the ionization state of the various constituents throughout the star. We have already seen, for example, that the particle pressure of a pure hydrogen gas that is fully ionized (HII) is twice the pressure of HI, all else being equal (Sect. 2.1.4),[4] and pressure as a function of radius, $P(r)$, is a fundamental property of stellar structure. Stellar opacity (Sect. 2.3), which depends on elemental ionization states, is also of critical importance to stellar structure and evolution.

In order for an atom to be ionized, sufficient energy must be present for this to occur via either radiation (*photoionization*) or particle collisions (*collisional ionization*). For hydrogen, the ionization energy from the bound ground state (principal quantum number n = 1) to a free state (n → ∞) is 13.6 eV. If an electron is in a higher bound state (i.e. it starts in an excited state), then a smaller amount of energy is needed to ionize the atom. An ionization energy, though, is usually considered to be from n = 1 because an electron has the highest probability of being in the ground state under most astrophysical conditions.

Figure 2.2 plots the ionization energy from the *ground state* as a function of atomic number for the first five ionization levels of 28 different elements. Recall from Box 1.3 on page 14 that ionization state I refers to neutral elements, ionization state II are ions with single electron gone, etc. Notice that it takes *more* energy to ionize an ion. For example, it takes 24.6 eV to remove one electron from the ground state of helium, but once that electron in gone, it then takes 54.4 eV to remove the second one from the ground state. HeII is no longer neutral, so it now has a stronger binding force between its remaining electron and the two protons in its nucleus.

Peaks for neutral atoms (red curve) occur at the locations of the noble gases, which are very stable elements because of their filled electronic shells: i.e. He (atomic number 2), Ne (10) and Ar (18). This would also be true of Kr (36), Xe (54) and Rn (86), as well, had the plot included them. It takes more energy to ionize these elements because of their stability. At progressively higher states of ionization (II, III, etc.) such stability shifts to higher atomic numbers.

---

[4] In fact, a pure HI gas would not have the same temperature as a pure HII gas, so all else would not, technically, be equal.

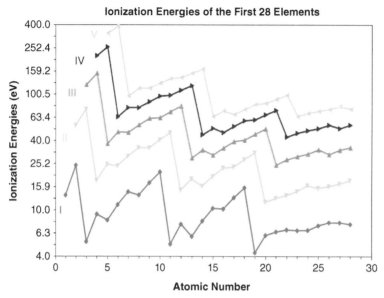

Figure 2.2 Ionization energies (from the ground state) for the first 28 elements (data from [190]). Neutral elements are designated I, singly ionized elements II, and so on, as described in Box 1.3 on page 14. Note the logarithmic stretch of the y-axis.

The *ionization fraction* of an element, f_i, is the number of particles in a gas that are in ionization state, i, compared to all particles. Writing this fraction in terms of number densities,

$$f_i = \frac{n_i}{n} = \frac{n_i}{n_1 + n_2 + \cdots n_{max}} \tag{2.22}$$

where n is the total number density of atoms plus ions of that element, n_1 would be the number density of neutral atoms, n_2 would be number density of particles in the first ionization state, etc., and n_{max} would be number density of particles that are completely ionized.

As an example, say we are interested in knowing how many lithium atoms are doubly ionized (LiIII) compared to the total number of lithium atoms. Eq. 2.22 would then look like

$$f_{LiIII} = \frac{n_{LiIII}}{n_{LiI} + n_{LiII} + n_{LiIII} + n_{LiIV}} \tag{2.23}$$

If we divide the numerator and denominator by the number density of neutral lithium, we find

$$f_{LiIII} = \frac{\dfrac{n_{LiIII}}{n_{LiI}}}{1 + \dfrac{n_{LiII}}{n_{LiI}} + \dfrac{n_{LiIII}}{n_{LiI}} + \dfrac{n_{LiIV}}{n_{LiI}}} \tag{2.24}$$

$$= \frac{\dfrac{n_{LiIII}}{n_{LiII}} \dfrac{n_{LiII}}{n_{LiI}}}{1 + \dfrac{n_{LiII}}{n_{LiI}} + \dfrac{n_{LiIII}}{n_{LiII}} \dfrac{n_{LiII}}{n_{LiI}} + \dfrac{n_{LiIV}}{n_{LiIII}} \dfrac{n_{LiIII}}{n_{LiII}} \dfrac{n_{LiII}}{n_{LiI}}} \tag{2.25}$$

Eq. 2.25 is a simple arithmetic manipulation of Eq. 2.24. The reason for this approach is that we now have a series of ratios that compare *adjacent* ionization states, and we know how adjacent states compare to each other. In LTE (see Box 1.2 on Page 12), the populations of two adjacent ionization states (e.g. between ionization state i and the more highly ionized state $i + 1$ are described by the *Saha equation*,

$$\frac{n_{i+1}}{n_i} = \frac{2U_{i+1}}{n_e U_i} \left(\frac{2\pi m_e \, kT}{h^2} \right)^{3/2} e^{-\frac{\chi_i}{k\,T}} \tag{2.26}$$

where m_e is the electron mass, T is the temperature, k and h are Boltzmann's constant and Planck's constant, respectively, n_e is the total number density of free electrons, U_i is the *partition function*[5] of the ionization state and χ_i is the ionization energy of state i from its ground state energy level. Notice the inverse dependence on n_e, implying that higher electron densities favour recombination.

The partition function is the sum of all possible states available to an electron in the ionization state of interest: that is, for any given ionization state (here dropping the subscript i or $i + 1$)

$$U = \sum_{n=1}^{n_{max}} g_n e^{-\left(\frac{\Delta E_n}{kT} \right)} \tag{2.27}$$

where g_n is called the *statistical weight* of the excitation level (principal quantum number), n, in the ionization state of interest, and ΔE_n is the energy difference between the ground state (n $=$ 1) and the excited state, n. The statistical weight is then the number of states available to an electron in a given excitation state, and the partition function is the total number of states available in all levels, n, for a particular ionization state.

For example, $U_{HI} = 2$ because if hydrogen is neutral, essentially all particles are in the ground state and there are two possible electron spins in the ground state (see also Sect. 5.4.5 of [154]). $U_{HII} = 1$ because a single proton has only one state. Notice that U, in general, is a function of temperature. Values of partition functions can be found online: for example, see Online Resource 2.1.[6]

We now have a way of determining the ionization fraction of an element for some ionization state for a given temperature and total electron density. Notice, however, that this process could become rather complicated for real situations because there are many possible ionization states and many elements. In turn, the total electron density itself depends on how many electrons are given up by the various elements in the gas; in other words, the ionization state depends on n_e, and n_e depends on the collective ionization states of the elements. Fortunately, some simplifications can

---

[5] Do not confuse the roman U used for the partition function with the math U, which is used for energy.

[6] For this Online Resource, note that the temperature must be specified in eV (e.g. 100 K $==>$ kT $= 1.38 \times 10^{-14}$ erg $= 0.0086$ eV), and the resulting partition function, called 'Z', for the specified temperature, can be found *below* the output table. For example, specifying H I at T $= 0.086$ eV yields Z $= 2$, as expected.

usually be made. Stars are, after all, about 90% hydrogen by number, so dealing with hydrogen alone is a good place to start (e.g. Prob. 2.8).

For the sake of completeness, there is a corresponding equation for the ratio of the populations of two states within any given atom under LTE conditions. This is called the *Boltzmann equation* and is

$$\frac{n_p}{n_q} = \left(\frac{g_p}{g_q}\right) e^{-\frac{\Delta E}{kT}} \tag{2.28}$$

where p and q refer to the quantum numbers of the states being compared, ΔE is the energy difference between the two states and g refers to the statistical weight of the state. Therefore this equation refers to the *excitation state* of atoms in a gas, whereas the Saha equation refers to the *ionization state* of the gas.

2.1.6 Pressure of a Partially Ionized Ideal Gas

For a calculation of the particle pressure, either μ would have to be computed for all species (meaning all ionization states of all elements plus free electrons) for use in Eq. 2.4, or one would have to sum the various contributions over the different species for use in Eq. 2.3. For example, the total particle pressure of a partially ionized gas would be the sum of the contributions from all ions plus the contribution from free electrons,

$$P = P_{\text{ion}} + P_e = \left[\Sigma_j \left(\Sigma_i f_{ij} n_j\right)\right] kT + n_e kT \tag{2.29}$$

where f_{ij} is the fractional ionization of ionization state, i, of element j, n_j is the total number density of element j, and n_e is the total electron density. For the case of a pure hydrogen gas that is partially ionized, Eq. 2.29 simplifies to

$$P = \left(n_{HI} + 2n_{HII}\right) kT \tag{2.30}$$

where n_{HI} is the total number density of neutral hydrogen particles and n_{HII} is the total number density of ionized hydrogen particles. The factor of 2 is because every hydrogen ion has a corresponding free electron.

2.1.7 Degrees of Freedom, Adiabatic Index and Specific Heats

The number of *degrees of freedom*, f, is the number of ways that energy can be distributed in a system. A single particle in a gas is free to move (*translate*) in three spatial directions. If we pump energy into such a gas, the number of degrees of freedom 'available to' each particle is at least f = 3 for an isotropic case. If these were the only modes possible, all of the input energy would go into heating the gas, because it is the random translational motions of particles that determine the gas temperature. However, a gas might store energy in other ways as well.

If the particles are atoms with bound electrons and some of the energy being pumped into the gas excites the electrons into higher energy states (even for a brief time), there will be more degrees of freedom, depending on which states and how much energy are available. Similarly, if energy goes into ionizing atoms, there will again be more degrees of freedom. Both *excitation* and *ionization*, then, are additional ways of storing energy, in which case $f > 3$.

Molecules are more complicated because they can *rotate* or *vibrate* in addition to excitation and translational motion. Thus, if energy is pumped into a gas of molecules, some of that energy may go into initiating rotation or vibration, depending on the amount of energy being introduced and details of the rotational and vibrational energy levels of the molecule. So, again, for molecules, f could be greater than 3. Although it may seem surprising, the atmospheres of cool stars contain a large variety of molecules (see e.g. [27]). Even the Sun, in the cooler regions of sunspots, contains molecules including gaseous water (H_2O, [316]).

Large, complex molecules can also experience modes that are not strictly vibrational or rotational, such as *bending*, *stretching*, *drumhead* or *twisting* modes (e.g. [272]), which occur in, for example, molecules like *polycyclic aromatic hydrocarbons* (PAHs) whose signatures can be seen in the interstellar medium (ISM) as well as in stellar envelopes (e.g. [52]). A visual example of a drumhead mode in PAHs is shown in Fig. 2.3. If energy is added to a gas containing PAHs or other complex molecules, some of it may go into such modes, rather than heat.

We now define the *adiabatic index*, γ_a, which is a measure of the number of degrees of freedom,

$$\gamma_a \equiv \frac{f + 2}{f} \tag{2.31}$$

For a gas, the minimum value of f is 3. If there are many ways to store energy aside from the three translational motions described earlier, f could theoretically become very high, so γ_a must range between approximately 1 and 5/3.

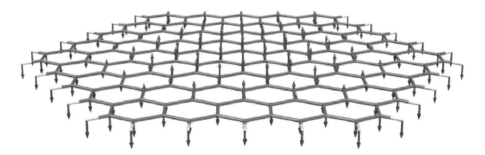

Figure 2.3 Model of PAH molecular structure, showing a drumhead mode, from [262] / With permission of IOP Publishing. Notice how the outer atoms move down while the inner atoms move up in this snapshot. The vibration is like a mode that is seen on a drum, resulting in infra-red spectral features. Credit: Alessandra Ricca.

Because the temperature of a gas is a measure of the random translational motions of its constituent particles, pumping energy into a gas that has no extra degrees of freedom (i.e. $f = 3$) will easily result in an increase in temperature for a given energy input. However, if a gas has many degrees of freedom, much more energy is required to increase its temperature because some of the input energy is going into those extra degrees of freedom. A way to measure this is by the *heat capacity* of the gas.

The heat capacity at constant volume, C_V (erg K^{-1}), and the heat capacity at constant pressure, C_p (same units), are the amount of energy required to raise the temperature of a substance by one degree Kelvin, holding either the volume or the pressure constant, respectively. If the energy is U (erg), we can write

$$C_V = \left.\frac{\partial U}{\partial T}\right|_V \tag{2.32}$$

$$C_p = \left.\frac{\partial U}{\partial T}\right|_P \tag{2.33}$$

The quantity C_p will be larger than C_V because when the pressure is kept constant, the gas must expand, doing work against its surroundings. It will take more energy to increase the temperature of the gas because some of that energy is going into the expansion. We could also define the *specific heat capacities*, or just 'specific heats', c_V and c_p, which are the heat capacities *per unit mass* (erg K^{-1} g^{-1})

$$c_p = \frac{C_p}{m}, \qquad c_v = \frac{C_V}{m} \tag{2.34}$$

From this discussion of both degrees of freedom and heat capacities, we are asking how much energy is required to raise the temperature of a gas, i.e. how much energy is available to go into random translational motion as opposed to some other way of storing energy. If the number of degrees of freedom is large, then the specific heats are also large, implying that more energy is required to raise the temperature in such cases. Thus, the specific heats should be related to the number of degrees of freedom.

For an ideal gas, these relationships are quite straightforward; it can be shown from standard thermodynamics that

$$\gamma_a = \frac{C_p}{C_V} = \frac{c_p}{c_v} = \frac{f+2}{f} \tag{2.35}$$

$$C_V = \frac{f}{2}\,\mathcal{N}\mathcal{R} = \frac{f}{2}\,N\,k \tag{2.36}$$

$$C_p = C_V + \mathcal{N}\mathcal{R} = C_V + Nk = \frac{\gamma_a}{\gamma_a - 1}\,N\,k \tag{2.37}$$

For example, if $f = 3$, then $\gamma_a = 5/3$, $C_V = 5/3\,N\,k$ and $C_p = 5/2\,N\,k$. As indicated earlier, these equations could be normalized by mass so that the heat capacities are instead expressed as specific heats.

2.1.8 Adiabatic and Isothermal Gases

When gases compress or expand, it is useful to consider two extreme cases: *adiabatic* and *isothermal* gases. An *adiabatic gas* is one in which heat does not transfer between the gas and its surroundings, whereas an *isothermal gas* is one in which heat transfer can occur such that the temperature of the gas, T, remains constant.

A helpful visualization is gas in a chamber with a piston that is compressing it (Fig. 2.4). As the piston does work on the gas, the gas heats up. If the piston is well-insulated, then no heat escapes (adiabatic, *Bottom*), so the temperature of the gas increases. If the piston is poorly insulated, then heat can escape. In a limiting case, heat escapes at such a rate that no temperature change occurs (isothermal, *Top*).

One could reverse these arguments for the case of expansion. For adiabatic expansion, the gas in the piston does work on its surroundings and therefore cools, even though no heat is lost directly from the gas through the piston walls to its surroundings. For isothermal expansion, some heat must be exchanged from the surroundings to the gas in order to keep its temperature constant during expansion.

The issue, then, is how well insulated the box is. Put another way, what is the timescale for heat transfer? For a pocket or layer of gas in a star, the surroundings are almost identical to the pocket or layer. We will see in Sect. 7.2, however, that the *dynamical timescale* is much shorter than the *thermal timescale*. If our pocket or layer is in motion (dynamical), then there is typically insufficient time for heat to transfer (thermal) into or out of it during the motion. This suggests that gas motions in stable stars are adiabatic. Similarly, sound waves, which are small perturbations of pressure, are adiabatic, as indicated in Box 2.1 on page 42.

For an ideal gas, any one of the ideal gas equations Eqs. 2.1 through 2.4 describes its equation of state. For the isothermal case, then,

$$P = C_1 \rho \qquad (2.38)$$

Compression of gas in a piston

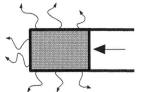

Isothermal (heat escapes, T = constant)

Figure 2.4 Diagram of a gas in a piston that is being compressed. **Top**: The isothermal case in which heat is escaping and the temperature of the gas remains constant. **Bottom**: No heat can escape, so the temperature rises.

Adiabatic (no heat escapes, T increases)

where $C_1 = kT/(\mu\, m_H)$, as in Eq. 2.4. For the adiabatic case, standard thermodynamics leads to

$$V\, T^{f/2} = C_2 \tag{2.39}$$
$$V^{\gamma_a} P = C_3 \tag{2.40}$$

Here, C_2 and C_3 are also constants. As long as the volume, V, and temperature, T, are known at any time during the adiabatic process, the constant C_2 is known. For example, if the initial values are known, then $V\, T^{f/2} = C_2 = V_i\, T_i^{f/2}$, where the subscript i indicates 'initial'. Similarly, if the volume, V, and pressure, P, are known at any point in the process (say, the 'final' point, f) then $V^{\gamma_a} P = C_3 = V_f^{\gamma_a} P_f$.

It is useful to express Eqs. 2.39 and 2.40 in a slightly different way. Since the density is $\rho = M/V$, then for a pocket of gas in which the mass is fixed, we could fold the mass of the particles into the constant to write

$$T = C_4 \rho^{\gamma_a - 1} \tag{2.41}$$
$$P = C_5 \rho^{\gamma_a} \tag{2.42}$$

Therefore, for an adiabatic pocket of gas, $T \propto \rho^{\gamma_a - 1}$ and $P \propto \rho^{\gamma_a}$.

Box 2.1

The Sound Speed

The speed of sound is an important parameter in stars, especially when we consider *convection* (Sect. 4.3) and asteroseismology (Sect. 9.2). Sound is a longitudinal pressure wave, meaning the variation in pressure is in the same direction as the propagation of the wave. Any material has a natural frequency or 'ringing' when it is perturbed, and the speed with which such a signal can travel through a medium is given by the speed of sound, c_s. Repeating Eq. 1.22,

$$c_s = \sqrt{\gamma_a \frac{P}{\rho}} \tag{2.43}$$

where γ_a is the adiabatic index, P is the pressure and ρ is the density.

Within stars, the pressure and density vary, so the sound speed also varies depending on location. Notice that the sound speed equation involves the adiabatic index, implying that sound travels adiabatically. This means the time it takes for a sound wave to propagate is less than the time it would take for heat to diffuse out of or into a region in which gas is compressed or rarefied, respectively, as described in Sect. 2.1.8.

When the gas is ideal (Eq. 2.4), we obtain

$$c_s = \sqrt{\frac{\gamma_a\, kT}{\mu\, m_H}} \tag{2.44}$$

2.1.9 Gas Motions and the Doppler Shift

If a feature of known rest wavelength λ_0 can be identified in a stellar spectrum, then its measured wavelength λ will be displaced from the rest wavelength by an amount $\Delta\lambda = \lambda - \lambda_0$ when the star has a motion *in the line of sight* with respect to the observer. Provided the motion is not relativistic,

$$\frac{\Delta\lambda}{\lambda_0} = \frac{v_r}{c} \quad (+v_r : -v_r \Rightarrow \text{star receding} : \text{advancing}) \tag{2.45}$$

where c is the speed of light and v_r is its line-of-sight or *radial velocity*. In practice, the identifiable feature is usually a spectral line. Figure 2.5 provides an illustration in which a star has a velocity v at an inclination i to the sky plane. Then

$$v_r = v \sin i \quad \text{(radial velocity)} \tag{2.46}$$

$$v_t = v \cos i \quad \text{(tangential velocity)} \tag{2.47}$$

Each spectral line has a different λ_0 and a different corresponding $\Delta\lambda$, resulting in one value of v_r. The Doppler shift leads only to v_r but not v_t. If a star or object is advancing, the spectral line is said to be *blueshifted*; and if it is receding, the line is *redshifted*.

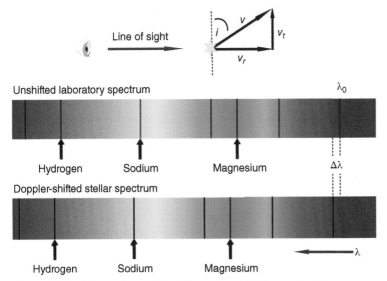

Figure 2.5 Illustration of the Doppler shift for a star. **Top**: A star is moving with a velocity v that has components v_r and v_t. **Bottom**: Each spectral line in the star is Doppler-shifted away from its rest wavelength because of the star's radial velocity.

2.2 THE RADIATIVE STAR

A star is a radiative body, and since it is also an opaque body in *local thermodynamic equilibrium* (LTE, Box 1.2 on page 12), it emits as a *black body* throughout its interior. A black body is an object that absorbs all radiation that falls upon it (no reflection from it or transmission through it). Note that a perfect absorber in thermal equilibrium is also a perfect emitter. In reality, stars depart somewhat from this condition, but the equations applicable to black bodies describe the radiation throughout stars to good accuracy. For the Sun, LTE breaks down in the chromosphere (Sect. 1.2.3) and higher in the atmosphere.

The radiation from a black body follows the *Planck function* expressed in its frequency-dependent or wavelength-dependent form, respectively,

$$B_\nu(T) = \frac{2h\nu^3}{c^2} \frac{1}{e^{\frac{h\nu}{kT}} - 1} \tag{2.48}$$

$$B_\lambda(T) = \frac{2hc^2}{\lambda^5} \frac{1}{e^{\frac{hc}{\lambda kT}} - 1} \tag{2.49}$$

where h is Planck's constant (6.626×10^{-27} erg s), c is the speed of light, k is Boltzmann's constant, and ν and λ are the frequency or wavelength being considered, respectively. It is important to note that $B_\nu \, d\nu = B_\lambda \, d\lambda$, which is why one cannot just substitute $\lambda = c/\nu$ into the first equation to get the second. The cgs units of $B_\nu(T)$ and $B_\lambda(T)$ are units of *specific intensity*, I_ν, namely erg s^{-1} cm^{-2} Hz^{-1} sr^{-1} and erg s^{-1} cm^{-2} cm^{-1} sr^{-1}, respectively. One should resist the temptation to combine the units (e.g. eliminate s^{-1} Hz^{-1}) because the first refers to time and the second to bandwidth. The form $B_\lambda(T)$ is not normally expressed in pure cgs units because 'per cm of bandwidth' is very large. Rather, a conversion is often made to 'per Angstrom' or other more practical units.

Any temperature that appears in an equation describing radiation is, by definition, a *radiation temperature*, and this is the case for Eqs. 2.48 and 2.49. For the Planck function, moreover, T is also the kinetic temperature of the gas because a black body is an opaque 'soup' of particles and radiation that are in equilibrium with each other. The functions have maxima at

$$\nu_{max} = 5.88 \times 10^{10} \, T \tag{2.50}$$

$$\lambda_{max} T = 0.29 \tag{2.51}$$

in cgs units, corresponding to the frequency or wavelength version of the function, respectively. Either of these relations is known as *Wien's displacement law*. Higher-temperature black bodies have peaks at higher frequencies. Thus, the black body curves of hotter stars are shifted to higher frequencies than cooler stars; they also have higher specific intensities than cooler stars, as an examination of Eqs. 2.48 and 2.49 reveals. In any given star, the same shift occurs as one goes deeper and deeper into the stellar interior.

For a source that subtends a small solid angle Ω (steradians or sr) in the sky, the specific intensity, I_ν, is related to the flux density, f_ν (erg s^{-1} cm^{-2} Hz^{-1}), via

$$f_\nu = \Omega I_\nu = \Omega B_\nu(T) \tag{2.52}$$

The specific intensity is quite general, whereas replacing it with the Planck function applies only to objects that approximate black bodies. Most stars are unresolved, so Ω is unknown, in which case the measurable quantity is f_ν (used in Eq. I.8 in Box I.2 on page xxiii).

For outward-directed flux (emission over 2π steradians),[7] such as at the surface of a star, the flux density (erg s^{-1} cm^{-2} Hz^{-1}) is

$$F_\nu = \pi I_\nu = \pi B_\nu(T) \tag{2.53}$$

where F_ν is used rather than f_ν to specify that flux density refers to the stellar surface in this equation, rather than at some distance away. It could also apply to outwards directed flux from a layer within a star. Again, the specific intensity, I_ν, applies to any type of radiation field, whereas the use of $B_\nu(T)$ applies only to sources that approximate black bodies.

Integration over frequency results in the well-known flux (erg s^{-1} cm^{-2}) relation for black body radiation

$$F_{BB} = \sigma\, T^4 \tag{2.54}$$

where σ is the *Stefan-Boltzmann constant* (5.67×10^{-5} erg s^{-1} cm^{-2} K^{-4}).

Equation 2.54 *defines* the *effective temperature*, T_{eff}, which can be thought of as the 'surface temperature'. Because stars are not perfect black bodies (for example, spectral lines in the atmosphere change the outgoing flux at different frequencies), an integration of Eq. 2.53 over frequency gives a flux, $F_\star$, that is not exactly the same as F_{BB} (see Fig. 2.6). Consequently T_{eff} is simply whatever temperature returns the measured value of $F_\star$, i.e.

$$F_\star = \sigma T_{eff}^{\,4} \tag{2.55}$$

Similarly, a star's luminosity (erg s^{-1}) is related to its flux by integrating the flux over the surface area to find the *Stefan-Boltzmann law* that relates the luminosity to the effective temperature:

$$L_\star = 4\pi R_\star^2 F_\star = 4\pi R_\star^2 \sigma\, T_{eff}^{\,4} \tag{2.56}$$

For isotropic radiation, such as in the interior of a star, the *energy density*, u (erg cm^{-3}), is related to the intensity, I ($I = \int I_\nu d\nu$), via

$$u = \frac{4\pi}{c}\, I = \frac{4\pi}{c}\, B = aT^4 \tag{2.57}$$

---

[7] The flux density f_ν or F_ν of Eqs. 2.52 and 2.53, respectively, results from $\int I_\nu \cos(\theta)\, d\Omega$. A distant source that is small in angular size observed with a detector that is perpendicular to the line of sight ($\theta = 0$) results in a measurement of f_ν as in Eq. 2.52. A source that is emitting over a large solid angle requires rewriting $d\Omega$ in a spherical coordinate system, giving a flux as in Eq. 2.53. See Chapter 3 of [154] for more details.

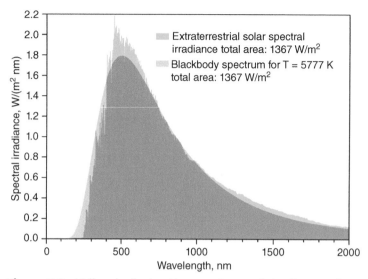

Figure 2.6 Yellow indicates the spectrum of the Sun, and grey shows the spectrum of a pure black body. The shaded regions (areas under the curves) represent an integration of the flux density over wavelength. Numerous absorption lines cut into the Sun's spectrum, resulting in a departure from a pure black body spectrum, but in this figure, the yellow-shaded region and the grey-shaded region have equal areas. The temperature of a black body for which these areas are equal is called the *effective temperature* (Eq. 2.55). Credit: https://commons.wikimedia.org/wiki/File:EffectiveTemperature_300dpi_e.png, last accessed October 08, 2022. Licensed under CC BY-SA 3.0

where $B = \int B_\nu \, d\nu$ replaces I in the stellar interior because this region closely resembles a black body. The constant a is the *radiation constant* (7.57×10^{-15} erg cm^{-3} K^{-4}).

Finally, also in an isotropic radiation field, the *radiation pressure*, P_{rad} (dyn cm^{-2}), is

$$P_{rad} = \frac{1}{3}u = \frac{1}{3}aT^4 \tag{2.58}$$

The radiation pressure would have to be added to the particle pressure to obtain the total pressure. However, often the radiation pressure can be neglected when it is a small fraction of the gas pressure, as is the case for the Sun.

2.3 STELLAR OPACITIES

Stellar opacity generally refers to how transparent or opaque a star is at a particular position within it. All scattering and absorption processes are included in determining the net opacity, so the optical depth (see Box 1.1 on page 7) must include any and all such processes that are significant contributors, as we outline here. The optical depth and mean free path of a photon depend on the type of particle with which a photon could interact, the density of the gas and the frequency. Any change in these quantities can make a significant difference in the path that a photon can travel through a substance (e.g. see Example 7.6 of [154]).

For stars, we are mostly concerned with how far a photon can travel before interacting, but the concept can be extended to particles as well. For example, a typical value for the mean free path, $\bar{l}$, of a photon in the Sun is about a centimetre (Box 1.2 on page 12). In reality, $\bar{l}$ varies from the center of the Sun outwards, but it is safe to say that the Sun is highly optically thick. In fact, using this mean free path, the optical depth of the Sun is $\tau \approx 10^{11}$! It is also true that $\bar{l} << R_\star$ for other stars, where $R_\star$ is the stellar radius.

However, the core of the Sun, where nuclear reactions take place, is a source of *neutrinos* (Box 5.2 on page 121). Solar neutrinos have a mean free path of more than a pc – using typical solar values and assuming that the gaseous region through which they travel were to extend that far! Consequently, a typical neutrino passes through the Sun effectively without any interaction at all. If your eyes were sensitive to neutrinos rather than optical light, the Sun would only appear to be the size of the small core shown in Fig. 1.9 and would be highly transparent.

Determining the opacity is an extremely important and challenging task, and opacity is a crucial input parameter to radiative transport that will be described in Sect. 4.2. The most common way to express the opacity in stars is to use the mass absorption coefficient, κ_ν (cm^2 g^{-1}).[8] The mass absorption coefficient (Eq. 1.24) is most useful because it is paired with the mass density, ρ (g cm^{-3}), which appears frequently in the equations of stellar structure (Sect. 6.1).

Let us now consider some of the most important photon-matter interactions in stars.

For example, consider a neutral atom. If a photon impinges on this atom at a frequency greater than or equal to the frequency required to *ionize* the atom, ionization may indeed occur; the photon's energy then goes into freeing the electron as well as giving the electron some kinetic energy. Since the photon is now lost to any given line of sight, its loss contributes to the opacity at the original photon frequency. Opacity that is a result of ionization is called *bound-free* opacity, $\kappa_{\nu_{bf}}$. Of course, energy levels in atoms and ions are quantized, so as long as a photon has sufficient energy to ionize any one of the energy levels and an electron is actually in the level, bound-free absorption can occur.

Another important source of bound-free opacity in cooler regions of stars is due to the hydrogen ion, H^-; this is hydrogen that has captured an extra electron. The extra electron is weakly bound, and any photon with energy greater than 0.754 eV can free it again, thus creating another source of opacity, $\kappa_{\nu_{H^-}}$. This ionization process (actually, the process is making hydrogen electrically neutral again) is very important

---

[8] By κ_ν, we mean the value *at* some frequency, not the value *per* frequency interval.

in any star cooler than spectral type F0, including our Sun, and it is usually considered separately from κ_{bf}.

Bound-free opacity is far from the only source of opacity in stars. The quantization of energy levels in atoms ensures that if a photon is at the frequency required for *excitation* from one bound level to another, then the photon is absorbed and the electron can make a transition to a higher-energy state. This is called *bound-bound* opacity, $\kappa_{\nu_{bb}}$; and again, there are many possibilities, especially for *metals*. Although metals are much less abundant than hydrogen or helium, the fact that there are many possibilities for line transitions makes them important sources of opacity in cooler stellar atmospheres.

Two more important sources of opacity, both related to free electrons, are important in stars.

One is *electron scattering*. When the energy of the photon, E_{ph}, is $E_{ph} << m_e c^2$, where m_e is the rest mass of the electron and c is the speed of light, this process is called *Thomson scattering*. It is frequency-independent and is therefore simpler than all of the other processes. Such scattering has a fixed *Thomson scattering cross-section* of $\sigma_T = 6.65 \times 10^{-25}$ cm^2. A conceptual way of looking at electron scattering is that a photon of any frequency will start a free electron oscillating at the same frequency. The oscillating electron then acts like an electric dipole and re-emits radiation at the same frequency. The net result is that the photon that would originally have emerged along a line of sight is instead *scattered* primarily out of the line of sight, thereby producing a source of opacity, κ_{es}. If the photon's energy exceeds $m_e c^2$, then the process is called *Compton scattering* and the cross-section, $\kappa_{\nu_{Cs}}$, depends on frequency. Compton scattering is only important in high-energy regions.

The second source of opacity is called *free-free absorption*. This occurs when a free electron is in the vicinity of an ion – for example, a free electron in the vicinity of a free proton. A photon that impinges upon such an electron will be absorbed, causing the electron to pick up speed. Again, since the photon is now gone, this is another source of opacity, $\kappa_{\nu_{ff}}$. It is perhaps easier to understand the inverse possibility: that is, an electron that decelerates (mainly changes direction) while it passes by an ion will emit radiation called free-free radiation, also known as Bremsstrahlung (braking) radiation. Both $\kappa_{\nu_{es}}$ and $\kappa_{\nu_{ff}}$ are due to free electrons, but $\kappa_{\nu_{ff}}$ requires that the electron is in the vicinity of an ion and the photon is absorbed rather than scattered.

The *total* opacity at any frequency is, finally,

$$\kappa_\nu = \kappa_{\nu_{bb}} + \kappa_{\nu_{bf}} + \kappa_{\nu_{ff}} + \kappa_{es} + \kappa_{\nu_{H^-}} \tag{2.59}$$

where the subscripts refer to each of the processes described. Note that electron scattering has no frequency subscript.

Although we have not included *molecular transitions* explicitly, they are indeed present in stars, as explained in Sect. 2.1.7, especially in cooler stellar atmospheres. The processes described previously can still occur. However, there are now more possibilities, including rotational, vibrational and other modal bound-bound transitions, or *photodissociations*. The latter is when the molecule splits apart (dissociates) due to the incoming photon. All of these have to be included as well, if opacity is to be accurately modeled in such regions.

Fortunately, there are regions in stars where one source (or at least not many sources) of opacity are actually important, in which case the others can be treated as negligible. An example is deep inside very hot stars where the temperature is so high that most atoms are completely ionized. In such a case, only opacities related to free electrons need be considered. In Sects. 4.2.1 and 4.2.2, we consider these opacities in a more quantitative fashion.

Online Resources

2.1 *Partition Functions, National Institute of Standards and Technology (NIST)*: https://physics.nist.gov/PhysRefData/ASD/levels_form.html
2.2 *R. Townsend's mad star EZ-Web stellar evolution code*: http://www.astro.wisc.edu/~townsend/static.php?ref=ez-web
2.3 *R. Townsend's mad star MESA-Web stellar evolution code (more input options)*: http://www.astro.wisc.edu/~townsend/static.php?ref=mesa-web-submit

PROBLEMS

2.1. (a) The mean molecular weight for a completely neutral atomic gas is

$$\mu_{neutral} = \frac{1}{X + \frac{1}{4}Y + \frac{1}{\overline{A}_m}Z} \tag{2.60}$$

where $\overline{A}_m$ is the mean atomic weight of metals. Derive this equation by starting with the definition of the mean molecular weight. Then find its value for the solar photospheric[9] composition, assuming that $\overline{A}_m \approx 16$ (oxygen).

(b) The mean molecular weight for a completely ionized gas is

$$\mu_{ionized} = \frac{1}{2X + \frac{3}{4}Y + \frac{1}{2}Z} \tag{2.61}$$

Derive this equation (note that it includes all free particles, both ions and electrons). Then find its value for the deep interior of the Sun.

2.2. Show that for a completely ionized gas with composition X, Y and Z:

(a) the electron density is

$$n_e = (1 + X) \frac{\rho}{2m_H} \tag{2.62}$$

where ρ is the mass density and m_H is the mass of a hydrogen atom.
[HINT: $n_e = N_e/V = N_e\, \rho/M$, where N_e is the number of electrons and M is the total mass.]

---

[9] Note that some metals in the photosphere will be ionized, but hydrogen (the most abundant atom) is neutral, justifying the use of this equation.

(b) the total number density (free electrons plus ions) is

$$n = \left(1 + 3X + \frac{1}{2}Y\right)\frac{\rho}{2m_H} \tag{2.63}$$

Equation 2.61 will be useful.

2.3. Equation 2.62 for the electron density in a completely ionized gas can be expressed as

$$n_e = \frac{\rho}{\mu_e m_H} \tag{2.64}$$

where μ_e is the mean molecular weight of electrons. Show that

$$\mu_e = \frac{1}{X + \frac{1}{2}Y + \frac{1}{2}Z} \tag{2.65}$$

2.4. Consider a position[10] that is 2.5 Mm below the surface of the Sun, corresponding to the lowest x-axis position in the plots of Appendix B. Compute the electron density, assuming complete ionization (Eq. 2.62), and compare your result to the value given in Fig. B.3. Is this location beneath the surface fully ionized or not? Explain.

2.5. (a) Find the most probable speed, v_{mp} (in km s^{-1}), for the atoms hydrogen, sodium and iron in the Sun at a radius of $R = 0.51\ R_\odot$ (see Table C.1). Assume typical non-isotopic atomic weights for each.

(b) Convert the temperature to an energy in eV. Will these atoms be neutral or ionized?

(c) What is the most probable speed of a free electron?

(d) Comment on the accuracy of the statement made by Eddington at the beginning of this chapter.

2.6. Consider a position at the center of the Sun, and refer to standard solar model (SSM, Sect. 1.3.1) data as needed.

(a) Find the mean molecular weight for the Sun's center. State any assumptions.

(b) For the density and temperature of the center, calculate the pressure using only the ideal gas law.

(c) Compare your result with the value of pressure given by the SSM and comment on whether the ideal gas law can be used throughout the Sun.

2.7. (a) Refer to the *mad star* stellar evolution model found at Online Resource 2.2, and obtain a 'summary output' for a 17 M$_\odot$ star evolved to an age of 5 million years with the default metallicity. Find the following quantities in the units specified: L ($L_\odot$), R ($R_\odot$), T$_{surface}$ (K), T$_{central}$ (K), $\rho_{central}$ (g cm^{-3}) and the particle pressure P$_{central}$ (dyn cm^{-2}).

---

[10] Be careful of units in this problem.

(b) For this star, determine the following:

    (i)   The central radiation pressure, P_{rad}, and the fraction $P_{rad}/P_{central}$.

    (ii) The average mass density of the star, $\bar{\rho}$ (g cm^{-3}). Is the average density of this massive star lower or higher than that of the Sun?

    (iii) The number density, n (cm^{-3}), at the average mass density, assuming $X = 0.71$ and complete ionization. [HINT: See Eq. 2.61.]

    (iv) The mean free path of a photon at the average mass density, assuming that the most important interaction is Thomson scattering. [HINT: See Eq. 2.62.]

2.8. At a certain location in the solar photosphere, the temperature is $T = 9000$ K and the electron density, from all ionized sources, is $n_e = 2.6 \times 10^{15}$ cm^{-3}.

(a) As a first approximation, consider that the gas consists of pure hydrogen, and compute the following quantities:

    (i)   the fraction of hydrogen that is ionized, f_{HII}

    (ii) the fraction of hydrogen that is neutral, f_{HI}

    (iii) the total number density of hydrogen particles, n_H, both neutral and ionized

(b) Compare your result for n_H to the modeled value of $n_H = 1.34 \times 10^{17}$ cm^{-3} [218], and comment on how well the assumption of pure hydrogen is sufficient for these conditions.

2.9. The pure He atmosphere of a white dwarf star has a temperature of 20,000 K and a density (ions plus neutral particles) of $n_{He} = 2 \times 10^{30}$ cm^{-3}.

(a) Determine the following quantities. [HINT: You will have a system of equations that can be solved with the help of computer algebra software.]

    (i)   The electron density, n_e

    (ii) The number fractions f_{HeI}, f_{HeII} and f_{HeIII}

    (iii) The number densities n_{HeI}, n_{HeII} and n_{HeIII}

(b) Is this He atmosphere largely neutral, singly ionized or doubly ionized?

Chapter 3

The Observed Star – Finding the Essential Parameters

♪ Good morning starshine
The earth says hello
You twinkle above us
We twinkle below

Writers: Gerome Ragni, James Rado, Galt MacDermot
Publisher: Sony/ATV Music Publishing LLC

As emphasized in the Introduction, it is important to match theory with observation. In this chapter, we focus on observations. This means that, for the most part, we are restricted to studying either the surfaces or the global properties of stars. Is it possible to 'observe' the interior of a star? In fact, there are clever ways of probing stellar interiors. For example, solar neutrinos that are generated in nuclear reactions pass through the Sun as if it were transparent, and measurement of these elusive particles provides information about conditions deep inside the Sun (Box 5.2 on page 121). Studies of solar and stellar pulsations also tell us about the interior structures of stars; this relatively new field of *asteroseismology*, or *helioseismology* when it refers to the Sun (Sect. 1.3.1), is now driving significant advances in our knowledge of stellar interiors (Sect. 9.2). Here, though, we will take a more classical approach, stepping through stellar properties like luminosity, temperature, chemical composition, mass, radius, rotation rate and winds. Our question is: how do we know these parameters?

As you will see, there are a variety of inventive methods, and often there is more than one way to find any given parameter. This is clearly a good thing! It is desirable to have redundancies so that the methods can be cross-checked against each other and we then gain confidence in the result. As we start to see relations between these parameters, we build up *calibration systems*. For example, the colour index (Box I.2 on page xxiii) correlates with stellar temperature (Sect. 3.1). If we then observe a star for which the colour index is known but not the temperature, we can use the known calibration to determine the temperature. Similarly, the data shown in Appendix F present calibrations for some other parameters. Let's see how we arrive at those values.

3.1 TEMPERATURE AND SPECTRAL TYPE

Numerous methods of finding a star's temperature have been used over the years (e.g. [124]), starting with the first realization about 100 years ago at Harvard Observatory that stellar *spectral types* were correlated with stellar *temperature*. Spectral classes were first organized by Henry Norris Russell, followed by Annie Jump Cannon. As shown in Appendix F, spectral types are designated from hot to cool stars as O, B, A, F, G, K and M, with subclasses numbered from 0 to 9 (hotter to cooler, respectively). Our Sun, for example, has a spectral type of G2. However, the temperature sequence was not confirmed until the work of Cecilia Payne-Gaposchkin. By eye, she examined the strengths of hydrogen lines and other elements, and she used Saha's equation (Eq. 2.26) to relate these to temperature [246]. Stellar spectra can be complex, but it is not actually necessary to measure every line in a spectrum to determine its temperature. A trained eye can pick out key lines and compare them to standard spectra to provide a good estimate of temperature.

Nowadays, reliance on the human eye has diminished. Grids of real standard stellar spectra as well as libraries of *synthetic* stellar spectra (e.g. Online Resource 3.4 or 3.6) have been amassed, the latter consisting of theoretical spectra based on model atmospheres and known atomic and molecular line lists. Observed spectra can then be digitally cross-correlated with known standards to obtain the temperature and many other stellar properties. A list of libraries of synthetic spectra can be found in [210].

Is there a more straightforward method of finding temperature, though, without carrying out time-consuming spectroscopic observations and poring over the resulting spectra?

In Sect. 2.2, we presented stars as black bodies whose spectrum can be described by the Planck function, $B_\nu(T)$, to first order. That is, the star's specific intensity, $I_\nu \approx B_\nu(T)$. Since almost all stars are *unresolved*, we do not actually measure a star's specific intensity, but rather its flux density, $f_\nu = I_\nu \, \Omega_\star$, where $\Omega_\star$ is the solid angle subtended by the star (Eq. 2.52). However, as long as $\Omega_\star$ does not vary with frequency[1], the *spectral shape* of f_ν should be the same as the shape of I_ν.

If stars were perfect black bodies, then, it would be quite easy to find their temperatures. One need only find the wavelength or frequency at which f_ν is a maximum and then use Wien's displacement law (Eq. 2.50 or 2.51) to determine the temperature. All that would be required is a few measurements of f_ν (or apparent magnitude) on either side of the Planck peak so that the frequency of the peak itself could be identified.

Stars are not perfect black bodies, however, as Fig. 3.1 illustrates. Here we see spectra of a variety of main sequence stars from cooler stars at the bottom to hotter stars at the top. The background Planck function is broadly outlined, with the peak shifting to shorter wavelengths (to the left) as the temperature increases. For example, the curve of the coolest star displayed (M5V at the bottom) shows a slant towards the right (longer wavelengths). By contrast, the curve of the O5V star (top curve) clearly increases to the left, and its peak would be to the left of the displayed wavelength range. However, it is obvious that numerous absorption lines make an accurate determination of the Planck peak difficult. Absorption lines are formed when background Planck photons are absorbed by atoms and molecules in stellar photospheres, as we saw for the Sun (e.g. Fig. 1.5). For the M5V star, for example, broad absorption lines due to molecules render the background Planck continuum barely recognizable. Cool stars are particularly susceptible to showing many absorption lines in their spectra, an effect called *line blanketing*, contributing to their opacity.

---

[1] In some cases, however, $\Omega_\star$ does vary with ν. For example, the Sun is larger at some radio wavelengths than at optical wavelengths because the radio 'surface' is in the corona [124].

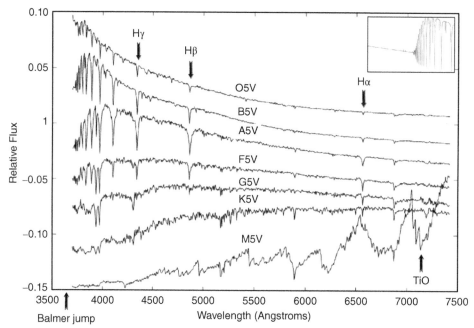

Figure 3.1 Sample stellar spectra for main sequence stars of different spectral type (marked), with a relative vertical scale. The coolest star (M5V) is at the bottom and hottest (O5V) at the top. Numerous absorption lines can be seen, with three Balmer lines (Hα, Hβ and Hγ indicated. Broad absorption lines from molecules are present in the M5V spectrum, of which one line of titanium oxide (TiO) is marked. The location of the Balmer jump at λ 3645 Å is also indicated. *Inset*: Cut-out of the spectrum (wavelength decreases to the left) from a 12,020 K star showing the Balmer jump – the point near the center at which all absorption lines converge into a single continuum (data from Online Resource 3.4).

Balmer absorption lines from hydrogen (see Box 1.3 on page 14) can also be seen in the hotter stars, when background photons cause bound electrons to jump from principal quantum number n = 2 to upper levels. Jumps to higher upper levels require higher-energy photons (shorter wavelengths). Eventually, when λ = 3645 Å, an electron can jump from n = 2 to a free state (hydrogen is ionized), a location called the *Balmer jump*. Consequently, at wavelengths shorter than this, there is a smooth *continuum* in the spectrum because any more-energetic photon will also ionize hydrogen from the n = 2 level (see figure *Inset*). Again, this masks the Planck peak. For example, a B5V star appears to have a peak near 4000 Å in Fig. 3.1. However, such a star has a temperature of 14,720 K (Table F.1), which, by Wien's displacement law, gives a Planck peak of λ = 1970 Å and is off the figure to the left. A plot showing a broader range of wavelengths would help, as shown for the Sun in Fig. 2.6 on page 46, but relying on a measurement of temperature using only the apparent Planck peak is subject to error.

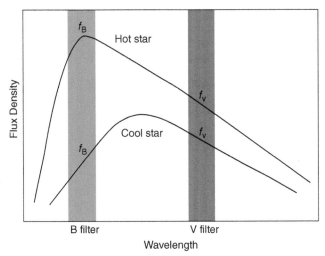

Figure 3.2 Stylized sketch of two different Planck curves for different temperatures showing B and V filters (centered at λ 438 nm and λ 545 nm, respectively) and illustrating how the ratios f_V/f_B are different for stars of different temperatures.

It is, first of all, helpful to identify a temperature that applies across all wavelengths and 'smooths over' details as to how the curves depart from pure black bodies. The choice is invariably the *effective temperature*, T_{eff}, as given in Eq. 2.55 and illustrated in Fig. 2.6. It is the effective temperature that is quoted in the tables of Appendix F, along with $B - V$ and $V - R$ colour indices. But we still need to link the imperfect stellar spectral shape to T_{eff}.

Stylized Planck curves at different temperatures are sketched in Fig. 3.2 as well as a rough placement of the B and V filters. Now we see that the *shapes* of the curves vary with temperature, and therefore so do the flux ratios, f_V/f_B. This links the flux ratio to the temperature of the black body. We know that the colour index $B - V = -2.5 \, log \left(\frac{f_B}{f_V} \right)$ (see Box I.2 on page xxiii), so knowledge of the colour index should be sufficient to find the temperature of a black body. With measurement at two wavelengths, the colour index method improves upon the method of finding the Planck peak, but the specifics of the flux at the two wavelengths (e.g. varying line blanketing in the two bands as a function of temperature) means that careful calibration is required.

Much effort has been expended to understand how colour indices relate to a star's effective temperature. There are many filter bands besides B and V, for example, and each pair of flux density measurements can, in principle, be used to find T_{eff}. Examples of filter bands can be found in Tables 3.1 and 3.3 of [154]. Information on the filters used by the Gaia satellite (Box 3.1 on page 60) is presented in Table 3.1.

More than just stellar absorption lines needs to be considered in such a calibration, including instellar reddening (footnote 9 on page 228), stellar metallicity

Table 3.1 Gaia satellite filters

	G (nm)	G_{BP}^{a} (nm)	G_{RP}^{a} (nm)
λ_m^{b}	639.74	516.47^{c}	783.05
$\Delta\lambda^{d}$	330 → 1050	330 → 680	630 → 1050

[a] Filters are sometimes abbreviated to *BP* and *RP*.
[b] Mean wavelength, weighted by the filter response function.
[c] For sources fainter than $G = 10.99$, the value is 511.78 nm.
[d] Wavelength range.
Credit: [154]/John Wiley & Sons.

and surface gravity. Once a reliable calibration is available, however, the colour index method is a standard for finding T_{eff}. It is also relatively straightforward because it requires only two measurements of flux density, rather than the more time-consuming process of obtaining spectra. It is also important to note that *no measurement of distance is required* to obtain a star's temperature! The result of all this effort is a calibrated system where temperature, colour index and spectral type are correlated. The colour index to temperature calibration from the Gaia satellite [46] is shown in Fig. 3.3.

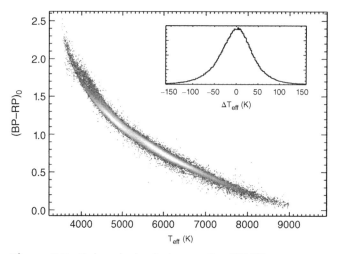

Figure 3.3 Colour index between the BP filter (λ 516.47 nm) and RP filter (λ 783.05 nm) used by the Gaia satellite, as a function of effective temperature. The calibration also used independent infrared data. The subscript 0 on the colour index indicates that the values have been corrected for interstellar reddening caused by dust. *Inset:* The distribution of residuals from the fit, indicating internal uncertainties of order 50 K. Credit: Luca Casagrande, [46] / With permission of Oxford University Press.

3.2 LUMINOSITY AND LUMINOSITY CLASS

The challenge in finding a star's luminosity is really a challenge in finding its distance, d. This is because what we measure is an *apparent magnitude*, m_ν, or a *flux density*, f_ν, at some frequency (Eq. I.8 in Box I.2 on page xxiii), either of which can be considered a 'measurable quantity'. However, what we need is an *absolute magnitude*, M_ν, which requires knowing the distance. Distances have, historically, been difficult to measure, but significant improvements are being made with new precise parallax measurements (see Box 3.1 on page 60).

In practise, m_ν is measured in some filter band, say the V-filter or the B-filter, etc., in which case it is more common to write the apparent magnitude as V, B, etc. With knowledge of d, the absolute magnitude in the corresponding band can be found (Eq. I.11).

In order to obtain the luminosity, we must convert from the absolute magnitude in some band, say M_V, to the absolute magnitude integrated over frequency, M_{bol}. To do so, a *bolometric correction* (BC) must be applied (e.g. Eq. 1.18). Values of BC are generally known for stars based on calibrations of standards; examples are in the tables of Appendix F. It is then straightforward to compute a luminosity (see Eq. I.12).

However, a bit more work needs to be done to determine which bolometric correction to use. Is the star a main sequence star, a red giant or some other type of star, for instance? This leads us to consider the second parameter to the spectral type, called the *luminosity class*. The main classes for normal stars, illustrated in Fig. 3.4, are

O: Extremely luminous supergiants (hypergiants)
Ia: Luminous supergiants
Ib: Less luminous supergiants
II: Bright giants
III: Giants
IV: Subgiants
V: Main-sequence stars, or 'dwarfs'
VI: Subdwarfs – stars that form the main sequence of Population II stars
(see Sect. 7.5)

The implication in these classes is that we have some information about the star's size. Stellar radius will be discussed in Sect. 3.5, but at the very least, the placement of a star on the H-R diagram helps to determine which luminosity class a star belongs to, as in Fig. 3.4. More accurately, though, the luminosity class can be determined from a star's spectrum. Specifically, spectral line *widths* provide us with the relevant information because line widths increase with luminosity class (I → V).

Consider the gravitational acceleration at the surface of a star, $g \propto M/R^2$ (Eq. 6.5), with the strongest effect from the (squared) radius R. Giant stars therefore have a low value of g in comparison to dwarf stars. For example, M0V and M0Ib stars have similar temperatures, differing by only about 10%. However, the ratio of their surface gravity, taking into account their different masses, is $g(\text{M0Ib})/g(\text{MV})) \sim 10^{-5}$! The

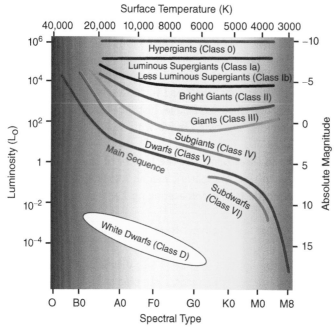

Figure 3.4 Luminosity classes in the H-R diagram (see text). Credit: Tim Cole, CC BY-SA 4.0.

dwarf will consequently have very high surface gas pressures and densities compared to the supergiant. When particles are close together, the electric fields of surrounding particles cause small shifts in the energy levels of a given line-emitting atom. For a collection of particles, these shifts cause an observed line to spread out in frequency so the line width increases. This effect is called *pressure broadening* or *collision broadening* (see also Sect. 11.3.2 of [154]). The observed line width, then, provides information on the luminosity class of a star and the consequent adoption of the right value of BC to determine the bolometric luminosity.

Between the previous section and this section, we now have sufficient information to transform an H-R diagram from the observational plane (i.e. a colour-magnitude diagram, bottom and right labelling in Fig. 3.4) into the theoretical plane (i.e. a temperature-luminosity diagram, top and left labelling in Fig. 3.4).

Box 3.1

Parallax and the Distance Ladder

A fundamental method of finding the distances to astronomical objects is *parallax*. A simple demonstration is holding up a finger at arm's length with an object like a tree in the far distance. As you close the right eye and then the left, the finger will appear to jump right and left in front of the tree. The *baseline* is the distance between your eyes, the tree is in the *far field* and the *parallactic angle* is the angle that the finger appears to jump. In astronomy, a nearby star at a distance d (center star in the picture) is seen against a backdrop of more distant stars in the far field (stars at right). The nearby star appears to shift by an angle $2\,\Pi$ when the Earth (blue dot) is on one side of the Sun (yellow circle) compared to 6 months later when the Earth is on the other side of the Sun. The parallactic angle is Π (exaggerated in the picture for clarity), and the baseline is the distance between the Earth and the Sun. By geometry, for small angles, $\tan \Pi \approx \Pi$ =baseline/distance. Therefore, a star's distance, d (pc), can be found from $d = 1/\Pi$, for Π in arcseconds. The *parsec* ('parallax-second') is *defined* by this relation.

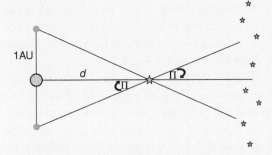

Gaia, a satellite of the European Space Agency, has taken parallax measurements to a whole new level. By the end of the mission, Gaia will have observed almost 2 billion stars and measured parallaxes as small as $\sim 10\,\mu$ arcsec corresponding to distances up to 100 kpc! Gaia is truly mapping out the Milky Way galaxy with unprecedented accuracy (for more information, see Online Resource 3.1 or 3.2).

Parallax represents the bottom rung of the astronomical *distance ladder*. Parallaxes are generally too small to measure for distances beyond the Milky Way, so there are numerous other methods available, such as finding objects whose intrinsic luminosities, L, are known. Such objects are called *standard candles*. L gives the absolute magnitude (e.g. Eq. I.12), and then a measurement of the apparent magnitude leads to the distance (Eq. I.11). Examples are Cepheid variable stars (Sect. 9.5), the tip of the red giant branch (Sect. 8.2.2.2) and Type Ia supernovae (Sect. 10.3.2). Stepping out to farther distances is like climbing up a ladder, with each rung representing a different kind of distance measurement. However, one doesn't take another step without first checking the new method against the previous one, and that is why the bottom rung is the most important.

3.3 CHEMICAL COMPOSITION

The chemical composition of stars is determined by studying spectral lines. For the Sun, optical spectral lines are formed in the photosphere and are seen in absorption because this region is cooler than the background continuum (Fig. 1.2). This is largely true for other stars as well, as Fig. 3.1 illustrates. There are some cases in which *emission lines* can also be seen in stellar spectra, but they are usually associated with stars that have ejected some material and/or have circumstellar disks or shells. Consequently, these lines come from regions in which there is no background continuum and so are seen in emission.

Spectral lines contain a wealth of information. Each element or molecule has its own characteristic spectral lines at known wavelengths, providing a 'fingerprint' for deciphering what elements are present. The line frequencies vary for an ion so the ionization state is also discernable. Also, the absorption line *strength* (depth and width, or integral over the line) is related to the abundance of the atom or molecule. All in all, the chemical composition or metallicity (recall *X, Y, Z* from Sect. 2.1.2) can be found from spectra. An example of the richness of the solar spectrum, with background continuum subtracted, is shown in Fig. 3.5. Once analysis is complete, a list of abundances can be drawn up, as was done for the Sun in Fig. 2.1.

An important proxy for metallicity is iron (Fe). This element shows many absorption lines in the optical part of the spectrum because of the excitation and ionization energies from the various quantum levels and the fact that there are so many electrons (26 in the neutral state). For example, 171 FeI and 13 FeII lines have been identified as unique in reference stars of spectral types F, G and K from the Gaia satellite [160]. By measuring only the iron lines in stellar spectra, it is standard to compare the iron abundance to the iron abundance of the Sun on a logarithmic scale. A measure of the metallicity of the star, ★, is then $[Fe/H]_\star$, defined by

$$\left[\frac{Fe}{H}\right]_\star \equiv log\left[\frac{n_{Fe}}{n_H}\right]_\star - log\left[\frac{n_{Fe}}{n_H}\right]_\odot \tag{3.1}$$

where n_{Fe} and n_H are number densities of iron and hydrogen, respectively. Then $[Fe/H]_\odot = 0$ by definition. A star with $[Fe/H]_\star = -1$ has a metallicity that is one-tenth solar, a star with $[Fe/H]_\star = -2$ is one-hundredth solar, and so on.

There are some shortcuts to obtaining metallicity. It is time consuming to obtain spectra but much quicker to do *photometry* in which the total light is measured in different filter bands. There are many absorption lines in the ultraviolet part of the spectrum compared to the optical, for example (cf. Fig. 3.1), and these are due mainly to metals. With careful calibration, a comparison can be made between the U − B colour index and the B − V colour index so that both temperature and metallicity can be taken into account [124]. The result is called a *photometric metallicity*. Different carefully chosen filter bands can also be used, depending on the sensitivity of these bands to the metallicity (e.g. those of Gaia [327]). There are no shortcuts in determining the calibration, however!

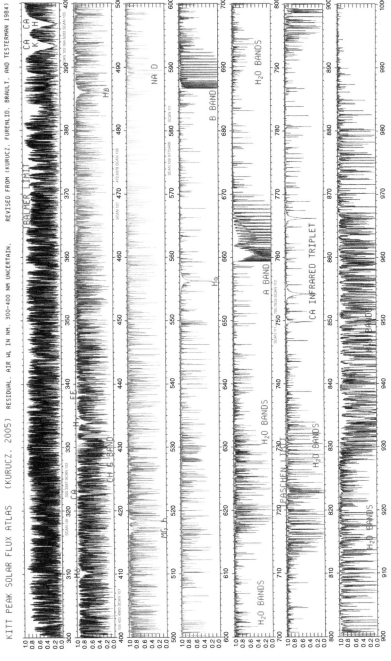

Figure 3.5 Sections of the solar spectrum with background continuum subtracted, from [182] / With permission of National Solar Observatory. Flux is represented on the y-axes and wavelength in nm on the x-axes. A number of lines in the Sun have been identified and labelled, as well as some terrestrial lines. Notice the strong Ca H and K lines in the top row at far right. Credit: Robert L. Kurucz.

3.4 MASS

The most fundamental property of a star is its mass, and to determine the mass of a star, we need to consider *dynamics*. It is fortunate, then, that many stars are not alone but are in binary or multiple systems. By one estimate, out of 100 solar-type stars in the disk of the Milky Way, there would be 62 companions [82]. This is a useful observational fact because binary star systems provide the most important and accurate route to mass determination. To this end, it is useful to summarize Kepler's laws, which were originally applied to the solar system.

Kepler's first law: The orbits of the planets are ellipses with the Sun at one focus. Since the mass of all the planets combined is much less than the mass of the Sun, the Sun can be considered at rest.[2] For any two stars, however, one is not necessarily much lower in mass than the other. That being the case, the two stars both orbit their common center of mass[3] (see small 'x' in Fig. 3.6 *Top*). Each orbit is an ellipse of different size but the same *eccentricity* or *ellipticity*, e ($e = 0$ is a circle, and higher values of e up to 1 are flatter and flatter ellipses). The center of mass of the system is the 'balance point', like a fulcrum, with the two stars on either side of it. The higher-mass star, m_1 or the *primary*, will always be closer to the center of mass, and the lower-mass star, m_2 or the *secondary*, will always be farther from it. If r_1 and r_2 are the distances of m_1 and m_2 from the center of mass, respectively, then at any time in the orbit,

$$m_1 \, r_1 = m_2 \, r_2 \qquad (3.2)$$

By the geometry of an ellipse, this also means that the semi-major axes of the two orbits, a_1 and a_2, which are measured to the centers of the ellipses, are related in the same way:

$$m_1 \, a_1 = m_2 \, a_2 \qquad (3.3)$$

Periastron is when the stars are closest together, and *apastron* is when they are farthest apart.

The *absolute orbit* is the orbit of both masses about the common center of mass, but it is also possible to construct a corresponding *relative orbit*, as shown in Fig. 3.6 *Bottom* and described mathematically in Box 3.2 on page 65. In the relative orbit, the secondary traces out a larger ellipse of the same eccentricity, but now the primary and secondary masses, plus semi-major axes, are, respectively,

$$M = m_1 + m_2 \qquad \text{fixed primary} \qquad (3.4)$$

$$\mu_m = \frac{m_1 \, m_2}{m_1 + m_2} \qquad \text{secondary} \qquad (3.5)$$

---

[2] The center of gravity of the solar system is called the *barycenter*, and it does wander somewhat with respect to the center of the Sun, mostly because of the gravitational pull of Jupiter. This movement is shown in Online Resource 3.7.

[3] For a simple binary star system in the Newtonian limit, the center of mass and the center of gravity are the same.

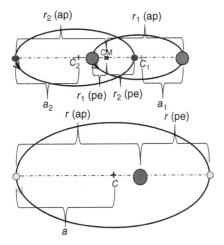

Figure 3.6 Illustration of a binary star orbit that is in the plane of the sky.
Top: The two components each orbit the center of mass, CM, marked with an x.
The larger red dot represents the larger mass, m_1, and the smaller blue dot
represents m_2. Arrows show the direction of the orbital motion. The picture
shows the stars when they are both at periastron (pe) and when they are both
at apastron (ap). The semi-major axes of the orbits are denoted a_1 and a_2, and
the centers of the ellipses are marked c_1 and c_2 with + signs. The distances of
the components from the center of mass are denoted r_1 and r_2. The ellipticities
of the two orbits are the same, but the sizes are different. **Bottom**: The same
orbital motion, but the sum of the masses, M (green), is taken to be fixed in
position at the focus of the ellipse. The secondary mass is now the reduced
mass μ_m (Eq. 3.5), the semi-major axis is a (Eq. 3.6) and the center of the
ellipse is c.

$$a = a_1 + a_2 \qquad \text{semi-major axis} \qquad (3.6)$$

where μ_m is the *reduced mass* (as in Eq. 2.18) and the total mass, M, is at the focus.
Note that if the absolute orbits about the center of mass are circular, so is the larger
relative orbit. If the orbit is inclined to the plane of the sky, then the primary star
shifts away from the focus of the projected ellipse. An observation of that shift helps
to determine the inclination of the orbital plane to the plane of the sky.

Box 3.2

From Absolute to Relative Orbits

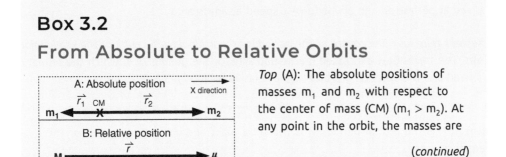

Top (A): The absolute positions of
masses m_1 and m_2 with respect to
the center of mass (CM) ($m_1 > m_2$). At
any point in the orbit, the masses are

(continued)

(continued)

separated by a distance r, and they are balanced about the CM like a fulcrum (Eq. 3.2). In vector form, respectively,

$$\vec{r} = \vec{r}_2 - \vec{r}_1, \quad m_1 \vec{r}_1 + m_2 \vec{r}_2 = 0 \tag{3.7}$$

where $\vec{r}_1$ is negative. These two equations lead to

$$\vec{r} = \vec{r}_2 \left(1 + \frac{m_2}{m_1}\right) \implies \ddot{\vec{r}} = \ddot{\vec{r}}_2 \left(1 + \frac{m_2}{m_1}\right) \tag{3.8}$$

where the double dots in the right expression represent the second derivative (i.e. acceleration). From Newton's second law and the universal law of gravitation, we know that the mass, m_2, times its acceleration, $\ddot{\vec{r}}_2$, is described by

$$m_2 \ddot{\vec{r}}_2 = -\frac{G\, m_1\, m_2}{r^2}\, \hat{\imath} \tag{3.9}$$

where $\hat{\imath}$ is the unit vector in the x direction. Inserting Eq. 3.8 (right) into Eq. 3.9 to eliminate $\ddot{\vec{r}}_2$ leads to

$$\left(\frac{m_1\, m_2}{m_1 + m_2}\right) \ddot{\vec{r}} = -\frac{G\, m_1\, m_2}{r^2}\, \hat{\imath} \tag{3.10}$$

In addition, from Eqs. 3.4 and 3.5, we have expressions for the reduced mass, μ_m, and the total mass, M. Making these substitutions in Eq. 3.10 gives

$$\mu_m \ddot{\vec{r}} = -\frac{G\, \mu_m\, M}{r^2}\, \hat{\imath} \tag{3.11}$$

Bottom (B): Equation 3.11 (cf. Eq. 3.9) now shows that a mass whose value is μ_m has an acceleration of $\ddot{\vec{r}}$ as it moves in a relative orbit about a mass M. The same equation would result if $m_1 < m_2$.

From the geometry of an ellipse (see Fig. 3.6), a relation for the apastron point is $r_{ap} = a\,(1 + e)$ (scalar form). The eccentricity, e, is fixed, so if $r_{ap} = r_{ap_1} + r_{ap_2}$, then $a = a_1 + a_2$, as given in Eq. 3.6.

Kepler's second law: Equal areas, A, of the ellipse are 'swept out' in equal units of time, t, i.e. dA/dt = constant. For binary stars, this means the two stars move at the highest speed at periastron and at the lowest speed at apastron.

Kepler's third law: The square of the period is proportional to the cube of the semi-major axis. This law is best expressed mathematically for the geometry in which the primary is fixed and the secondary orbits it (the relative orbit, Fig. 3.6 *Bottom*),

$$P^2 = \frac{4\pi^2}{GM}\, a^3 \tag{3.12}$$

where P is the orbital period, M is the total mass and we use cgs units for this equation.

If a is expressed in astronomical units, M in solar masses and P in years, then the constants become 1 so that

$$P^2 = \frac{a^3}{M}$$ (3.13)

For the solar system, it is straightforward to compute the mass of the Sun because, essentially, all of the mass is concentrated in the Sun so $M \sim M_\odot$. All that is necessary to find $M_\odot$ is to measure the period and semi-major axis of any planet. For Mars, for example, we have $P_{Mars} = 1.88$ years and $a = 1.524$ AU, leading to $M = 1\,M_\odot$, as expected. The same process holds for moons around planets or any object orbiting another that is much more massive, provided gravity is the dominant force. It is true of exoplanets as well, although planets around other stars are faint so such measurements are difficult.

For binary star systems, however, usually m_2 is *not* much less than m_1, and we will now see how these relations can help us pin down the mass of each component.

3.4.1 Visual Binaries

Visual binaries are ones in which each star can be seen individually so the movements of each can be tracked with time. Corrections are made for the inclination of the orbit with respect to the plane of the sky, and the elliptical paths (e.g. Fig. 3.6 *Top*) of each star can be traced as each star orbits the center of mass. The individual semi-major axes a_1 and a_2 can be measured and Eq. 3.3 used to find the ratio m_1/m_2.

It is straightforward to measure the period, P, of a visual binary, provided a sufficient fraction of the orbit has been observed. To find the relative semi-major axis, $a = a_1 + a_2$ (Eq. 3.6), the *distance, d, must be known* so that an angular measure of a can be converted to a linear size. The simple *small angle formula* can be used,

$$a = d\,\theta, \; a_1 = d\,\theta_1, \; a_2 = d\,\theta_2$$ (3.14)

where θ (radians) is the angular measurement on the sky, d is the distance to the object and a is the linear distance on the sky corresponding to θ (and similarly for a_1 and a_2). Units for a match the units of d. With P and a known, we can use Kepler's third law (Eq. 3.12) to find the total mass, M. We now know both the ratio of masses as well as the sum of masses, and this is sufficient to obtain the individual masses m_1 and m_2.

Figure 3.7 shows an example of a visual binary system. The bright star shown at *Left* is Sirius, a main sequence star of spectral type A. Sirius is the brightest star in the night sky and is known as the 'Dog Star' because it is in the constellation of Canis Major or the Big Dog. It is also called Sirius A to distinguish it from its companion, Sirius B, which can be seen as a faint 'dot' at ≈ 7 o'clock in the picture. The companion is a white dwarf (Sect. 10.2), hotter than Sirius but much smaller. At *Right* is the *apparent orbit* of Sirius B around Sirius A, i.e. the orbit as projected onto the plane of the sky.

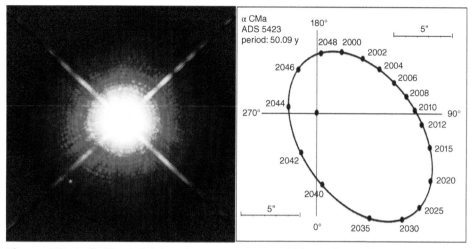

Figure 3.7 Left: The bright main sequence star Sirius A (α Canis Majoris A), with its very faint white dwarf companion, Sirius B (α Canis Majoris B) barely visible at about 7 o'clock in the picture. Credit: NASA, ESA, H. Bond (STScI) and M. Barstow (University of Leicester). **Right**: Apparent orbit of Sirius B about Sirius A, with different years marked. Sirius A is fixed where the vertical and horizontal lines cross in this diagram. The orbital plane is tilted with respect to the plane of the sky, so Sirius B traces out an apparent ellipse rather than a true ellipse. Sirius A is at the focus of the true ellipse but is offset from the focus of the apparent ellipse. The highest speed of Sirius B occurs when it is closest to Sirius A in the true orbit. Credit: FrancescoA at Wikimedia Commons.

Visual binaries are the most abundant type of binary. Online Resource 3.8 is a sample catalogue containing more than 125,000 entries, although some of these could be chance alignments rather than physically associated stars. An observational challenge for this type of binary is that because the stars are resolved, they are statistically more likely to have large separations and therefore long periods, which require much observational patience to track. The orbit of Sirius B is $\sim$ 50 years, for example, but other visual binaries can be much longer. By one measure, the distribution of periods for solar type stars peaks at P $\approx$ 250 yr with a dispersion of $\sigma_{log\ P} = 2.3$ and an average semi-major axis of $a \approx 45$ AU [82]. By Eq. 3.13, the corresponding average total mass is M $\approx$ 1.5 M$_\odot$.

In summary, visual binaries give us the masses of each component: $\Rightarrow$ m$_1$, m$_2$.

3.4.2 Spectroscopic Binaries

In many cases, the two stars of a binary system are *unresolved*, so they look like a single star and cannot be separately measured in the sky. If the system is studied via spectroscopy, however, the presence of a second star can be detected or inferred from the spectrum. When spectra from both stars are seen superimposed, the system is called a *double-line spectroscopic binary* and the radial motions of each star can be

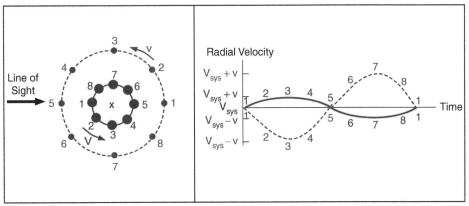

Figure 3.8 Illustration as to how the motions of two stars (red = more massive primary, blue = less massive secondary) orbiting a center of mass (marked X), produce shifting radial velocities. The simple case is shown in which the orbital plane is in the line of sight and the orbits are circular. The center of mass is moving at the *systemic velocity* (velocity of the system), V_{sys}, the primary is moving at $V_{sys} \pm V$ and the secondary is moving at $V_{sys} \pm v$. **Left**: The orbits of the two stars are shown as a function of time. When the primary (red) is at position 1, the secondary (blue) is also at position 1, and so on. **Right**: Corresponding shift in radial velocity for the two stars.

determined from *Doppler shifts* of their spectral lines (Sect. 2.1.9). If the spectrum only shows one star, but the motion of that star implies that it is moving in an orbit about an unseen companion, then the system is called a *single-line spectroscopic binary*.

Figure 3.8 shows a simplified system in which the orbit is circular and is in the line of sight, i.e. it has an inclination of $i = 90°$ to the plane of the sky. As the two stars orbit their common center of mass, their spectral lines shift with time, as shown on the right. The radial velocity curves are sinusoidal because the orbits are circular; if the orbits were elliptical, non-sinusoidal curves would be seen. The maximum radial velocity displacements are seen at times 3 and 7, when the orbital velocity equals the radial velocity. At times 1 and 5, there is no radial velocity component and therefore no velocity displacement from the systemic velocity, V_{sys}. From the maxima of the curves, we know the orbital speeds of the two stars, V (the primary) and v (the secondary).

For circular orbits, the semi-major axis is the same as the radius of the circle, and the speeds are constant and equal to the perimeter of the circle divided by the period, P:

$$V = \frac{2 \pi a_1}{P} \tag{3.15}$$

$$v = \frac{2 \pi a_2}{P} \tag{3.16}$$

These relations allow us to find a_1 and a_2 since the period is known and the same for both stars. From Eq. 3.2, we now obtain the *ratio of the masses*, m_1/m_2. From Eq. 3.6, we obtain a and, as before, use Kepler's third law (Eq. 3.12) to find the *sum of the masses*, M. Finally, since both the ratio and sum of the masses are known, each mass can individually be found.

If our circular orbit is inclined by some angle i with respect to the sky plane, then we do not measure V or v, but rather $V_r = V \sin i$ and $v_r = v \sin i$ (Eq. 2.46). In that case, from Eqs. 3.15 and 3.16, we find $a_1 \sin i$ and $a_2 \sin i$, leading to $a \sin i$. That is, only a *lower limit* to the semi-major axis is found. Equation 3.3 still leads to the ratio of masses, m_1/m_2, but Kepler's third law then leads to M $\sin^3 i = m_1 \sin^3 i + m_2 \sin^3 i$. Using the mass ratio, we find $m_1 \sin^3 i$ and $m_2 \sin^3 i$, i.e. a *lower limit* to the masses. It is beneficial in that case to have additional information to help pin down the masses. Since the spectrum is observed, the spectral type is known, for example, so appeals to prior calibrations that link spectral type to mass could be made.

Elliptical orbits are more complex, but the shapes of the radial velocity curves (Fig. 3.8 *Right*) depart from sinusoids in that case, and modelling the shape is an aid to understanding the orbit. In addition, statistically, unresolved binary systems tend to have small separations and shorter orbital periods: for example, hours or days rather than years. This is simply an observational selection effect because longer-period, more widely separated binaries are more likely to be seen as visual binaries (Sect. 3.4.1). But smaller orbits also tend to dynamically circularize from tidal effects, which can simplify the analysis. For example, [31] suggest that spectroscopic binaries with periods less than 30 days have eccentricities of zero.

If only a single-line spectroscopic binary is observed, finding the masses is more challenging because Fig. 3.8 would be modified to show only a single set of curves with, say, the blue secondary missing. Such would be the case if a star were perturbed by any companion whose emission was too faint to be observed, such as a very faint star or a planet. It would also be the case if the companion were a black hole without observable emission. For a single-line system, only the *mass function, MF*, can be obtained,

$$MF = \frac{m_2^3 \sin^3 i}{(m_1 + m_2)^2} \qquad (3.17)$$

where m_1 is the observed star and m_2 is the unseen companion. Such a function is of limited use on its own but can provide useful information on masses in a larger statistical sample.

Spectroscopic binaries are useful tools for calibrating the masses of stars, as given in Appendix F. Very high radial velocity precisions have been achieved: for example, internal uncertainties of order ± 1 m/s [158]. Moreover, Doppler shifts are *independent of distance*, so spectroscopic binaries do not suffer from sometimes problematic distance determinations. This approach is also used as a way of detecting planetary systems around other stars. The first exoplanet found around a normal star like our Sun used the radial velocity method [214].

In summary, double-line spectroscopic binaries give us: $\Rightarrow m_1 \sin^3 i$ and $m_2 \sin^3 i$. Single-line spectroscopic binaries give us: $\Rightarrow MF$ (Eq. 3.17).

3.5 RADIUS

Assuming that we now have knowledge of well-calibrated stellar luminosities (Sect. 3.2) and effective temperatures (Sect. 3.1), the simplest way to obtain a stellar radius is to use Eq. 2.56, which relates these quantities according to $L \propto R^2 T_{eff}^4$. The accuracy of the result depends on the accuracy of the calibration of T_{eff} and L, since $R \propto L^{0.5} T_{eff}^{-2}$. Moreover, uncertainties in distance also factor into the determination of L, and sometimes distances are not known to the accuracy that one would like. For example, the red giant star Betelgeuse in the constellation of Orion is too bright for a parallax distance measurement using the Gaia satellite (Box 3.1 on page 60), but it does have a measurement of parallax from an earlier satellite called Hipparcos. With a measurement of $d = 197 \pm 45$ pc, a luminosity of $log\,(L/L_{\odot}) = 5.10 \pm 0.22$ [137] and an effective temperature of $T_{eff} = 3600 \pm 25$ K [193], the resulting (impressively large) radius is $R = 914\,R_{\odot}$, but with a hefty uncertainty of about 26%. It would clearly be preferable to have an independent and more direct method of determining R, which could then feed back into existing calibration schemes and improve upon them.

Aside from our Sun, stars are mostly unresolved objects, so direct measurements of stellar angular sizes are not possible. In a few cases, though, if the stars are very large and/or nearby, their angular sizes can be measured via optical or infrared interferometry,[4] and such measurements have indeed been made where possible [195, 301]. A good example is our friend, Betelgeuse, whose angular diameter has been measured to high precision, resulting in $\theta = 42.36 \pm 0.05$ milliarcseconds [223], an uncertainty of only 0.1%! The distance uncertainty dominates by far, being about 18% in an updated distance of $d = 222$ pc [136]. The resulting updated radius, from Eq. 3.14, is $R = 1010\,R_{\odot}$ with the same percentage uncertainty. Betelgeuse is a giant indeed.

Interferometric observations push the limits of angular measurements that are possible today. The requirement of needing close and/or giant stars so that the angular size of the star is big enough to be measurable limits the number of stars that are accessible. As of 2016, more than 400 stellar angular sizes have been measured directly using this technique out to a distance of about 2 kpc [231]. Online Resource 3.9 provides a more comprehensive list [85].

Clever techniques continue to be used. For example *lunar occultations* provide another direct measurement of radius. When the Moon passes in front of a star, the starlight blinks out, and the time it takes to disappear is directly related to its size. Since we know the position and velocity of the Moon very accurately, a careful

---

[4] Interferometry is a technique that uses the superimposition of light in an interference pattern from mirrors that have a wider separation than typical telescopic single mirrors. The spatial resolution of a signal from a single mirror is approximately $\theta = \lambda/D$, where θ is the resolution in radians, λ is the wavelength of the observation and D is the diameter of the mirror. However, by using two mirrors separated by a larger distance, d, $\theta = \lambda/d$, and the spatial resolution is much improved. Variants of the approach also exist: for example using multiple mirrors or segmented mirrors. See also Sect. 4.2.5 of [154].

measurement of the time it takes to blink out yields the star's radius. The 'blink-out' time is very short, less than a millisecond for a star at 10 pc, requiring very fast cameras. And short exposures mean only bright stars register on the equipment. With the additional restriction that the Moon only occults a small fraction of stars in its orbital path around the Earth, fewer than 200 stars have so far been measured this way. Nevertheless, typical measurements of a star's angular size in this fashion have an accuracy of ∼ 1% [264].

Is there another way? The answer is yes, and the approach continues to exploit the fact that stars are often in binary systems. Just as we saw that binaries can lead to accurate masses, now we will see how binaries can lead to radii, provided these systems are *eclipsing binaries*.

3.5.1 Eclipsing Binaries

An eclipsing binary is a system in which one star goes in front of the other, causing an eclipse, so that the light is diminished for a time. Over the course of an orbital period, each star is eclipsed by the other, once. In order to see an eclipse, the orbit must be highly inclined from the sky plane: that is, the orbital plane should be roughly along the line of sight ($i \approx 90°$). If the orbital plane is tilted from that angle by a small amount, it may still be possible to see an eclipse, but only if the orbits have small semi-major axes. This is because widely separated stars whose orbits are inclined by $i < 90°$ will not pass in front of each other along the observer's line of sight. Statistically, then, eclipsing binaries tend to be close binaries with short periods. Of the ∼ 2,900 eclipsing binaries detected by NASA's Kepler space telescope, the majority have periods of less than a day [167]. Including other observations from a variety of telescopes, more than several hundred thousand eclipsing binaries are known [328].

Close binaries are *unresolved*. We do not see each star individually, but we can observe dips in the *light curve*, which is a plot of the collective light as a function of time. As we saw for spectroscopic binaries (Sect. 3.4.2), short-period systems tend to circularize from dynamical effects, and this can simplify the analysis. On the other hand, if the stars are close enough to each other, mass transfer can occur between the two stars, or gravitational tidal effects can distort one or both into ellipsoidal shapes, as we saw earlier for the star Regulus (Box I.1 on page xxi). Sophisticated modelling is required to take such effects into account.

An example of an eclipsing binary that is perfectly edge-on (inclination, $i = 90°$) with a circular orbit is shown in Fig. 3.9. In this example, the larger star (the primary, red in colour) is the cooler one, presumably having evolved off of the main sequence, and the smaller star (the secondary, yellow in colour) is the hotter one. The orbit is shown at the *Top*. On the *Left*, the hotter star goes in front of the larger, cooler star, causing a dip in the light, shown at the bottom. On the *Right*, the hotter star then continues in its orbit, eventually passing behind the cooler primary, again causing a dip in the light (*Bottom*). The dip on the right is the deeper one because the smaller

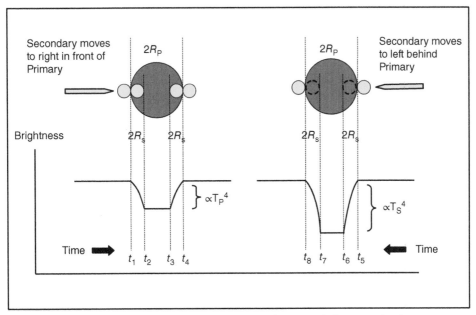

Figure 3.9 Diagram showing the light curve (**Bottom**) that results from an eclipsing binary system (**Top**) in which a small hot star (yellow, secondary) and a giant cool star (red, primary) are orbiting their common center of mass. In this case, the orbit is circular and in the line of sight. Radii R and times t are labelled. The depths of the eclipses depend on the temperatures T.

star, whose light has been blocked, is the hotter star. We can see that the stellar *area* that is blocked is the same for both eclipse minima, so from Eq. 2.56, the ratio of the eclipse depths, ratio$_{ecl}$, depends on the relative temperatures of the stars,

$$\text{ratio}_{ecl} = \left(\frac{T_P}{T_S}\right)^4 \tag{3.18}$$

From this, the ratio of stellar temperatures can be found.

From Fig. 3.9, we can also see that the radius of the secondary, R_S, and primary, R_P, are related to the times of various contacts and the *relative* velocity, v, between the two stars,

$$2\,R_S = v\,(t_2 - t_1) = v\,(t_4 - t_3) \tag{3.19}$$

$$2\,R_S + 2\,R_P = v\,(t_4 - t_1) \tag{3.20}$$

Similar equations could be written for the second eclipse. We also know that the semi-major axis of the relative orbit, a, is related to the relative velocity via,

$$\frac{2\,\pi\,a}{P} = v \tag{3.21}$$

for a circular orbit (i.e. a is just the radius of the circle), where P is the period. Combining Eqs. 3.19, 3.20 and 3.21 leads to

$$\frac{R_S}{a} = \frac{\pi (t_2 - t_1)}{P} \tag{3.22}$$

$$\frac{R_P}{a} = \frac{\pi (t_4 - t_2)}{P} \tag{3.23}$$

Measurement of the orbital period and other times as shown in the figure is straightforward, and those times are all that are needed to find the radii of the two stars in terms of the orbital radius, a.

How can we obtain stellar temperature and radii in absolute terms? For this, we need to add in information from spectra. For example, if the eclipsing binaries are also spectroscopic binaries, then Eqs. 3.15, 3.16 and 3.6 give us a, and the stellar radii can then be found explicitly *without the need to know the distance to the system*. Uncertainties in the radius of less than 1% have been achieved this way [213], a considerable improvement over the results given for Betelgeuse at the beginning of this section. And if the temperature of at least one star can also be found from the spectrum (Sect. 3.1), then the temperatures of both stars can be found from Eq. 3.18. *Eclipsing spectroscopic binaries* are like a 'gold standard' for deriving stellar parameters and setting the calibrations for stars for which some of these parameters are not known. A sample catalogue can be found in [328].

In summary, eclipsing binaries give us: ⇒ R_S/a and R_P/a, or their ratio, R_S/R_P. When combined with spectroscopy (Sect. 3.4.2): ⇒ R_S and R_P.

Box 3.3

Starring ... P Cygni

Distance: $d = 1.56 \pm 0.25$ kpc [266]; Variable Type: LBV; V = 4.8 [263]; Effective Temperature: $T_{eff} = 18,700$ K, Mass: M = 37 $M_\odot$ [266].

P Cygni (alternate name 34 Cygni) is in the constellation of Cygnus, the Swan. It is in a class of stars called *luminous blue variables* (LBVs). These are massive post-main-sequence stars with strong stellar winds, although the causes and evolutionary status of these stars are still debated [91]. The first documented account of the brightening of P Cygni was made in the year 1600 by a student of the Danish astronomer Tycho Brahe.

P Cygni's variability is stochastic (i.e. brightening at unpredictable times), rather than periodic. It is currently in a relatively stable period with only minor magnitude variations. This star is the prototype of a class whose spectrum clearly reveals the presence of outflowing winds. A 'P Cygni' profile has a classic shape, as shown in Fig. 3.11. P Cygni is near the limit of naked-eye visibility. Try finding this star on a clear dark night.

Credit: Adapted from theskylive.com.

3.6 ROTATION AND WINDS

When it comes to *motions*, such as rotation or winds, we can rely on Doppler shifts (Sect. 2.1.9) to help disentangle the kinematics.

Figure 3.10 *Left* shows a diagram for a simple case in which a star is at rest with respect to the observer and is rotating around a polar axis that is in the plane of the sky (inclination $i = 0$). A spectral line produced at the right side of the star in the figure will be *blueshifted* with respect to the line center because the light is approaching with respect to the observer. A spectral line produced on the left side will be *redshifted* because the light is receding with respect to the observer. Light from the star that is transverse (black arrows at the surface) will show no Doppler shift and therefore contributes to the center of the line. Because the star is unresolved, any measurement detects light from all parts of the star's surface that is facing the observer, so redshifted, blueshifted and unshifted light are collected together. The net result is a *broadening* of the spectral line.

Examples of rotationally broadened lines are seen on the *Right* of the figure. Faster rotation results in broader, shallower lines. If the star has some systemic velocity as it moves away or towards the observer, the lines are broadened in the same way, but the line center is shifted to the systemic velocity of the star. A measurement of half of the full line width, converted to velocity via the Doppler formula (Eq. 2.45), yields the rotational velocity, v_{rot}. If the radius of the star is known (Sect. 3.5), then the period can also be found from $P = 2 \pi R / v_{rot}$.

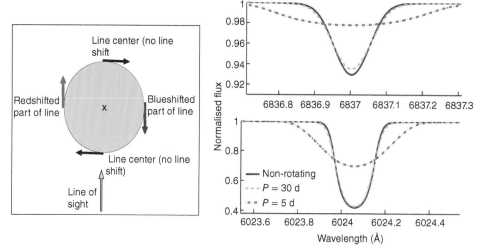

Figure 3.10 **Left**: Top view of a star at rest with respect to the observer. The star's rotation is around a polar axis that is vertical to the plane of the paper going through the x. Arrows represent the star's velocity at the surface. On the right, any spectral line will be blueshifted; and on the left, spectral lines will be redshifted. Lines coming from positions in which the motion is transverse (black arrows at the surface) will not be shifted. **Right**: Two lines centered at 6837 Å (**Top**) and 6024 Å (**Bottom**) from a synthetic spectrum, showing how they widen depending on stellar rotation, from [289] / With permission of Cambridge University Press. As rotation speed increases (period, P, decreases), the lines widen. Credit: Adam Stevens.

More problematic is the fact that the star's polar axis is unlikely to be in the plane of the sky but will be tilted with some inclination $i > 0$. Just as we saw for spectroscopic binaries (Sect. 3.4.2), we do not then measure v_{rot} but rather $v_{rot} \sin i$. For any given star, this inclination is unknown. However, for a statistically large sample in which stellar inclinations are randomly oriented, one expects an average[5] of $< \sin i > = \pi/4$. Therefore, in a sample of stars in which an average $< v_{rot} \sin i >$ is measured, the average rotational velocity, $< v_{rot} >$, can be estimated (e.g. [228]) from the correction

$$< v_{rot} > = \frac{4}{\pi} < v_{rot} \sin i > \tag{3.24}$$

Is there any other way to obtain a star's rotational period? Surprisingly, yes, thanks to impressive improvements in photometric sensitivity. Just as Christoph Scheiner determined the rotation of the Sun by observing sunspots as they shifted across the solar surface (Fig. 1.10), we can do something similar today. Stars have *starspots*, and although we cannot see the spots as they travel across the face of the unresolved star, small changes in brightness result as the spots move across. The Kepler space

---

[5] Medians are sometimes preferred instead of averages.

telescope that has been so useful in obtaining eclipsing binary data (Sect. 3.5), for example, has also been scoured for small changes in light and the timing of that change due to starspot motion. More than 30,000 rotation periods have been determined by this method [216]. The sophistication of such techniques is impressive. Even starspot cycles (analogous to sunspot cycles, Sect. 1.4) are being studied to understand magnetic field variations on distant stars [229].

It is also possible to tell if winds are present, and for this, we return to Doppler shifts.

When there are winds, there is typically *circumstellar material* present that is expanding away from the star. The circumstellar region is usually much larger than the star (see Fig. 3.11). Lines produced in regions where there is no background light are seen in emission (large red arrows denoting the line of sight in the figure). Lines that come from a region immediately in front of the star are seen in absorption (large blue arrow) because the background continuum from the star is hotter

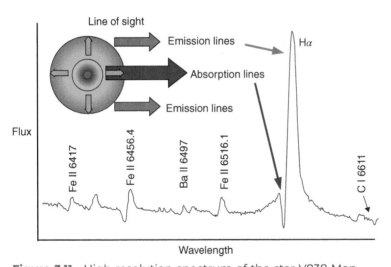

Figure 3.11 High-resolution spectrum of the star V838 Mon, with P Cygni profiles evident on almost every line, some of which are labelled. The *Top* diagram at left shows the geometry. The star is the red dot at the center, the green arrows indicate the direction of the wind outflow and the observer is towards the right. The larger, thick arrows indicate the line of sight from the various regions in which lines are produced. Red shows where emission lines are seen, and blue shows where absorption lines are seen. The absorption lines are blueshifted because of a higher velocity towards the observer. One absorption line and one emission line are pointed out in the spectrum. Credit: Adapted from Paolo Valisa, Società Astronomica G.V. Schiaparelli Centro Popolare Divulgativo Di Scienze Naturali, Varese, Italy.

than the cooler expanding material. The result is a characteristic line profile in which each spectral line has an absorption component that is *blueshifted* next to a stronger emission component (see the Hα line in the figure). The blueshift is because the wind is advancing towards the observer in comparison to the emission line region. This line shape is called a *P Cygni* profile, named after the star in which the feature was originally identified (see Box 3.3 on page 74). The separation between the emission and absorption feature gives the wind velocity.

Online Resources

3.1 *The Gaia mission of the European Space Agency*:
     https://www.esa.int/Science_Exploration/Space_Science/Gaia
3.2 *Gaia technical performance*:
     https://www.cosmos.esa.int/web/gaia/science-performance
3.3 *Gaia archive search*:
     https://gea.esac.esa.int/archive
3.4 *The POLLUX database of stellar spectra*:
     https://npollux.lupm.univ-montp2.fr
3.5 *The VizieR online database of the Centre de Données astronomiques de Strasbourg*:
     https://vizier.u-strasbg.fr/viz-bin/VizieR
3.6 *Bruzual-Persson-Gunn-Stryker atlas of stellar spectra*:
     https://www.stsci.edu/hst/instrumentation/reference-data-for-calibration-and-tools/astronomical-catalogs/bruzual-persson-gunn-stryker-atlas-list
3.7 *Wandering of the solar system barycenter*:
     https://www.youtube.com/watch?v=oauf6W3Uz04
3.8 *The Washington Double Star Catalog*:
     http://www.astro.gsu.edu/wds
3.9 *The JMMC (Jean-Marie Mariotti Center) Stellar Diameter Catalogue*:
     https://www.jmmc.fr/english/tools/data-bases
3.10 *Stellar classification table with data*:
     http://www.isthe.com/chongo/tech/astro/HR-temp-mass-table-byhrclass.html

PROBLEMS

3.1. Beginning with Eq. 3.12 in cgs units, convert the units to find Eq. 3.13.
3.2. Suppose that the velocity curves of a double-line spectroscopic binary are sinusoidal and the inclination of the orbital plane is $i = 90°$. Star A and star B have measured amplitudes of $v_A = 15$ km s^{-1} and $v_B = 45$ km s^{-1}, respectively. The period is $P = 2.5$ years.

   (a) What is the orbital eccentricity, e?
   (b) Find the ratio of stellar masses, M_A/M_B. Which star is more massive?

(c) Find the relative semi-major axis, a (AU).

(d) Find the mass of each star ($M_\odot$).

(e) Suppose the *same binary system* were inclined at an angle $i \neq 90°$. Which of the quantities v_A, v_B, P, e, M_A and M_B would change and how (i.e. increase, decrease or stay the same)?

3.3. (a) Write the Stefan-Boltzmann equation (Eq. 2.56) as $R = R(L, T)$ and, making use of the general expression for derivatives (Eq. 2.7), show that

$$\frac{dR}{R} = \frac{1}{2}\frac{dL}{L} - 2\frac{dT}{T} \qquad (3.25)$$

(b) Assuming that L and T are independent, let us estimate the fractional error in radius to be

$$\frac{dR}{R} = \sqrt{\left(\frac{1}{2}\frac{dL}{L}\right)^2 + \left(2\frac{dT}{T}\right)^2} \qquad (3.26)$$

In the second data release of the Gaia satellite (Box 3.1 on page 60), typical errors in temperature and luminosity are, respectively, $dT = \pm 324$ K and $dL = \pm 15\%$ [9]. If a star is measured by Gaia to have $L = 6.31\ L_\odot$ and $T = 7178$ K, find the radius, R, and its uncertainty.

(c) How does your result compare with the 1% uncertainty that has been achieved for some eclipsing spectroscopic binaries (Sect. 3.5)?

3.4. A star is observed with the Gaia satellite. Its measured parameters are parallax $\Pi = 70.737$ milliarcsec; apparent magnitude in G-band $G = 5.3276$; dust-corrected colour index BP − RP = 0.8171 (see Fig. 3.3), bolometric correction, BC $= M_{bol} - M_G = -0.05$.

(a) Determine (or estimate) the following quantities in the units indicated. [HINT: The information in Box I.2 on page xxiii will be useful.]

(i) The distance, d (pc)

(ii) The absolute G magnitude, M_G

(iii) The luminosity ($L_\odot$)

(iv) The effective temperature (K).

(v) The radius ($R_\odot$)

(b) What are the spectral type and luminosity class of this star (see Appendix F)?

3.5. From the geometry of Fig. 3.9 and Eq. 2.56, derive Eq. 3.18.

3.6. Visit Online Resource 3.5, search for *stellar spectra* catalogues, and then go to the *MUSE*[6] *library of stellar spectra*. Use this resource to plot the spectra of two stars: HD209290 and HD306799. Identify the strongest/deepest line between 5500 and 7000 Å. Zoom in on this line, and extract the results for both stars over the same wavelength limits. Which star is the main sequence star, and which is the giant? Explain. Identify the lines you have just plotted (try an internet search).

---

[6] Multi-Unit Spectroscopic Explorer.

Chapter 4
The Shining Star – Interiors

Pumbaa: Timon, ever wonder what those sparkly dots are up there? ...
Timon: They're fireflies. Fireflies that, uh ... got stuck up on that big bluish-black thing.
Pumbaa: Oh, gee. I always thought they were balls of gas burning billions of miles away.

The Lion King [6]

We take the **stellar interior** to be the region of a star from the edge of the core to the surface. Recall that, for the Sun, the core constitutes the inner 25% of the radius, as drawn in Fig. 1.9. The core and its nuclear reactions will be discussed separately in Chapter 5. The gas surrounding the core that takes up the remaining 75% of the solar radius is essentially an inert envelope, like an atmosphere, without any energy generation. How, then, does the generated energy get out to the surface? In other words, how does the star shine? That is the focus of this chapter.

Astrophysics: Decoding the Stars, First Edition. Judith Irwin.
© 2023 John Wiley & Sons Ltd. Published 2023 by John Wiley & Sons Ltd.

4.1 ENERGY TRANSPORT IN STARS

There are three ways in which energy can be transported from the deep interior where energy is generated to the surface to be radiated away: i) *radiation*, ii) *(electrical) conduction*, and iii) *convection*.[1] It is possible that more than one process is occurring within the same star at different locations. For example, the Sun transports energy largely through radiation, but its outer layers are convective. Conduction is generally not important in main sequence stars. It is most important in stellar remnants such as *white dwarfs* and also in the very dense cores of stars that have evolved off of the main sequence. Most of this chapter is dedicated to radiative and convective transport. However, we also briefly discuss conduction in Sect. 4.4.

It is worth remembering that by the second law of thermodynamics, energy is transported from hotter to cooler regions. Of course, this means outwards energy transport in general. However, there are some conditions in which temperature inversions do occur. For example, in the cores of stars in late stages of stellar evolution, neutrino losses can cool the core from the center outwards, causing a temperature *inversion*. In such a case, energy would flow inwards from the region around the core towards the center (e.g. [113, 325], Sect. 8.2.1)! In this section, however, we deal with the more standard situation of outwards energy transport.

4.2 RADIATIVE TRANSPORT

Radiative transport is arguably the most important transport mechanism in stars. It is a *diffusive* mechanism: that is, photons diffuse outwards from where they are produced in the core. This is a direct consequence of the fact that the mean free path of a photon in a stellar interior is much smaller than the stellar radius, i.e. $\bar{l} << R_\star$. Rather than having a photon travel freely from the core to the surface, it can be thought of as taking a *random walk* to the surface. The photon will have high energy when it leaves the core, but interactions with intervening particles will cause it to give up some of its energy to those particles en route. The photon that emerges from the surface (not the same one!) will have an energy typical of the surface temperature of the star.

Since photons are trapped, or rather 'leaking through' the star, any stellar interior will be in local thermodynamic equilibrium (LTE, see Box 1.2 on page 12). This means there are many interactions between photons and matter, allowing sufficient time for an equilibrium to be established between radiation and matter at any location. The result is that the *radiation temperature* and the *gas kinetic temperature* are the same 'locally' and the equations summarized for black body radiation (Sect. 2.2) are applicable.

What does 'local' actually mean? Generally, this can be determined via physical insight. For example, if $\bar{l}$ is small for a photon, then the photon will interact with a

---

[1] A fourth well-known energy transport mechanism is thermal conduction, such as occurs in solids, but this mechanism is negligible in stars [251].

particle in a region whose physical properties are about the same as the properties of the region in which the photon was emitted. The Sun is again a good example. As we saw in Box 1.2 on page 12, over a typical mean free path, the temperature change is only about 10^{-4} K. This difference in temperature is so small that thermodynamic equilibrium is an adequate description of the radiation/matter mix at any interior location. On the other hand, this small temperature difference is also what *drives* the outwards radiative flux, a point to which we will return in Sect. 4.2.3.

4.2.1 The Rosseland Mean Opacity

As outlined in Sect. 2.3, there can be many contributors to stellar opacity, so it can be a challenge to calculate accurate opacities for all frequencies. Ideally, these opacities (at least, those that are dominant) should be summed frequency by frequency (see Eq. 2.59) prior to any frequency-averaging. This is desirable in stellar atmospheres where LTE conditions break down. However, such a detailed approach is not necessary (or easy) throughout the stellar interior because the *structure* of a star depends only on the *bolometric* (i.e. integrated over all frequencies) flux, F (erg cm^{-2} s^{-1}), or luminosity, L (erg s^{-1}). The specifics of F_ν (erg cm^{-2} s^{-1} Hz^{-1}) or L_ν (erg s^{-1} Hz^{-1}) can then be taken into account in a simpler fashion.

For this, we define the *Rosseland mean opacity*, κ_R (cm^2 g^{-1}), after its originator, Svein Rosseland. The Rosseland mean opacity is the *flux-weighted mean opacity*. By 'mean', we mean a weighted average over *frequency*, i.e.

$$\kappa_R \equiv \frac{\int_0^\infty \kappa_\nu F_\nu \, d\nu}{\int_0^\infty F_\nu \, d\nu} \tag{4.1}$$

The denominator is just the bolometric flux. Eq. 4.1 would have to be evaluated at every radius within a star.

Let us approach the Rosseland mean opacity in another way, by beginning with the radiation pressure.

From Eqs. 2.57 and 2.58, the bolometric and frequency-specific radiation pressures in LTE are, respectively,

$$P_{rad} = \frac{4\pi}{3c} B(T) \tag{4.2}$$

$$P_{rad_\nu} = \frac{4\pi}{3c} B_\nu(T) \tag{4.3}$$

where the first equation has cgs units of dyn cm^{-2} and the second is in dyn cm^{-2} Hz^{-1}.

There is a small change in flux (and intensity) with radius and hence a small change in pressure as well. From basic radiation theory,

$$dP_{rad_\nu} = dF_\nu/c \tag{4.4}$$

for *outwards-directed* flux density and pressure (e.g. Sect. 3.5 of [154]). Also, the equation of radiative transfer for an absorbing, but not emitting, cloud is

$$dF_\nu = -\kappa_\nu \rho F_\nu \, dr \tag{4.5}$$

(e.g. Sect. 8.3.2 from the same reference), where dr is a small change in radius. Then, from Eqs. 4.4 and 4.5,

$$dP_{rad_\nu} = -\frac{\kappa_\nu \rho}{c} F_\nu \, dr \qquad (4.6)$$

The unitless quantity $-\kappa_\nu \rho \, dr$ is just $d\tau_\nu$. The negative sign enters because P_{rad_ν} and F_ν decrease with increasing radius r. Notice that we do *not* need to consider an *emission term* as long as we look outside of the core and therefore outside of the region in which energy is generated (i.e. there is no need for an emitting or 'source term').

Let us rearrange Eq. 4.6 and equate it to the differentiation of Eq. 4.3 with respect to r, followed by a change of variable:

$$\frac{dP_{rad_\nu}}{dr} = -\frac{\kappa_\nu \rho}{c} F_\nu = \frac{4\pi}{3c} \frac{d[B_\nu(T)]}{dr} = \frac{4\pi}{3c}\left(\frac{d[B_\nu(T)]}{dT}\right)\left(\frac{dT}{dr}\right) \qquad (4.7)$$

From Eq. 4.7, we can write

$$\kappa_\nu F_\nu = -\frac{4\pi}{3\rho}\left(\frac{d[B_\nu(T)]}{dT}\right)\left(\frac{dT}{dr}\right) \qquad (4.8)$$

$$F_\nu = -\frac{4\pi}{3\rho}\left(\frac{1}{\kappa_\nu}\frac{d[B_\nu(T)]}{dT}\right)\left(\frac{dT}{dr}\right) \qquad (4.9)$$

Notice that the left-hand sides of Eqs. 4.8 and 4.9 appear in the numerator and denominator, respectively, of the definition of the Rosseland mean opacity (Eq. 4.1). We can therefore integrate Eqs. 4.8 and 4.9 over frequency to obtain another representation of the Rosseland mean opacity. Note that the quantity dT/dr is not frequency-dependent and so does not need to enter into the integral. Substituting those integrals into the numerator and denominator, respectively, of Eq. 4.1 and *inverting* gives us our second representation of the Rosseland mean opacity:

$$\frac{1}{\kappa_R} = \frac{\int_0^\infty \frac{1}{\kappa_\nu}\frac{d[B_\nu(T)]}{dT}\,d\nu}{\int_0^\infty \frac{d[B_\nu(T)]}{dT}\,d\nu} = \frac{\int_0^\infty \frac{1}{\kappa_\nu}\frac{d[B_\nu(T)]}{dT}\,d\nu}{\frac{dB(T)}{dT}} \qquad (4.10)$$

Equation 4.10 is the inverse of Eq. 4.1. The resulting κ_R is still a flux-weighted mean but is now weighted in terms of the temperature gradient of the Planck curve $(d[B_\nu(T)]/dT)$. This may seem somewhat 'convoluted' except for two useful results.

The first is that this definition of κ_R can be used together with Eq. 4.9 (and some minor substitutions) to produce the equation of radiative transport (Eq. 4.21 and Prob. 4.3), which is one of the fundamental equations of stellar structure, to be discussed in Sect. 4.2.3.

The second is that the Rosseland mean opacity is usefully viewed as a *harmonic mean* quantity. The largest contributors to κ_R come from frequency ranges where the flux density is *highest* (Eq. 4.1); but this is precisely the region over which κ_ν is *smallest*, i.e. more flux gets through when the opacity is small. Thus, the Rosseland mean gives *highest weight* to the *lowest opacities*. Equation 4.10, which reveals the harmonic mean property of κ_R, is perhaps easier to understand as a 'transparency' $(1/\kappa_R)$ rather than an opacity. If the temperature gradient of the Planck curve is highest, which is close to the frequency at which the Planck curve itself is highest (Prob 4.1), then the 'transparency' is weighted higher. The situation is entirely analogous to electrical resistance in parallel circuits. When there are more routes for current to take, the effective resistance is low. Rosseland mean opacities are complicated to calculate and are typically not calculated at the time when a stellar model is computed. Instead, they are calculated in advance, as a function of density and temperature, and stored in tables that can be accessed when a model is 'built'. Examples are provided in the Online Resources at the end of this chapter.

4.2.2 Analytical Forms for the Mean Opacities

It is not a simple task to evaluate κ_R, given the many potential contributions to opacities as described in Sect. 2.3 and the variations that can occur for different densities and temperatures. However, approximate analytical forms have been found for some frequency-averaged quantities.

The first are the bound-free opacity, $\overline{\kappa_{bf}}$, and free-free opacity, $\overline{\kappa_{ff}}$, whose dependences on density and temperature are such that $\kappa \propto \rho/T^{3.5}$. Such a relation is referred to as a *Kramers law* (e.g. [201]). In units of cm^2 g^{-1},

$$\boxed{\overline{\kappa_{bf}} = 4.34 \times 10^{25} \left(\frac{\overline{g_{bf}}}{t} \right) Z(1+X) \frac{\rho}{T^{3.5}}} \tag{4.11}$$

$$\boxed{\overline{\kappa_{ff}} = 3.68 \times 10^{22} \overline{g_{ff}} (X+Y)(1+X) \frac{\rho}{T^{3.5}}} \tag{4.12}$$

where X, Y and Z are the usual mass fractions of hydrogen, helium and metals, respectively. The quantities $\overline{g_{bf}}$ and $\overline{g_{ff}}$ are the frequency-averaged bound-free and free-free *gaunt factors*, respectively. These are quantum mechanical correction factors and are typically $\mathcal{O}(1)$. The quantity t is ominously referred to as the *guillotine factor* and is a factor that takes into account a cutoff in the contribution of an atom to the opacity once it has been fully ionized. The guillotine factor varies between 1 and 100.

Both ρ and T increase with depth in stars, but the strong dependence on T shows that Kramers law bound-free and free-free opacities decrease towards the interiors of stars. The appearance of Z in Eq. 4.11 reveals the importance of metals for the bound-free opacity, especially in the deep interior of stars where hydrogen and helium have already been stripped of their electrons. The free-free opacity, on the other hand, is dominated by free electrons from hydrogen and helium because these are the most abundant elements. The magnitudes of these two opacities tend to be similar.

We now consider the electron scattering opacity, κ_{es}, which for Thomson scattering (i.e. when photon energies are less than the electron rest mass energy) is frequency-independent and therefore easier to compute. For complete ionization, the expression is simple,

$$\kappa_{es} = \frac{\sigma_{es}\,n_e}{\rho} = 0.19\,(1 + X) \tag{4.13}$$

The first equivalence makes use of $\kappa\rho = \sigma n$ (cf. Box 1.1 on page 7). The second applies the expression for electron density (Eq. 2.62) and evaluates using the Thomson scattering cross-section. This term is small compared to the bound-free and free-free opacities, all else being equal. In high-temperature regions, however, electron scattering can dominate the other terms. This is because Kramers law opacities continue to decline with increasing temperature, whereas the electron scattering opacity remains constant with T. In fact, κ_{es} can be thought of as a lower limit to the total absorption coefficient in stellar interiors.

As outlined in Sect. 2.3, the opacity due to the H^- ion can be an extremely important contributor to the overall opacity and is, in fact, the *dominant* source of opacity in the solar atmosphere. The weakly bound extra electron can be ejected with a photon of energy > 0.754 eV (8,700 K). For this ion, there is an analytical form applicable to restrictive ranges of density, temperature and abundance. For the temperature range $3000\,\mathrm{K} \le T \le 6000\,\mathrm{K}$, density range $10^{-10}\,\mathrm{g\,cm^{-3}} \le \rho \le 10^{-5}\,\mathrm{g\,cm^{-3}}$ and abundance ranges $X \sim 0.7$ and $0.001 < Z < 0.03$,

$$\overline{\kappa_{H^-}} \approx 2.5 \times 10^{-31} \left(\frac{Z}{0.02}\right) \rho^{1/2}\,T^9 \tag{4.14}$$

Clearly, this relation is not a Kramers law and has an extremely steep positive dependence on temperature. The free electrons from which H^- is formed are supplied largely by metals. The calculation of $\overline{\kappa_{H^-}}$ is complex, requiring quantum mechanical considerations.

As for bound-bound opacities, as can be imagined, analytical forms are not generally possible because of the quantum nature of the many transitions of interest, so the calculation must be done numerically. Bound-bound transitions only occur when photons have frequencies corresponding to the energy difference between bound states. Bound-bound opacities are generally not as important as the bound-free, free-free and electron scattering opacities described earlier in stellar interiors. However, especially in the atmospheres of cool stars where there are many molecules, and therefore many more possibilities for bound-bound transitions, this opacity can become very important.

Figure 4.1 illustrates how the important opacities vary with temperature for a *fixed* density, taking into account the ionization conditions, and shows which terms dominate for different temperatures. Here, the *Kramers law* regime includes both *bf* and *ff* opacities.

In Fig. 4.2, we show a plot of the Rosseland mean opacity taken from extensive numerical calculations, including detailed quantum effects. Here, the Rosseland mean is plotted against temperature for a variety of densities. At any temperature, scanning from the bottom to the top of the graph, it is clear that higher densities result

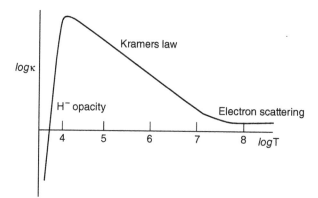

Figure 4.1 The opacity as a function of temperature, showing the various regimes in which each dominates. *No* variation of density has been taken into account. This is a logarithmic plot, so power laws are seen as straight lines.

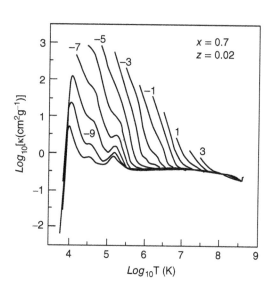

Figure 4.2 Rosseland mean opacity for a composition of $X = 0.70$, $Y = 0.28$ and $Z = 0.02$. The x-axis is the logarithm of the temperature, T, in K, and the y-axis is the logarithm of the opacity, κ, in $cm^2\ g^{-1}$. Each curve is labelled by the logarithmic value of the density, ρ, in $g\ cm^{-3}$. Credit: Adapted from [44].

in higher opacities, as would be expected. Scanning from left (low temperatures) to right (high temperatures) shows more complex behaviour for any given density. Between about 3,000 K ($log T = 3.5$) and 6,000 K ($log T = 3.8$), there is a steep rise from removing the extra electron from the H^- ion (Eq. 4.14). The curve continues to climb, though, as hydrogen and helium both become ionized, providing increasing numbers of free electrons from these two species. Once the temperature has reached $\sim 10,000$ K ($log T = 4$), hydrogen has become fully ionized (HII) and helium has become singly ionized (HeII). As the temperature continues to climb, bound-free and free-free opacities are important and decline following Kramers laws ($\propto T^{-3.5}$, Eqs. 4.11

and 4.12). By 40,000 K ($logT = 4.6$), helium becomes fully ionized (HeIII), and the excess electrons produce a small bump at this temperature. Around 10^5 K ($logT = 5$), ionization of iron and other metals produces another bump. Bumps in the curve tend to be associated with bound-free absorptions from specific elements and levels. Finally, at the highest temperatures, almost all atoms become ionized, and the dominant opacity is from electron scattering (Thomson scattering). Thomson scattering has no dependence on density or temperature (Eq. 4.13), so all curves converge to the same 'floor', as can be seen at the right of the graph.

Online opacity tables can be found at Online Resource 4.1 or 4.2. For more information, see [16].

4.2.3 The Equation of Radiative Transport

The *equation of radiative transport* indicates how the flux, F (erg cm^{-2} s^{-1}), or luminosity, L (erg s^{-1}), at any position within a star relates to the star's temperature gradient, dT/dr.

Consider a 'surface' within a star at a distance r from the center, as shown in Fig. 4.3. Photons are being emitted isotropically at points above and below the surface, but we initially restrict this development to the *radial direction*. Then only those photons that are within a distance $\bar{l}$ above and below the surface will pass through it.

The differential flux can be expressed as

$$dF \equiv c\,du \tag{4.15}$$

where du (erg cm^{-3}) is a differential energy density and c (cm s^{-1}) is the speed of light. To determine the net flux at r, restricting the limits to a mean free path above and below the surface in the radial direction, we can integrate Eq. 4.15 as

$$F = c\Delta u \tag{4.16}$$

where Δu is the difference in energy density between $r + \bar{l}$ and $r - \bar{l}$. This value is

$$\Delta u = u(r + \bar{l}) - u(r - \bar{l}) = \left[u(r - \bar{l}) + 2\bar{l}\frac{du}{dr} \right] - u(r - \bar{l}) = 2\bar{l}\frac{du}{dr} \tag{4.17}$$

Note that du/dr is the gradient in the energy density and is negative because u declines as r increases. Then the flux going through the surface at r from above and below (radial direction only) is

$$F = -2c\bar{l}\frac{du}{dr} \tag{4.18}$$

The negative sign enters because du/dr is negative and the flux must be positive outwards (the direction of increasing r). A more accurate result that takes into account the fact that photons are emitted *isotropically* (in all directions) gives

$$F = -\frac{1}{3}c\bar{l}\frac{du}{dr} \tag{4.19}$$

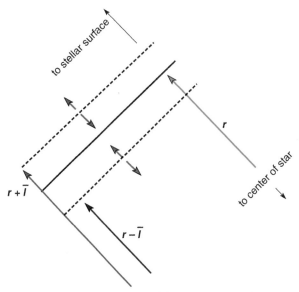

Figure 4.3 A 'surface' within a star at a distance r (green arrow) from the center is shown as a solid black line. Photons (red arrows) from the layers near this surface will reach the surface as long as they are not more than a mean free path, $\bar{l}$, away. These nearby surfaces are shown as dashed black lines with positional vectors $r + \bar{l}$ (blue) and $r - \bar{l}$ (dark blue) from the center.

For this, we require a consideration of geometry that we will not carry out here.[2]

Because we are considering values that have been integrated over frequency, we now use the Rosseland mean opacity in the equation for the mean free path (Eq. 1.24), i.e. $\bar{l} = 1/(\kappa_R \rho)$. We also change variables by writing $du/dr = (du/dT)(dT/dr)$, and from Eq. 2.57, we set $du = 4aT^3 dT$. Performing these actions on Eq. 4.19 gives

$$F = -\frac{4}{3}\left(\frac{c}{\kappa_R \rho}\right) aT^3 \frac{dT}{dr} \tag{4.20}$$

It is straightforward, by dimensional analysis, to verify that the units on either side are those of flux (erg s^{-1} cm^{-2}). This is the net flux passing through a surface at a distance r within the star.

Let us now rearrange Eq. 4.20 and express the temperature gradient in terms of the *luminosity* at some radius; it is customary to express the luminosity as L_r (see

---

[2] This development is similar to what was discussed in Sect. 4.2.1 for the Rosseland mean opacity, except that we looked at pressure in that case, instead of energy density. (The two differ by the factor 1/3; cf. Eqs. 2.57, 2.58.) Problem 4.3 reveals how these two developments compare.

comments in Sect. 6.1). Since $L_r = 4\pi r^2 F$ (Eq. 2.56), we find the *equation of radiative transport*,

$$\frac{dT}{dr} = -\left(\frac{3}{16\pi a c}\right)\left(\frac{\kappa_R \rho L_r}{r^2 T^3}\right) \qquad (4.21)$$

where κ_R is defined by Eq. 4.1 or 4.10.

Equation 4.21 is a description as to how radiation is transported (diffuses) outwards through a star. It is not a simple equation because it depends on parameters that also depend on r: for example, temperature and density, as well as chemical composition (via κ_R).

Recall that the temperature gradient *drives* the outwards radiative flux. From Eq. 4.21, if the temperature at a position is *lower* (all else being equal), then the temperature gradient will be steeper. When the opacity is high, the temperature gradient will also be steeper. This has relevance for whether or not *convection* occurs which we will consider next.

4.3 CONVECTIVE TRANSPORT

Convection is similar to a boiling pot of water, with pockets of gas rising and pockets falling. A good example can be seen on the surface of the Sun and is called *granulation*, as shown in Fig. 4.4. The Sun's outer $\approx 30\%$ in radius is convective. Rising pockets of hot gas transfer their heat to surrounding material, and then the cooler (and denser) regions descend again, heating up. A net transfer of energy upwards occurs, with subsequent radiative release of energy. The brightness of the inner part of a granule can be $\approx 20\%$ brighter than the descending cooler material around it. Recall, however, that $L \propto F \propto T^4$ (Eq. 2.54), so the temperature difference may be only $\approx 4\%$.

4.3.1 Condition for Convective Instability

Consider a small pocket of gas a distance r from the stellar center; this pocket is at the local stellar ambient values of pressure, P, and density, ρ. We will use the subscript a to refer to the pocket of gas and the subscript ★ to refer to the ambient medium wherever that pocket might be. The subscript a stands for *adiabatic* (explained shortly).

Suppose that this pocket is perturbed slightly upwards to a new position, $r + dr$. Since stellar pressure decreases upwards, the pocket will expand until its pressure balances the ambient pressure again. However, expansion means that the density in the pocket decreases. If the density of the pocket, ρ_a, is then *less than* the density of the surrounding ambient gas at the new radius, $\rho_\star$, *buoyancy* will drive the pocket upwards even more, and we say that system is *unstable to convection*. However, if the pocket of gas is more dense than the new ambient density, it will descend again. This is illustrated in Fig. 4.5.

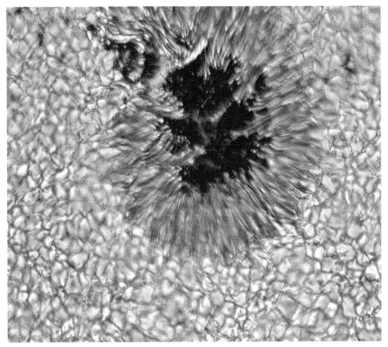

Figure 4.4 This sunspot is surrounded by a *granulation pattern* that illustrates convection in the outer layers of the Sun. Brighter regions are hotter (ascending gas), and darker regions are cooler (descending gas). A typical granule is about 1000 km across (the image resolution is 100 km). Credit: National Optical-Infrared Astronomy Research Laboratory, CC BY 4.0.

Since we are considering upward motion of the pocket, in the following, keep in mind that small changes $d\rho$, dT, dP are negative quantities, while dr is positive.

The pocket of gas initially began at the ambient density at location r, so the criterion for *convective instability* is (remembering that the density gradient is negative)

$$\left(\frac{d\rho}{dr}\right)_a < \left(\frac{d\rho}{dr}\right)_\star \qquad (4.22)$$

That is, if the density gradient of the pocket of gas is steeper (more negative) than the density gradient of the star, we have convective instability. We can now change variables,

$$\left(\frac{d\rho}{dP}\right)_a \left(\frac{dP}{dr}\right)_a < \left(\frac{d\rho}{dP}\right)_\star \left(\frac{dP}{dr}\right)_\star \qquad (4.23)$$

and note that the pressure gradient for the pocket as well as the star are the same, because pressure equalization is very fast. The pressure gradients are negative, so if

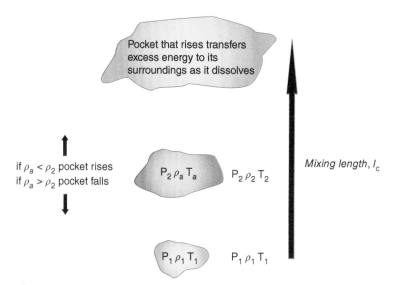

Figure 4.5 Diagram showing the process of convection. **Bottom**: A pocket of gas is initially at the same pressure, P_1, density, ρ_1, and temperature, T_1, as its surroundings. **Middle**: If the pocket is perturbed upwards, it has a new density, ρ_a, and temperature, T_a, which are different from its surroundings at ρ_2 and T_2, although its pressure equalizes to its surroundings as it rises. If the pocket's density is greater than its surroundings, it will fall down again, implying stability against convection. If the pocket's density is less than its surroundings, it will continue to rise by buoyancy, a situation of instability to convection. **Top**: In the case of instability, the pocket continues to rise until it dissipates, giving up its energy. The rise distance is called the *mixing length*.

we cancel the negative quantities, we have $d\rho/dP|_a > d\rho/dP|_*$. Inverting the ratios, we find *instability* if

$$\left(\frac{dP}{d\rho}\right)_a < \left(\frac{dP}{d\rho}\right)_*$$

(4.24)

We have emphasized that the expanding pocket is adiabatic. This is because, for such a motion, the *thermal timescale* (Sect. 7.2.2) is much greater than the *dynamical timescale* (Sect. 7.2.1), so no heat is exchanged between the pocket and its surroundings over a short rise time. However, a temperature change can occur because the expanding pocket is doing work against its surroundings (Sect. 2.1.8). Recall from thermodynamics that for an ideal gas in an adiabatic process (and assuming negligible radiation pressure), the relation between pressure and density follows Eq. 2.42, i.e.

$$P = const\, \rho^{\gamma_a}$$

(4.25)

where γ_a is the *adiabatic index* introduced in Sect. 2.1.7 and *const* incorporates the mass of the pocket, m. Differentiating Eq. 4.25 with a minor manipulation leads, for an adiabatic pocket of gas, to

$$\left(\frac{dP}{d\rho}\right)_a = \gamma_a \frac{P_a}{\rho_a} \approx \gamma_a \frac{P_\star}{\rho_\star} \tag{4.26}$$

For the approximation on the right-hand side of this equation, we have let $P_a/\rho_a \approx P_\star/\rho_\star$ since we are considering only small perturbations. Notice that $dP/d\rho$ in the pocket of gas differs from the gross slope in the ambient medium (over the rise length) only by a factor γ_a. We will now drop the subscripts that apply to the ambient medium and write simply P/ρ. Then from Eqs. 4.26 and 4.24 (and noting that even though $dP/d\rho$ is positive, both dP and $d\rho$ are negative),

$$\frac{dP}{P} < \gamma_a \frac{d\rho}{\rho} \tag{4.27}$$

We now have a criterion for instability that involves only the properties of the star, rather than a specific pocket of gas.

For an ideal gas, we know how dP/P and $d\rho/\rho$ are related to each other (Eq. 2.9), in which case our instability criterion Eq. 4.27 becomes, after eliminating ρ and dividing by dr,

$$\frac{dT}{dr} < \left(\frac{\gamma_a - 1}{\gamma_a}\right) \frac{T}{P} \frac{dP}{dr} \tag{4.28}$$

This instability criterion is called the *Schwarzschild criterion* after Karl Schwarzschild,[3] who originally looked at the stability question in the year 1906.

We now define the *adiabatic temperature gradient*, $dT/dr|_a$, to be

$$\boxed{\left.\frac{dT}{dr}\right|_a \equiv \left(\frac{\gamma_a - 1}{\gamma_a}\right) \frac{T}{P} \frac{dP}{dr}} \tag{4.29}$$

Since both the temperature and pressure gradients are negative, we will rewrite our final instability criterion

$$\left|\frac{dT}{dr}\right| > \left|\frac{dT}{dr}\right|_a \tag{4.30}$$

In other words, if the star's temperature gradient in the region is steeper than the adiabatic temperature gradient, the star is unstable to convection in the region. The region is said to be *superadiabatic*, and convection will proceed.

There will always be some perturbations in a star. After all, the star is rotating, and there is plenty of internal and surface activity related to magnetic fields. Suppose then that the instability criterion (Eq. 4.28) has been met. It will be met *as soon as* the adiabatic temperature gradient (Eq. 4.29) has been achieved. That is, the temperature gradient of a pocket of gas need only be marginally steeper than the adiabatic

---

[3] This is the same Schwarzschild after whom the *Schwarzschild radius* (Eq. 10.42) is named.

temperature gradient in order for convection to proceed. Our conclusion is that if convection is occurring at all, the temperature gradient of a pocket of gas is just *given* by Eq. 4.29.

For *stability* against convection,

$$\left|\frac{dT}{dr}\right| < \left|\frac{dT}{dr}\right|_a \tag{4.31}$$

The adiabatic temperature gradient is the maximum temperature gradient for stability above which convection will set in. Locations within stars with steep temperature gradients are therefore more likely to show convection. This is the case in the outer third of the Sun, as illustrated by the temperature curve in Appendix C (see Fig. C2 Inset).

Figure 4.6 shows an illustration of where convection is important in stars of different mass on the main sequence. Convection occurs near the surfaces of low-mass stars and occurs in the centers of stars more massive than the Sun. The switch from the surface to the core occurs at $M \approx 1.3 M_\odot$, which corresponds to an F-type star, the latter spanning the range from 1.1 to 1.6 $M_\odot$. Rather than having no convection,

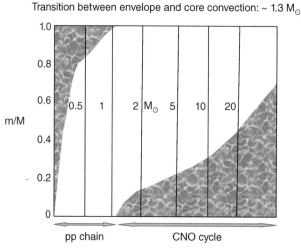

Transition between envelope and core convection: ~ 1.3 $M_\odot$

pp chain   CNO cycle

Figure 4.6 An indication of where convection (blue mottled regions) occurs in main sequence stars of different mass. The left vertical axis shows m(r)/M, where M is the total mass of the star (center of star at bottom and surface of star at top). The horizontal axis labels the type of energy generation within the star. The vertical lines designate the mass of the star in solar mass units. Above $M \approx 1.3 M_\odot$, convection shifts from the surface to the core. For the Sun, note that $r = 0.7 R_\odot$ (approximately where convection begins in the outer layers) corresponds to $m = 0.97 M_\odot$, according to the SSM. The bottom labelling shows approximately where nuclear energy generation is predominantly via the PP-chain or the CNO cycle (see Sec. 5.4). Adapted from [166].

such stars show narrow convection zones in *both* the surface layers and core regions [13].

Off the main sequence, red giant branch (RGB) and asymptotic giant branch (AGB) stars (Sect. 8.2) have very extended convective envelopes. Pre-main-sequence stars and main sequence stars less massive than about 0.3 $M_\odot$ are fully convective (Fig. 4.6); recall that lower temperatures favour a steeper temperature gradient (Eq. 4.21). Finally, convection is also present in the pre-supernova stages of both Type I and Type II supernovae as well as the collapse phase of Type II supernovae.

4.3.2 Mixing Length Theory

How can convection be handled mathematically? As can be imagined, convection is a difficult-to-model process. However, a reasonable approximation is called *mixing length theory*, originally proposed by Ludwig Prandtl in 1925 and developed for stars by Ludwig Biermann in the 1930s. It is not a complete theory in the classical sense, but it is a development that helps in understanding the physical processes involved in convection.

Consider that a pocket of gas is rising adiabatically as described in the previous subsection. The density and temperature of the pocket change as it rises. We assume that the pocket travels a distance l_c, called the *mixing length*, which is the radial distance a pocket can travel before it 'mixes' with its surroundings and releases its energy. One can think of the mixing length as a kind of 'mean free path' for a convective pocket. The density and temperature of the pocket change as it rises, so there are differences in density and temperature, $\delta\rho$ and δT, respectively, between the pocket and the ambient medium of the star around it. However, as we will justify shortly, the velocity of the pocket is much less than the sound speed. Since pressures should equalize on a sound speed timescale, pressure equilibrium is assumed ($\delta P = 0$) throughout the rise time.

The mixing length is generally unknown and must enter into models as a free parameter. However, it is taken to be proportional to the *pressure scale height*, h_P, of the star, possibly $l_c/h_P \approx 1 \rightarrow 2$ (see Box 4.1 on page 95). That is, the pocket likely releases its heat after travelling one to two pressure scale heights.

Box 4.1

The Pressure Scale Height

The pressure scale height, h_P, is defined by a distribution that declines exponentially with increasing radius, $P = P_0 \, exp(-r/h_P)$, where P_0 is the reference pressure at $r = 0$. Differentiating this expression with some minor manipulation

(continued)

(*continued*)

gives

$$\frac{dP}{dr} = -\frac{P}{h_P} \tag{4.32}$$

Anticipating a result from Chapter 6 in which *hydrostatic equilibrium* is adopted (Eq. 6.6), we can also write

$$\frac{dP}{dr} = -g\rho$$

where ρ is the density at position r and g is the value of the local acceleration due to gravity at the same position, i.e. $g = GM_r/r^2$ (Eq. 6.5). Then equating these differentials and solving for h_P yields

$$h_P = \frac{P}{g\rho} \tag{4.33}$$

Finally, we use the *perfect gas law* in the form given in Eq. 2.4 to find

$$h_P = \frac{kT}{\mu m_H g}$$

This tells us that the pressure scale height will be large when g is small, assuming that variations in g are greater than variations in T. Stars that are very large (small g) should therefore have large scale heights. In the solar photosphere ($g = 2.7 \times 10^4$ cm s^{-2}, P $= 1.0 \times 10^5$ dyn cm^{-2}, $\rho = 2.6 \times 10^{-7}$ g cm^{-3}), $h_P = 1.4 \times 10^7$ cm (140 km). At $r = R_\odot/2$ ($g = 9.8 \times 10^4$ cm s^{-2}, P $= 7.3 \times 10^{14}$ dyn cm^{-2}, $\rho = 1.4$ g cm^{-3}), the scale height is much larger, i.e. $h_P = 5.5 \times 10^9$ cm (55,000 km) [166].

How much energy is transferred outwards in the star by the motion of such a pocket? The mass flux (g cm^{-2} s^{-1}) will be $\rho_a v_c$, where ρ_a is the density and v_c is the velocity of the pocket.[4] The mass flux must be multiplied by the energy contained in the pocket; that energy can be represented by the specific heat at constant pressure c_P (erg g^{-1} K^{-1}, Sect. 2.1.7) times a temperature difference, δT. Here, δT is the difference between the temperature of the pocket and the temperature of the surrounding ambient medium once the pocket has risen l_c. Thus the *energy flux due to convection* (erg cm^{-2} s^{-1}) will be

$$F_c = \rho_a v_c c_P \delta T \tag{4.34}$$

Our goal now is to obtain an expression for F_c that is dependent not on the specifics of an average pocket, but rather on the properties of the star.

---

[4] We use the subscript a as a reminder that the pocket is adiabatic and the subscript c to specify that the velocity is due to convection. Both refer to the pocket of gas.

For δT, we can write

$$\delta T = \left(\left| \frac{dT}{dr} \right|_\star - \left| \frac{dT}{dr} \right|_a \right) dr = \left(\left| \frac{dT}{dr} \right|_\star - \left| \frac{dT}{dr} \right|_a \right) l_c \tag{4.35}$$

The term $|dT/dr|_a$ is the adiabatic temperature gradient, as seen in the previous subsection. The subscript $\star$ represents the surrounding ambient medium. One could write a similar expression for $\delta \rho$, the difference in density between the pocket and its surroundings,

$$\delta \rho = \left(\left| \frac{d\rho}{dr} \right|_\star - \left| \frac{d\rho}{dr} \right|_a \right) l_c \tag{4.36}$$

and recall that $\delta P = 0$.

Let us first consider the adiabatic term in Eq. 4.35. We will use a change of variable and note that for any variable x, $d(ln(x)) = dx/x$. Then

$$\left(\frac{dT}{dr} \right)_a = \left(\frac{dT}{dP} \right)_a \left(\frac{dP}{dr} \right)_a => \left(\frac{dT}{dr} \right)_a = \left(\frac{T}{P} \right)_\star \left(\frac{d\,ln\,T}{d\,ln\,P} \right)_a \left(\frac{dP}{dr} \right)_\star \tag{4.37}$$

Note that for the expression on the right, we have exchanged the subscript a for the subscript $\star$ in the T/P term and dP/dr. This is because small perturbations on T or P make little difference to the former term and because the dP/dr of the star is the same as that of the pocket of gas for the latter term. For the stellar term in Eq. 4.35, $(dT/dr)_\star$, we can express it in a fashion similar to $(dT/dr)_a$. We then substitute both the adiabatic and stellar terms into Eq. 4.35, to find

$$\delta T = T \left[\left(\frac{d\,ln\,T}{d\,ln\,P} \right)_\star - \left(\frac{d\,ln\,T}{d\,ln\,P} \right)_a \right] \left[l_c \left(\frac{1}{P} \frac{dP}{dr} \right)_\star \right] \tag{4.38}$$

We can now see the usefulness of Eq. 4.38 because the term in the last set of brackets is related to the pressure scale height (Eq. 4.32). Thus, one can fold together the quantities contained in these brackets into a single free parameter (now dropping the $\star$ subscript since pressures and pressure gradients are the same for both the pocket of gas and the ambient medium) called the *mixing length parameter*, α,

$$\alpha \equiv -\frac{l_c}{P} \left(\frac{dP}{dr} \right) = \frac{l_c}{h_P} = \frac{l_c g \rho}{P} \tag{4.39}$$

where we have used Eq. 4.32. The right-hand side has also made use of Eq. 6.6 in Sect. 6.1.2, where g is the local acceleration due to gravity. The mixing length parameter is therefore just the mixing length, l_c, in units of the scale height, h_P. Note that α is $\mathcal{O}(1)$ if indeed a pocket rises $\mathcal{O}(1)$ pressure-scale heights before releasing its heat.

Let us now look at the convective velocity, v_c (Eq. 4.34). Since the buoyancy force is moving the pocket of gas upwards, the pocket is actually accelerating, so we will compute its average velocity, which is given from standard mechanics by $v_c = \sqrt{\frac{1}{2} a_B l_c}$, where a_B is the acceleration due to buoyancy.[5] The buoyancy force is

---

[5] Recall from kinematics that the average velocity is $\bar{v} = v_f/2$ for an initial velocity of zero and a final velocity of v_f. Also recall that $v_f{}^2 = 2ax$, where a is the acceleration and x is the displacement. Therefore, combining these equations gives $\bar{v} = \sqrt{ax/2}$, which is what has been written for the average velocity of a convective pocket.

given by Archimedes' principle as $F_B = V_a g \delta\rho$, where V_a is the volume of the pocket, $\delta\rho$ is the difference between the density of the pocket and the density of the surrounding medium (expressed in Eq. 4.36) and g is again the local acceleration due to gravity. Then $a_B = F_B/m_a = (\delta\rho/\rho_a)g$, where we have expressed the mass of a pocket $m_a = \rho_a V_a$. Thus the average velocity due to buoyancy is

$$v_c = \sqrt{\frac{1}{2}\frac{|\delta\rho|}{\rho}g\,l_c}$$

(4.40)

Again, we have dropped the subscript from ρ since $\delta\rho/\rho_a \approx \delta\rho/\rho$, and we have used an absolute value sign for $\delta\rho$ to ensure that the square root can be taken as real.

We can now make use of Eq. 2.9 and the fact that $\delta P = 0$ so that $\delta\rho/\rho = \delta T/T$. Also, we can eliminate g by using the right side of Eq. 4.39 to find

$$v_c = \sqrt{\frac{1}{2}\frac{|\delta T|}{T}\alpha\left(\frac{P}{\rho}\right)}$$

(4.41)

Again, we have ensured that δT is positive.

It is now very interesting to compare Eq. 4.41 with the equation for the speed of sound, i.e. $c_s = \sqrt{\gamma_a \frac{P}{\rho}}$ (Eq. 2.43 in Box 2.1 on page 42). Since α and γ are both $\mathcal{O}(1)$, and $\delta T/T$ should be $\ll 1$, then v_c should be $\ll c_s$. That is, a pocket of gas rises much more slowly than the speed of sound. When a pocket of gas is perturbed, its pressure should adjust to that of its surroundings at the sound speed. This, in fact, supports our assumption of constant pressure ($\delta P = 0$).

As an example, at the base of the convection zone in the Sun, an estimate of $v_c \approx 5000$ cm s$^{-1} \approx 10^{-4} c_s$ [44]. If $\alpha = 1$ ($l_c = h_P$), it would take 13 days for a pocket of gas to rise over a scale height and give up its energy. Also for these conditions, $\delta T/T \approx 10^{-7}$ and $\delta\rho/\rho$ is $\approx 10^{-7}$ (Eq. 2.9 with $\delta P = 0$). This means an adiabatically rising pocket of gas is only marginally different in temperature and density from its surroundings, e.g. $\rho_a \approx \rho$. As indicated in Box 4.1 on page 95, the scale height decreases with increasing radius in the Sun, so the timescale for convection also decreases. At the surface, granulation pattern timescales vary, but cells may last for a few hours [74] or less (e.g. Online Resource 4.3).

The conclusion that $\delta T/T$ is extremely small is a very important result! This tiny value is what drives the flux outwards due to convection and is a measure of the (difficult to pronounce) 'superadiabaticity' of the gas. If we wish to model the temperature gradient due to convection inside a star, and since we know that the temperature of any pocket of gas is virtually the same as the surrounding temperature, *we can simply use the adiabatic temperature gradient (Eq. 4.29) as the temperature gradient of the star for convection zones*. We will return to this point in Sect. 6.1.

We finally have a more tractable form for Eq. 4.34. We use Eq. 4.38 together with the first expression of Eq. 4.39 for δT, and use Eq. 4.41 again, substituting for δT to find the final result for the flux transport by convection:

$$F_c = \rho\, c_P\, T \sqrt{\frac{P}{\rho}}\left[\left(\frac{d\ln T}{d\ln P}\right)_\star - \left(\frac{d\ln T}{d\ln P}\right)_a\right]^{3/2}\alpha^2$$

(4.42)

The unsubscripted temperature, pressure and density refer to the ambient medium in the star. In Eq. 4.42, we have absorbed a term $\sqrt{\frac{1}{2}}$ into the unknown free parameter α. If some fraction of the work done goes into turbulence, for example, instead of upwards acceleration of the pocket of gas, a turbulent parameter could also be absorbed into α.

Equation 4.42 informs us as to the energy transport due to convection. Rather than applying to a single pocket of gas, it is actually a general expression. Equation 4.34, for example, does not specify the size or mass of any given pocket, so the pocket can represent the average energy flux due to convection at any location r. The *falling* pockets of gas also obey Eq. 4.34 [275]. This is because v_c becomes negative and so does dr in Eq. 4.35. Yet by the second law of thermodynamics, heat still transfers from hot to cold. The net *mass flux* (g cm^{-2} s^{-1}) is zero as pockets rise and fall, but this flux acts like a conveyor belt transporting energy upwards. Just how efficiently the flux is transported depends on the degree of superadiabaticity of the gas, as given by the term in square brackets in Eq. 4.42: a small quantity, but the driver of outward-moving energy flux by convection.

This development considers the net energy transport due to the convective process alone without any additional process at play. If *all* of the energy flux is a result of convection (there is no additional radiative term, for example), we can express the luminosity at some r as

$$L_r = 4\pi r^2 F_c \qquad (4.43)$$

Provided the convective zone is external to the region within which nuclear reactions are taking place, $L_r = L_\star$, where $L_\star$ is the total luminosity of the star. Using Eqs. 4.42 and 4.43 together is then a way of solving for the bracketed term in Eq. 4.42 and determining the magnitude of this tiny quantity. However, a situation in which essentially all energy flux is by convection is really only valid for convection zones in deep stellar interiors. Near stellar surfaces where the density is lower, even if convection is the dominant process, there can also be a small radiative energy transport component that needs to be included.

4.3.3 Real Convection

Mixing length theory can only be considered approximate and introduces the unknown free parameter α into the mix of the equations of stellar structure (Sect. 6.1). Turbulent pressure and viscosity are ignored, as is a varying velocity during the buoyant motion of a pocket of gas. The theory is also a one-dimensional (r-dependent only) approximation. At present, such an approach is still widely employed, although 3-D numerical simulations are improving (e.g. [157, 242, 302] and Fig. 4.7). Differential solar rotation (Fig. 1.11) also plays a part in the evolution of convective cells.

Convection is clearly the most important energy transport mechanism in low-mass stars, even to the point that the entire star is convective [280] (Fig. 4.6). Other

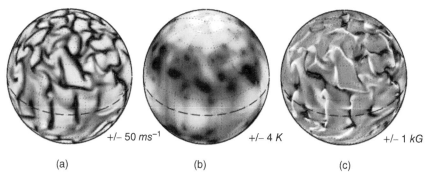

$+/- 50\ ms^{-1}$ $+/- 4\ K$ $+/- 1\ kG$

(a)   (b)   (c)

Figure 4.7 Three-dimensional simulations of the solar convection zone. **(a)** Radial velocity, **(b)** temperature and **(c)** radial magnetic field. Credit: Brun et al. (2004), American Astronomical Society.

convective-related effects can be extremely important in stars, even when they are not fully convective. For example, *convective overshoot* may occur, in which a pocket travels beyond the convective regions into a radiative region, due to its momentum. Convection tends to mix material. If convection is occurring in a nuclear burning core, the core may be homogenous rather than chemically differentiated as it would be if it were radiative. If convection is occurring near the surface and extends down into a region in which nuclear burning is occurring, it may dredge up material from the 'ashes' of the nuclear burning process. A number of *dredge-up* events can occur in stars that are off the main sequence, for instance (Sect. 8.2.1).

Convection plays an important role in *magnetic dynamo* generation, as we have already seen for the Sun (Sect. 1.4). A more esoteric consideration is the search for planets around stars by observing potential radial velocity variations in the parent star as a result of unseen planets. Because these effects are small, it is important to know surface radial velocities that result from convection (only about $2\ km\ s^{-1}$ for the Sun, for example) in order to properly correct for them when searching for planets. Details related to the magnetic field and field reversals, differential rotation and winds are all subjects of active research.

It is also interesting to examine the sizes of convective cells on other stars, at least of those that can be measured. Only the largest stars can be directly resolved by *interferometric* techniques to reveal cellular patterns on their surfaces. However, π^1Gruis is one such example (see Box 4.2 on page 101). The horizontal sizes of cells are known to be typically 3 to 10 times the pressure scale height, h_p [303]. We know that lower values of g lead to larger scale heights, as outlined in Box 4.1 on page 95. Therefore, the horizontal sizes of cells are also much larger in stars that have low values of g. π^1Gruis is one such example, with a stellar radius of 658 times the Sun and correspondingly low g. Consequently, its convective cells can be as large as 27% of its radius [239].

Box 4.2

Starring ... π^1Gruis

Distance: $d = 160$ pc; Spectral Type: S5,7

Effective Temperature: $T_{eff} = 3{,}200$ K; $V = 5.31 - 7.01$

Mass, M = 1.5 $M_\odot$; Radius, $R = 658\ R_\odot$ [239]

π^1Gruis is an *asymptotic giant branch* (AGB) star in the southern constellation of Grus (the crane). The 'S' spectral type denotes cool giants with carbon and oxygen in the atmosphere. π^1Gruis is very faint but discernable with the naked eye in a dark sky. It is also a close visual double with a fainter companion, π^2Gruis. π^1Gruis has been observed *interferometrically* (e.g. Sect. 4.2.5 of [154]), providing extremely high-resolution images, as shown in the accompanying figure with a milliarcsecond (mas) scale. This star is huge, leading to a very low value of g at its surface ($g = 0.36$ cm s^{-2}). Consequently, its *convective cells*, as revealed by the light and dark shadings in the image, are also very large, with a characteristic horizontal size of 1.2×10^8 km (cf. ≈ 1000 km for the Sun).

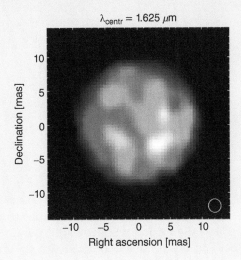

Convection also occurs in the mantles of planets and moons. Slow sea-floor spreading and subduction in the mantle of the Earth, driven by heat in the deep interior, is a convective process (Fig. 4.8 **Top**). The timescale for convection in the deep interior of the Earth, for example, is believed to be ≈ 200 million years ([173] and references therein), though can be shorter (≈ 50 million years) closer to the surface. Even everyday 'stove-top physics' can easily illustrate convection (Fig. 4.8 **Bottom**).

It is interesting to compare the small-scale convective cells of the Fig. 4.8 frying pan to the granulation patterns on the Sun (Fig. 4.4). Simulations of the convective

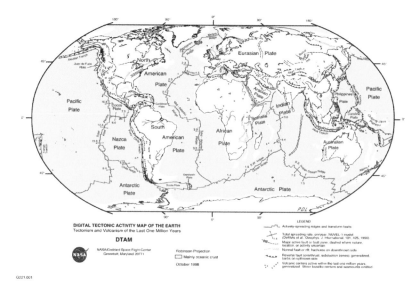

Figure 4.8 Top: Map of the Earth's surface showing the structure of tectonic plates. Credit: NASA/Goddard Space Flight Center. **Bottom**: Convective cells on the surface of hot oil (with small seeds added for visualization) in a frying pan. The structure resembles solar granulation patterns as shown in Fig. 4.4.

cells in the Earth (Fig. 4.9) help to explain the 'cell-like' nature that is observed at the surface of some convective regions. When material in convective cells reaches the surface, the material expands horizontally as its heat is given up. Material from adjacent cells then causes compression at the edges so that the cooler falling material appears line-like around the rising cells. For details on the numerical modelling of convection in stars and planets, see [120].

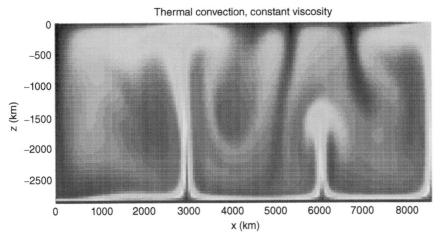

Figure 4.9 Simulation of the convective cells in the Earth. Red indicates hotter rising areas, and blue indicates cooler descending regions. Credit: Wikimedia Commons.

4.4 CONDUCTIVE ENERGY TRANSPORT IN DENSE REGIONS

As indicated in Sect. 4.1, thermal conduction, such as occurs in solids on Earth, is not important in gaseous stars. However, if the density is high enough, *electrical conduction* might be.[6]

Such conditions are not seen for stars along the main sequence. However, in brown dwarfs (Sect. 7.4.1) or in the cores of evolved stars (Sect. 8.2), densities can become so great that the space between particles is very small. In such a case, the Heisenberg uncertainty principle (HUP) states that $\Delta x \Delta p \geq h/4\pi$, where x is a distance, $p = m\upsilon$ is the momentum and h is Planck's constant. As x becomes very small, the velocity, υ, of the particles becomes very large. The high velocities of the particles then produce the pressure, the *electron degeneracy pressure*, that holds up the object. For the highest-density cores, the electrons can even be boosted to relativistic speeds. In the late stages of stellar evolution, the expanded low-density envelope that surrounds such cores is ejected (the result is called a *planetary nebula*), leaving behind only the degenerate core. We call these remnants *white dwarfs*. A discussion of such stars will be left until Sect. 10.2, but since the physics of a dense degenerate core and a white dwarf are the same, we can discuss the energy transport in such regions in this section.

---

[6] Recall that we saw that electrical conduction could transfer energy from the solar corona to the transition region on the Sun (Sect. 1.2.4). However, this was in a narrow layer and did not apply to a star or stellar remnant as a whole.

It can be shown that in high-density cores (typically $\approx 10^6$ g cm^{-3} or higher), the mean free path for electrons becomes very large (because most lower-energy states are already occupied), and therefore the probability of interaction is low. The electrons still diffuse outwards, just as we saw for photons; and for diffusion of photons or electrons, one can consider the *diffusion coefficient* (cm^2 s^{-1}),

$$D = \frac{1}{3} v \bar{l} \tag{4.44}$$

where v is the speed of the particle/photon and $\bar{l}$ is its mean free path. For dense cores, a combination of high (sometimes relativistic) velocities for electrons and a long mean free path mean electron transport (conduction) becomes the dominant energy transport mechanism, rather than radiative transport.

Since the process of conductive energy transport is analogous to that of radiative transport, the 'particle' being either an electron or photon, it is possible to write the energy flux equation in the same form as radiative transport (Eq. 4.20), i.e.

$$F = -\frac{4}{3} \left(\frac{ac}{\kappa_{ec} \rho} \right) T^3 \frac{dT}{dr} \tag{4.45}$$

The only difference is that the absorption coefficient is now the electron conduction absorption coefficient, κ_{ec}, rather than the Rosseland mean opacity, κ_R, which was used for photons.

To order of magnitude [201],

$$\kappa_{ec} \sim 10^2 \frac{T^{1/2}}{\rho} \text{ cm}^2/\text{g} \tag{4.46}$$

If both radiation and conduction are important, then,

$$F = -\frac{4}{3} \left(\frac{ac}{\rho} \right) \left(\frac{1}{\kappa_{ec}} + \frac{1}{\kappa_R} \right) T^3 \frac{dT}{dr} \tag{4.47}$$

This equation again illustrates the harmonic mean property of the opacity discussed in Sect. 4.2.1. The total opacity is then

$$\frac{1}{\kappa_{tot}} = \frac{1}{\kappa_{ec}} + \frac{1}{\kappa_R} \tag{4.48}$$

Even at the high densities at the center of the Sun, the efficiency of energy transport by electron conduction is only $\approx 10^{-5}$ that of radiative transport.

Online Resources

4.1   *OPAL opacities*: https://opalopacity.llnl.gov/existing.html
4.2   *The Opacity Project (OP) opacities*: http://opacity-cs.obspm.fr/opacity/
4.3   *Video showing solar granulation*: https://www.youtube.com/watch?v=CCzl0quTDHw

PROBLEMS

4.1. Consider a region in a star with a temperature $T = 5600$ K.

[HINT: Computer algebra will help with this problem. Making the substitution, $x \equiv (h\nu)/(kT)$ may also help with the math.]

(a) Plot the Planck function $B_\nu(5600)$ over a frequency from $\nu_1 = 10^{14}$ Hz to $\nu_2 = 10^{15}$ Hz. Verify that the frequency of the maximum, $\nu_{max}(B)$, is given by Eq. 2.50.

(b) Find the *derivative* of the Planck curve with respect to *temperature*, and verify that the result (cgs units assumed) is given by

$$\frac{dB_\nu(T)}{dT} = \frac{2\,h^2\,\nu^4\,e^{\frac{h\nu}{kT}}}{c^2\,k\left(e^{\frac{h\nu}{kT}} - 1\right)^2 T^2} \tag{4.49}$$

(c) Plot Eq. 4.49 for a temperature of 5600 K over the same frequency range as in part (a), and find the frequency of the maximum, $\nu_{max}(dB/dT)$.

(d) In Sect. 4.2.1, we indicated that the Rosseland mean opacity is weighted most strongly where the *derivative* of the Planck curve with respect to temperature is a maximum. How does $\nu_{max}(dB/dT)$ compare with the simple result of $\nu_{max}(B)$? Is it a fair statement to say that the two frequencies are close? Would this be true for any temperature?

4.2. Visit the home page of The Opacity Project (see Online Resource 4.2), enter your email address, and let all metals be zero except for carbon, which can be left at its default value. Once you receive the tables, identify and extract the opacity tables corresponding to

i. A: all hydrogen plus carbon only ($X = 0.9$, $Y = 0$, $Z = 0.1$)

ii. B: no hydrogen plus carbon only ($X = 0$, $Y = 0.9$, $Z = 0.1$)

Adopt a fixed (and typical) value of $log(R) \equiv log(\rho/T_6^3) = -3$, where T_6 is the temperature in units of 10^6 K and ρ is the density (g/cm^3).

(a) Plot $log(\kappa_R)$ against $log(T)$ on the same plot for the two cases.

(b) Offer a very *brief* qualitative discussion of the shapes and differences between these curves. As a start, for A, you should see two peaks between $log(T)$ of 3.5 and 5. The lower-temperature peak is due to bound-free absorption contributions from Balmer ($n = 2$) electrons, and the higher-temperature peak is due to bound-free absorptions from Lyman ($n = 1$) electrons.

4.3. In Sect. 4.2.1, we indicated that the development for the Rosseland mean opacity could lead to the equation of radiative transport. Let's see this link more explicitly. Beginning with Eq. 4.9 in Sect. 4.2.1, derive the equation of radiative transport, Eq. 4.21. [HINT: You will find Eqs. 4.10 and Eq. 2.57 to be useful in the process.]

4.4. The information in Sect. 2.1.7 as well as the ideal gas law will be useful for this problem.

(a) Show that the adiabatic temperature gradient in a star (Eq. 4.29) can be written simply as

$$\left.\frac{dT}{dr}\right|_a = -\frac{g}{c_P} \tag{4.50}$$

where g is the local acceleration due to gravity and c_P is the specific heat at constant pressure. [HINT: Eq. 6.5 will also be useful for this part.] Verify by dimensional analysis that the units on both sides of the equation agree.

(b) Show that

$$c_P = \left(\frac{\gamma_a}{\gamma_a - 1}\right)\frac{P}{\rho T} \tag{4.51}$$

where γ_a is the adiabatic index, P is the pressure, ρ is the density and T is the temperature.

4.5. (a) For a location in the convection zone of the Sun at $r = 0.8\,R_\odot$, find g, c_P and $\left.\frac{dT}{dr}\right|_a$. Use the standard solar model (SSM, Online Resource 1.12) to obtain the data, and use Eqs. 4.50 and 4.51.

(b) Find the pressure scale height (Eq. 4.33) at this location, expressed in units of solar radii, $R_\odot$.

(c) Suppose that the mixing length is $l_c \approx h_P$ and that a pocket of gas starting at $r = 0.8\,R_\odot$ rises to $r + \Delta r = r + l_c$ before its heat dissipates. Find $\left.\frac{\Delta T}{\Delta r}\right|_\star$ by reading the temperature change over this mixing length from SSM data.

(d) Compare $\left.\frac{dT}{dr}\right|_a$ from part *(a)* to $\left.\frac{\Delta T}{\Delta r}\right|_\star$ from part *(c)*, and comment on the agreement or disagreement between the two gradients.

4.6. Consider the following cases, and indicate *whether or not* convection is *more likely* to occur. Provide your reasons.

(a) The region is an *ionization zone*, i.e. photons are removing electrons from atoms in this zone.

(b) The opacity is high in the region.

(c) There is nuclear energy generation in the region that is highly sensitive to temperature.

4.7. The criteria for (in)stability against convection and the adiabatic temperature gradient are outlined in Eqs. 4.28 through 4.31. If one were modelling a star, it would be necessary to check whether the condition for convection was satisfied so that your models could switch from radiative transport to convection as the energy transport mechanism. Show that this check can be simply carried out via

$$\frac{d\ln P}{d\ln T} < \frac{\gamma_a}{\gamma_a - 1} \tag{4.52}$$

Does Eq. 4.52 imply instability or stability?

4.8. Consider a star whose density is $\rho = 10^6$ g cm^{-3} and temperature is T = 10^7 K. Ignore any variations in these quantities. Assume that the Rosseland mean opacity is dominated by the bound-free opacity, κ_{bf}, given by a Kramers law and that the electron conduction opacity is given by Eq. 4.46.

(a) Evaluate κ_R and κ_{ec}. Adopt reasonable parameters, as required. Which opacity is higher?

(b) What is the total opacity, κ_{tot}, and which is the dominant contributor? Is the energy flux by electron conduction or by radiation?

Chapter 5
The Burning Star – Cores

The Guide and I into that hidden road Now entered, to return to the bright world ... Till I beheld through a round aperture Some of the beauteous things that Heaven doth bear; Thence we came forth to rebehold the stars.

Dante Alighieri, Inferno [5]

By definition, *stellar cores* are the central regions within which nuclear reactions are taking place, referred to as nuclear 'burning'. This is where the star's energy/luminosity is generated. Figure 1.9 shows approximately the size of this region for the Sun, and Fig. C.3 in Appendix C illustrates how the luminosity increases in the core and then is constant at larger radii. There are different types of nuclear reactions, with different dependences on density and temperature and therefore different energy generation efficiencies. We explore these issues in this chapter.

Astrophysics: Decoding the Stars, First Edition. Judith Irwin.
© 2023 John Wiley & Sons Ltd. Published 2023 by John Wiley & Sons Ltd.

5.1 CLASSICAL AND QUANTUM APPROACHES

It is now well understood that nuclear reactions power stars. However, it wasn't always so, especially since any classical approach to the problem of energy generation does not result in temperatures that are high enough to overcome the *Coulomb potential energy barrier*.

The Coulomb barrier for proton-proton interactions is shown as the peak in Fig. 5.1. To the left of the peak is the size of the nucleus itself. Any particles within this region are those within the nucleus where the *strong force* dominates. This is the close-range force that binds together protons in nuclei and keeps them from repelling each other. To the right of the peak, two protons would experience a classical electrostatic positive-positive repulsion force that keeps the two charges separate.

Clearly, for two nuclei to combine, the energy of a (relatively) moving nucleus to the right of the barrier must be high enough that it can overcome the Coulomb barrier of the target nucleus. It is straightforward to estimate what temperature is required for two nuclei to overcome that barrier and fuse together, making a new heavier nucleus. We can do this by finding the potential energy of the barrier between any two protons, u_p, and equating it to the average kinetic energy of the moving protons, u_{th}, which is given in Eq. 2.17,

$$u_p = u_{th} \tag{5.1}$$

$$\frac{Z_1 Z_2 e^2}{r} = \frac{3}{2} kT \tag{5.2}$$

where r is the separation between the particles, Z_1 and Z_2 are the atomic numbers of the two particles (nuclei) and e is the atomic unit of charge. For two protons $Z_1 = Z_2 = 1$, $r = 10^{-13}$ cm (Fig. 5.1) and $e = 4.8 \times 10^{-10}$ esu, we find T $= 10^{10}$ K. This is much higher than the central temperature of the Sun, which is T $= 1.56 \times 10^7$ K

Figure 5.1 Plot of potential energy as a function of distance between two protons. As one proton approaches the other, it moves from right to left in the figure and encounters a potential energy barrier that it must overcome for the two protons to fuse. Credit: Adapted from [44].

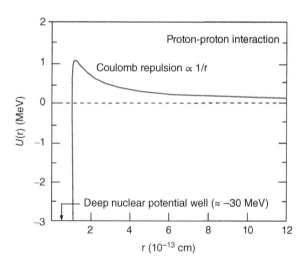

(Table C.1). It is clear that a classical approach, as shown by this simple comparison, is insufficient to explain the nuclear reactions in the core of the Sun.

The solution came only by employing a quantum mechanical approach, as discovered by George Gamov in 1928. There is some probability that a particle whose energy is lower than the Coulomb barrier, will 'tunnel through' the barrier, allowing for the possibility of nuclear reactions to occur. Arthur Eddington (see the Introduction) was convinced that the Sun was indeed hot enough for nuclear reactions to proceed, although not everyone at the time agreed. In one famous quote from 1930, Eddington said, 'If any physicist tells you that the temperature of 40 million degrees[1] is not enough for the generation of stellar energy, tell him to go and find a hotter place' [309].

The details of this *quantum mechanical tunnelling* are beyond the scope of this text (e.g. see [201]), but it can be shown that the probability of tunnelling is of order

$$P_r \approx exp\left[-\left(\frac{E_G}{E}\right)^{1/2}\right] \tag{5.3}$$

Here, $E = (3/2)kT$ (Eq. 2.17) represents the kinetic energy of a particle, and E_G is called the *Gamov energy* given by

$$E_G = \left(\pi\alpha_f Z_1 Z_2\right)^2 2\mu_m c^2 \tag{5.4}$$

where α_f is the *fine structure constant* given by

$$\alpha_f = \frac{e^2}{\hbar c} \sim \frac{1}{137} \tag{5.5}$$

Z_1 and Z_2 are the atomic numbers of the two particles, μ_m is the reduced mass of the two particles (Eq. 2.18), e is the electronic charge and c is the speed of light. Notice that E_G is independent of the gas temperature and only depends on the reduced mass, which is fixed for a given type of reaction. A reaction involving more massive nuclei (higher Z) results in a lower probability of tunnelling.

For two protons, $Z_1 = Z_2 = 1$ and $\mu_m = m_p/2 = 8.35 \times 10^{-25}$ g, leading to $E_G = 7.9 \times 10^{-7}$ erg = 493 keV. For a gas temperature typical of the center of the Sun ($T \sim 10^7$ K), $E = 2 \times 10^{-9}$ erg. Using these values of E and E_G in Eq. 5.3 results in a probability of $P_r = 2 \times 10^{-9}$. It is clear that the probability of quantum mechanical tunnelling is quite low. But it is not zero.

Not every particle that penetrates the Coulomb barrier is involved in the expected nuclear reaction. Whether or not the reaction proceeds depends on the appropriate cross-section of the interaction. The *reaction rate* (see the next section), however, will certainly be proportional to the probability given in Eq. 5.3. This low probability helps to explain why nuclear reactions tend to proceed slowly and lifetimes on the main sequence, where hydrogen is being converted into helium, tend to be long. We

---

[1] This is likely in °F, which corresponds to 22 million K, or about 40% high compared to modern values.

also see from Eq. 5.3 that more massive nuclei require higher temperatures for the reaction to proceed. For example, if one shifts from hydrogen burning (two protons) to helium burning (two α particles), then $E_G \propto (Z_1 Z_2)^2 \, \mu_m$ increases by a factor of 64, which requires that the temperature must increase by the same factor for an equivalent probability. This helps to explain why there are distinct, well-separated phases of nuclear burning as a star evolves. That is, once a temperature reaches a value at which a certain type of nuclear burning can occur with a reasonable probability, we can say that the nuclear burning 'ignites' or 'turns on'.

5.2 ENERGY GENERATION RATE

In order to obtain the energy generation rate per gram of material (units of erg s^{-1} g^{-1}), we need to look at the details of any particular reaction. These can become rather complicated, but it is possible to parameterize the equations, representing them as power laws that are centered about some temperature.

For a two-body interaction, the *reaction rate per unit volume* (number of reactions s^{-1} cm^{-3}) is

$$r_{i,j} = r_0 \, X_i \, X_j \, \rho^{\phi} \, T^{\beta} \tag{5.6}$$

where i and j represent the two species being considered (e.g. H and H, C and N, etc.), r_0 is a constant specific to the reaction of interest, X_i is the mass fraction of species i, X_j is the mass fraction of species j, and ρ and T have their usual meanings of mass density and temperature, respectively. Density and temperature are each raised to some power (ϕ and β, respectively) that is not yet specified.

We need to know the amount of energy released per reaction, ϵ_r (erg per reaction; see Sect. 5.3) in order to calculate the energy generation rate, i.e.

$$\epsilon_{i,j} = \frac{\epsilon_r}{\rho} r_{i,j} \tag{5.7}$$

$$\epsilon_{i,j} = \epsilon_{0_{i,j}} \, X_i \, X_j \, \rho^{\alpha} \, T^{\beta} \tag{5.8}$$

Equation 5.7 can be thought of as [(energy released per reaction) times (number of reactions per second per unit volume) divided by (amount of mass per unit volume)], giving the desired units of erg s^{-1} g^{-1}. Equation 5.6 has been inserted into Equation 5.7 to form Eq. 5.8, along with combining the constants, ϵ_r and r_0 to form a new constant, $\epsilon_{0_{i,j}}$, and the power on the density has become $\alpha = \phi - 1$.

Equation 5.8 tells us the energy generation rate for any particular reaction, and the power law representation is very useful because it allows different reactions to be represented in the same way except for the powers, α, β, and the constant, $\epsilon_{0_{i,j}}$. The power law form, however, results from mathematical expansions of more accurate calculations about specific temperatures. Therefore, they apply to temperatures that are *of order* the specified expansion temperature. Adding up $\epsilon_{i,j}$ over all mass within which the specific reactions are taking place gives the total energy generation rate in erg s^{-1}.

Equation 5.8 neglects several other effects. The most important is a *screening factor*, $f_{i,j}$. This factor results from the fact that at such high temperatures in stellar cores, atoms are completely ionized. Thus, a 'soup' of electrons forms a screen around the positive nuclei, reducing the effectiveness of the positive charge. This makes the effective charge *lower* than it otherwise would be. As a result, the Coulomb barrier is also lower, thereby *enhancing* the effectiveness of nuclear reactions. Screening can actually be significant, depending on the reaction type, sometimes enhancing helium-producing reactions by 10 to 50% [44]. Another effect is the neglect of higher-order correction terms in the expansions, which could be re-introduced as a multiplicative correction term, $C_{i,j}$. Finally, reactions can branch into different routes (e.g. Fig. 5.3), and a multiplicative term, $\Psi_{i,j}$, could be introduced to take that into account. By introducing each of these effects as unitless correction factors, we finally have a more general power law expression for the energy generation rate,

$$\epsilon_{i,j} = \epsilon_{0_{i,j}} \, f_{i,j} \, C_{i,j} \, \Psi_{i,j} \, X_i \, X_j \, \rho^\alpha \, T^\beta \quad \text{erg s}^{-1} \text{ g}^{-1} \tag{5.9}$$

Representing the energy generation rate in this form helps to reveal how different reactions depend on density and temperature.

5.3 ENERGY RELEASE AND BINDING ENERGY

En route to Eq. 5.9, we needed to know the amount of energy that is generated per reaction, ϵ_r, because it entered into Eq. 5.7. For this, we need to consider the *binding energy* of a nucleus, E_b. The binding energy is the difference in energy between nucleons (protons and neutrons) when they are bound within the nucleus, and the same nucleons when they are free:

$$E_b = \Delta m_{nuc} \, c^2 = \left[(Z \, m_p + (A - Z) \, m_n) - m_{nucleus} \right] c^2 \tag{5.10}$$

where $m_{nucleus}$ is the mass of the bound nucleus, m_p is the mass of the proton, m_n is the mass of the neutron, Z is the atomic number (number of protons), A is the atomic weight (total number of nucleons) and $(A - Z)$ is the number of neutrons. The quantity in the square brackets is the difference in mass, Δm_{nuc}, between the free and bound states, referred to as the *mass decrement* (or *mass deficit* or *mass defect*). The mass of free nucleons is greater than the mass of the same nucleons when they are bound in a nucleus, and it is this mass decrement that is converted into energy via Einstein's famous equation, $E = \Delta m_{nuc} \, c^2$. 'Collecting' free nucleons and fusing them into nuclei, or collecting lighter nuclei and fusing them into heavier nuclei – *nuclear fusion* – is what produces energy in stars.

For example, for helium, $Z = 2$ and $A - Z = 2$, so from Eq. 5.10,

$$E_b = \left[(2m_p + 2m_n) - m_{He} \right] c^2 \tag{5.11}$$

$$= [2 \, (1.00727647 \, u) + 2 \, (1.00866492 \, u) - 4.00153 \, u] \, c^2 \tag{5.12}$$

$$= 0.03035 \, u \, c^2 = 5.52983 \times 10^{-5} = 28.27 \text{ MeV} \tag{5.13}$$

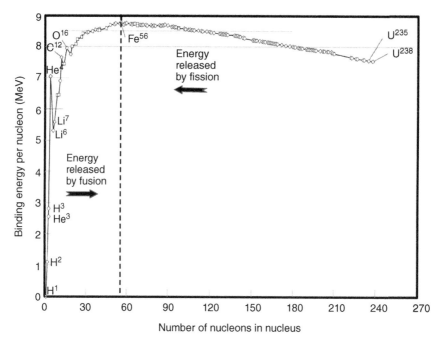

Figure 5.2 Binding energy per nucleon, as a function of the number of nucleons in the nucleus. The superscript numbers on the element indicate the number of nucleons in the nucleus. The maximum occurs at iron. Exothermic reactions are labelled with arrows; to the **left** of the vertical dashed line, nuclear fusion releases energy; and to the **right** of the vertical dashed line, nuclear fission releases energy. See also Online Resource 5.1 for data. Credit: Adapted from Wikimedia Commons.

where $u = 1.66053886 \times 10^{-24}$ g is the atomic mass unit (Appendix A) and m_{He} is the mass of a helium nucleus. Therefore, 28.27 MeV is released when two neutrons and two protons are converted into one helium nucleus. Put another way, it would take 28.27 MeV of energy to break apart the helium nucleus into its constituent parts. Then the average binding energy *per nucleon* for helium (there are four nucleons) is 7.07 MeV.

Figure 5.2 illustrates this concept further, showing the binding energy per nucleon plotted against the number of nucleons in a nucleus for the various elements. Notice that it increases, with a peak at $^{56}_{26}$Fe that is the most tightly bound nucleus. The shape of this curve is determined by a battle between the short-range, strong nuclear force that holds nuclear particles together, and the long-range Coulomb force, which is repulsive between positively charged protons. For light elements, as the number of nucleons increases, the effective strong force also increases. For heavy elements, however, protons are farther apart near the surface where the Coulomb force plays a relatively more important role, and the curve declines again (e.g. see [166]). This behaviour has fundamental consequences for natural nuclear reactions in stars and man-made nuclear reactions on Earth.

To the left of the peak, denoted by a vertical dashed line, *fusion* reactions are *exothermic*, i.e. they *release energy*. This corresponds to motion from left to right, e.g. hydrogen fusing into helium or helium fusing into carbon. Such energy produces pressure that holds the star up. To the right of the peak, fusion reactions are *endothermic*, i.e. they *require energy*. If such endothermic reactions are eventually the main reactions that occur in a star, energy production ceases, and the pressure required to hold the star up is no longer present. We will see the dramatic consequences of this in Sect. 10.3.1. Fusion reactions in stars, therefore, correspond to the region to the left of the peak. As such reactions proceed (left to right in the figure), notice that E_b increases.

Fission reactions, in which a nucleus is split up into lighter particles (motion from right to left to the right of the peak in the diagram), are also exothermic: for example, the fissioning of uranium into smaller daughter products. Man-made nuclear reactors belong to the region to the right in the diagram. Attempts are being made, however, to build nuclear reactors that use fusion energy rather than fission energy (see Box 5.1 on page 116).

Let us now express Eq. 5.11 as a reaction in which the free nucleons fuse together to form the helium nucleus:

$$2n + 2p \rightarrow {}^4He + \gamma \tag{5.14}$$

The quantity γ is the energy released, and because of our definition of binding energy, $\gamma = E_b = 28.27$ MeV. However, suppose we examine a reaction involving heavier nuclei. In that case, knowledge of the binding energy per nucleon is very helpful, as given by Fig. 5.2 or Online Resource 5.1. The nuclei that will fuse can be imagined to be broken up into their individual free particles and then reassembled. For example,

$$ {}^8Be + {}^4He \rightarrow {}^{12}C + \gamma \tag{5.15}$$

$$ 8(7.062 \text{ MeV}) + 4(7.074 \text{ MeV})) \rightarrow 12(7.680 \text{ MeV})) + \gamma \tag{5.16}$$

$$ 84.792 \text{ MeV} \rightarrow 92.160 \text{ MeV} + \gamma \tag{5.17}$$

$$ \gamma = -7.37 \text{ MeV} \tag{5.18}$$

The negative sign is simply a consequence of the fact that the binding energy was defined to be a positive quantity. Here, a negative γ can be thought of as the energy that is 'leaving the system' as a result of the reaction. As noted earlier, the binding energy of the product of nuclear fusion increases for light elements, and in this case, 7.37 MeV is produced as a result.

Reactions that convert hydrogen into helium in stellar cores involve a number of steps (cf. Figs. 5.3, 5.4) so the 'energy produced per reaction', or ϵ_r in Eq. 5.7, represents a net energy from all steps. We will consider this in more detail in Sects. 5.4.1 and 5.4.2.

Box 5.1
Star in a Box – Fusion Energy

Man-made nuclear reactors use *fission* to generate energy. High-mass particles, such as naturally occurring ^{235}U, are hit by neutrons that split them apart into smaller elements. Numerous products can result, including isotopes of iodine, cesium, strontium, xenon, barium and krypton, among others (e.g. Prob. 5.2). From Fig. 5.2, however, the slope of the curve of E_b per nucleon is much steeper for the lower-mass elements to the left of the peak than the higher-mass elements to the right of the peak. This means that, gram for gram, a typical low-mass fusion reaction should produce more energy than a fission reaction. With no carbon footprint or numerous radioactive products, the advantages are clear. Why not, then, make fusion nuclear reactors to generate power?

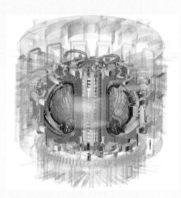

A number of different fusion experiments are taking place on Earth, but the largest of these is ITER (International Thermonuclear Experimental Reactor), which means 'the way' in Latin. An international project located in France, ITER uses a 'tokamak' (see figure), which is a cylindrical chamber containing strong magnetic fields that confine a hot plasma in a toroidal shape (purple in the figure). It is designed to mimic the fusion reactions in the Sun, except that isotopes of hydrogen (deuterium and tritium) are used instead of pure hydrogen. These isotopes start to fuse at temperatures lower than pure hydrogen would. ITER is expected to reach a pressure of $\gtrsim 2.5$ atmospheres and temperature of $\gtrsim 50$ million K in order to achieve conditions that are favourable for fusion. Such values require significant energy input, and confining and stabilizing the plasma are also challenges. See https://www.iter.org for more information.

5.4 MAIN SEQUENCE REACTIONS

In the H-R diagram of Fig. I.2, we saw the *main sequence* from an observational perspective. We can now explicitly address the physics of such stars. The *definition* of the main sequence is that it is the locus of points on the H-R diagram in which stars are converting hydrogen into helium in their cores, i.e. are undergoing 'core hydrogen burning'. As we will see in Sect. 7.2, a star spends the longest time of its nuclear burning lifetime on the main sequence. All other nuclear burning stages are shorter. Consequently, the highest density of stars on the H-R diagram occurs on the main sequence, as Fig. I.2 illustrates.

There are two ways of converting hydrogen into helium, namely via the *PP-chain* and the *CNO cycle*. Consequently, we need to understand these two important reactions in stars. An example of two visible main sequence stars, one in which the PP-chain is occurring and one in which the CNO cycle is occurring, are highlighted in Box 5.3 on Page 124.

5.4.1 The PP-chain

In our Sun and also the lower-mass stars along the main sequence, the dominant energy generation is by the PP-chain. 'PP' means 'proton-proton', equivalent to two hydrogen nuclei. In our Sun, 99.1% of the luminosity results from PP-chain reactions [234].

Figure 5.3 shows the most important reactions in this chain and also gives the probabilities that the reactions will follow a given path. In the figure, we include both the atomic weight (number of protons plus neutrons) as a superscript on the left as well as the atomic number (number of protons) as a subscript on the left. The subscripts are not really necessary since the number of protons is also given by the name of the element (e.g. hydrogen has one proton, helium has two, etc.), but they are included in the figure for additional clarity.[2] As before, when energy is produced, it is written as γ.

The three possible reaction chains are designated PPI, PPII and PPIII, with the PPI chain being the most important since it occurs 85% of the time. Of the remaining 15% of reactions, PPII occurs 99.9% of the time and PPIII only 0.1% of the time, as the figure illustrates. Notice that the first two reactions (*Top Center*) must occur twice in order to form the two ^{3}He that are required for the PPI route (*TOP-LEFT BOX*). Similarly, some ^{4}He must already have been formed in order for the reaction that forms ^{7}Be to proceed (*Top Right*), followed by the PPII and PPIII chains. For all chains, the net result is the conversion of four protons into one helium nucleus. The PPI chain can be summarized as

$$4\,^1\text{H} \rightarrow \,^4\text{He} + 2\,\text{e}^+ + 2\,\nu_e + 26.73 \text{ MeV} \tag{5.19}$$

---

[2] An alternate and somewhat more succinct way of writing a reaction is, for example, $^3\text{He}(\alpha, \gamma)^7\text{Be}$, which means $^3\text{He} + \alpha \rightarrow \gamma + ^7\text{Be}$. Here, the α particle is ^4_2He.

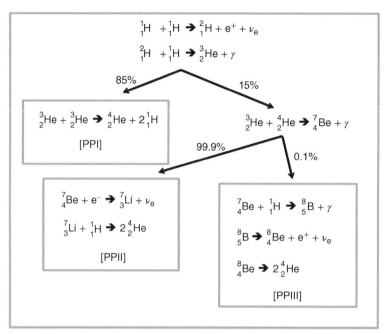

Figure 5.3 Most important reactions in the PP-chain in the Sun. The percentages specify the probability that a reaction will follow the route shown by the arrow. Nuclei are specified by total number of nucleons (**left** superscript) and number of protons (**left** subscript). γ represents energy generation in the form of photons, e^+ is a positron and ν_e is an electron neutrino. Credit: Adapted from [35].

where e^+ is a positron and ν_e is an electron neutrino (Box 5.2 on page 121). The energy released includes the energy from the annihilation of the positrons with surrounding electrons, but it does not include the energy carried away by neutrinos, because these do not react with other material in the star. The PPII and PPIII chains can be summarized similarly with slightly modified energies depending on how much energy is carried away by neutrinos [251].

Another branch of the PP-chain, not shown in the figure, is the PEP-ν branch. The first reaction in the top line of Fig. 5.3 has a 99.6% probability of occurring, in comparison to the PEP-ν reaction (not shown), which occurs 0.4% of the time. The PEP-ν reaction is $^1H + e^- + {}^1H \rightarrow {}^2H + \nu_e$. The reaction then continues to the second line in the figure. Neutrinos from these reactions are discussed in Box 5.2 on page 121.

The timescale for the PP-chain sequence of reactions is dictated by the slowest step. For the most probable PPI chain, this is the first step (hydrogen to deuterium), which takes about 10^{10} yr. The second step takes only 6 seconds, and the third (^{3}He to ^{4}He plus two protons) takes 10^6 yr [201]. The relative slowness of the first step is due to the fact that a proton must decay into a neutron, which is a process with a low probability.

The energy generation rate is given by

$$\epsilon_{pp} \sim 2.4 \times 10^6 \, f_{pp} \, \Psi \, C_{pp} \, X^2 \, \rho \, T_6^{-2/3} \, exp\left(-33.8/T_6^{1/3}\right) \text{ erg s}^{-1} \text{ g}^{-1} \qquad (5.20)$$

where

$$C_{pp} = 1 + 0.0123 \, T_6^{1/3} + 0.0109 \, T_6^{2/3} + 0.00095 \, T_6, \qquad (5.21)$$

X is the mass fraction of hydrogen, ρ is the mass density (g cm^{-3}) and T_6 is the temperature in units of 10^6 K. The values of f_{pp}, Ψ_{pp} and C_{pp}, which were introduced in Sect. 5.2, are all of order ~ 1, but we have included an expression for C_{pp} for additional accuracy.

This energy generation rate can also be represented in power law form following Eq. 5.8, with i, j denoted by PP [44],

$$\epsilon_{pp} \sim \epsilon_{0_{pp}} \, \rho \, X^2 \, f_{pp} \Psi_{pp} \, C_{pp} \, T_6^4 \text{ erg s}^{-1} \text{ g}^{-1} \qquad (5.22)$$

for temperatures near $T = 1.5 \times 10^7$ K, the correction factors set to 1 and $\epsilon_{0_{pp}} = 1.08 \times 10^{-5}$ erg cm^3 g^{-2} s^{-1}. Therefore, simply, $\epsilon_{pp} \propto T^4$.

5.4.2 The CNO Cycle

On the main sequence, the CNO cycle dominates nuclear reactions in more massive stars ($\gtrsim 1.3$ M$_\odot$). In the Sun, however, only 0.9% of the energy generation comes from the CNO cycle [234]. Of this percentage, the main CNO cycle, which is shown in Fig. 5.4, occurs 99.96% of the time, and another minor cycle involving flourine (not shown) occurs only 0.04% of the time [35]. Other minor CNO cycles also occur and become more important in more massive stars.

It is important to note that the CNO cycle occurs in main sequence stars, and by definition, this reaction must be converting hydrogen into helium. Carbon, nitrogen and oxygen are therefore *catalysts* in the CNO cycle. As Fig. 5.4 shows, as soon as one of these elements is produced, it is destroyed, so that the net result is a conversion from hydrogen (boxed in blue) to helium (boxed in red). Of course, some carbon, nitrogen and oxygen must be present originally for this cycle to proceed, but stars do have some initial metallicity derived from the cloud from which they were formed (Sect. 8.1).

CNO cycle reaction rates are faster than those of the PP-chain. The slowest timescale in this cycle belongs to the bottommost reaction in the figure (^{14}N plus a proton to ^{15}O), taking 10^8 yr. The second slowest is the topmost reaction (^{12}C plus a proton to ^{13}N), at 10^6 yr [201]. This means the bottleneck in the cycle occurs with ^{14}N 'waiting' to be converted into ^{15}O. Thus, the initial mix of C, N and O becomes mostly ^{14}N as this cycle progresses [166].

An examination of Fig. 5.4 indicates that the net reaction for this CNO cycle is equivalent to the PP-chain as given in Eq. 5.19. The energy produced is also the same. However, the neutrino energies are different, so the fractional amount of energy that is carried away by neutrinos and does not contribute to the solar luminosity is also different.

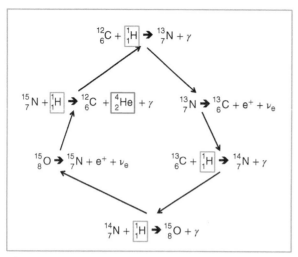

Figure 5.4 Most important reactions in the CNO cycle in the Sun. Notice that C, N and O are simply catalysts. The net reaction is a conversion from hydrogen (boxed in blue) into helium (boxed in red). The reactions involving radioactive decay that produce positrons also contribute a small amount of energy.

The energy generation rate is given by [201]

$$\epsilon_{CNO} \sim 8.7 \times 10^{27} \, f_{CNO} \, C_{CNO} \, X \, X_{CN} \, \rho \, T_6^{-2/3} \, exp\left(-152.3/T_6^{1/3}\right) \text{ erg s}^{-1} \text{ g}^{-1} \quad (5.23)$$

where

$$C_{CNO} = 1 + 0.0027 \, T_6^{1/3} - 0.0078 \, T_6^{2/3} - 0.00015 \, T_6, \quad (5.24)$$

where $f_{CNO} \sim 1$ for temperatures $T \geq 10^7$ and X_{CN} is the sum of the mass fractions for ^{12}C and ^{14}N.

As we did for the PP-chain, the CNO energy generation rate can be represented as a power law (cf. Eq. 5.8) with all correction factors set to 1 [44]

$$\epsilon_{CNO} \sim \epsilon_{0_{CNO}} \, \rho \, X \, X_{CN} \, T_6^{19.9} \quad \text{erg s}^{-1} \text{ g}^{-1} \quad (5.25)$$

for temperatures near $T = 1.5 \times 10^7$ K. Here, $\epsilon_{0_{CNO}} = 8.24 \times 10^{-24}$ erg cm^3 g^{-2} s^{-1}. The power-law form[3] makes it easier to see how much stronger the temperature dependence is for hydrogen burning by the CNO cycle, compared to the PP-chain, i.e. $\epsilon_{CNO} \propto T^{19.9}$. It is clear that in more massive stars with higher core temperatures, hydrogen burning by the CNO-cycle will quickly dominate over hydrogen burning by the PP-chain. The CNO energy generation rate climbs rapidly with increasing stellar

---

[3] The power varies strongly with temperature range (e.g. Fig. 12.3 of [201]).

temperature (and stellar mass), so the hydrogen fuel is used up more quickly. This helps to explain why massive stars have such short lifetimes compared to low-mass stars (see Fig. 7.1).

Box 5.2
Solar Neutrinos

Neutrinos (the 'little neutral ones') are the least-massive particles known ($\lesssim 0.3$ eV, [111]). Together with the fact that they have no charge, these particles have very small interaction cross-sections and long mean free paths, making them difficult to detect. Nevertheless, specialized detectors placed underground can indeed measure a tiny fraction of them when a low-probability interaction with matter occurs. Historically, there was a deficit in Earth-based fluxes compared to the number of neutrinos expected from the PP-chain, until it was understood that neutrinos *oscillate*, converting between different *flavours* as they travel from the solar core to the Earth. The neutrinos generated in the Sun are *electron neutrinos*, ν_e, but the other flavours are muon, ν_μ, and tau, ν_τ, neutrinos. Measurements by the KamiokaNDE/Super-Kamiokande project in Japan and the Sudbury Neutrino Observatory (SNO) in Canada showed definitively that these oscillations do occur, and the discrepancy was resolved.

At first, only the higher-energy electron neutrinos from the PPIII branch ($_5^8B \rightarrow {}_4^8Be + e^+ + \nu_e$; see Fig. 5.3) were detected. But now almost all solar neutrinos have been measured, including one from the minor PEP-ν branch (Sect. 5.4.1) as well as neutrinos from the CNO cycle [34] that produces only 0.9% of the solar luminosity. Neutrinos may have little mass, but their energy is not insignificant. In the effective reaction of the PP-chain given in Eq. 5.19, neutrinos in the MeV range carry away 2.3% of the Sun's nuclear energy production [314]. This is energy that is lost and does not contribute to the Sun's luminosity or internal pressure because neutrinos pass through the Sun as if it were transparent.

Measurements of solar neutrino fluxes have the potential to resolve the solar abundance problem (Sect. 1.3.1), can improve our knowledge of the core temperature profile and can aid in transitioning to developing high-precision solar models.

5.5 REACTIONS AFTER THE MAIN SEQUENCE

As stars evolve, they 'meander off' the main sequence. This wandering, sometimes back and forth on the H-R diagram, will be described in Sect. 8.2. The key point here, however, is that once the hydrogen fuel is used up in the core, hydrogen burning

Table 5.1 Nuclear burning examples

Process	Fuel	Products	Approx. ignition temperature (K)
Hydrogen burning	Hydrogen	Helium	1×10^7
Helium burning	Helium	Carbon, oxygen	1×10^8
Carbon burning	Carbon	Oxygen, neon, sodium, magnesium	5×10^8
Neon burning	Neon	Oxygen, magnesium	1×10^9
Oxygen burning	Oxygen	Magnesium to sulphur	2×10^9
Silicon burning	Silicon	Iron and nearby elements	3×10^9

Credit: [251] / John Wiley & Sons.

ceases, and the core contracts. The contraction leads to more heating (Sect. 7.1) and an increase in temperature that can lead to new reactions starting. In this section, we summarize some of the important nuclear reactions that can occur 'off the main sequence'. Table 5.1 indicates the approximate temperature required to ignite a variety of reactions, the fuel that is required and some of the products. Note that some of these temperatures are only achieved for massive stars. The Sun, for example, is not massive enough to burn elements heavier than helium, so only the first two rows of Table 5.1 are relevant, the first row applying to the main sequence and the second row applying post-main-sequence. (Some carbon is burned to oxygen, however, in late stages.)

5.5.1 The Triple-α Process – Helium Burning

The triple-α process, or 'helium burning', requires temperatures of order 10^8 K to proceed (Table 5.1). It is so-named because it involves helium nuclei, which are also called *alpha particles*, and three of them are needed, as follows

$$^4\text{He} + {}^4\text{He} \rightarrow {}^8\text{Be} + \gamma \tag{5.26}$$

$$^4\text{He} + {}^4\text{Be} \rightarrow {}^{12}\text{C} + \gamma \tag{5.27}$$

The first step is actually *endothermic*, requiring a small amount of energy (0.092 MeV) to proceed. This implies that the reaction 'wants' to go the other way, with berillium rapidly ($\sim 10^{-16}$ s) decaying back into helium again. However, if the berillium nucleus is struck by another alpha particle, it produces carbon, as the second step shows. Since the second step must have occurred within such a short time, the two steps can be considered a 'three-body interaction'. The net energy gain is $\gamma = 7.27$ MeV, which is much less than for hydrogen burning (cf. Eq. 5.19). This is consistent with the shallower slope between helium and carbon (compared to hydrogen to helium) in Fig. 5.2.

The energy generation rate is given by

$$\epsilon_{3\alpha} \sim 5.09 \times 10^{11} \, f_{3\alpha} \, Y^3 \, \rho^2 \, T_8^{-3} \, exp\left(-44.027/T_8\right) \text{ erg s}^{-1} \text{ g}^{-1} \tag{5.28}$$

where T_8 is the temperature in units of 10^8 K, Y is the mass fraction of helium and $f_{3\alpha}$ is the screening factor. Expressed as a power law centered about 10^8 K, and taking $f_{3\alpha} \sim 1$ [201] for temperatures about the same value,

$$\epsilon_{3\alpha} \sim \epsilon_{0_{3\alpha}} \, \rho^2 \, Y^3 \, T_8^{41.0} \tag{5.29}$$

We now see the extraordinarily strong temperature dependence of the 3 α energy generation rate. The Sun is currently not hot enough to burn helium, but from the standard solar model (SSM), we know that the temperature at the center of the Sun is a factor of two higher than at the outer part of the core (Fig. C.2). If a hotter stellar core capable of burning helium rose by the same factor, its energy generation rate would increase by a factor of 10^{12}! One can see how a nuclear reaction like this can be thought of as 'igniting', or being on/off, like a step function.

5.5.2 Additional and Higher Temperature Reactions

If the 3 α process has begun and sufficient carbon has been formed, carbon can combine with an α particle, followed by other α particle-related reactions (Table 5.1),

$$^{12}C + {}^4He \rightarrow {}^{16}O + \gamma \tag{5.30}$$

$$^{16}O + {}^4He \rightarrow {}^{20}Ne + \gamma \tag{5.31}$$

$$^{20}Ne + {}^4He \rightarrow {}^{24}Mg + \gamma \tag{5.32}$$

Of these reactions, the first is most important, with neon and magnesium production being minor by comparison [166]. This means once the 3 α process progresses, the result is a core primarily of carbon and oxygen (C/O), a result with important implications for the end-points of stellar evolution (Sect. 8.2).

At higher temperatures, numerous nuclear reactions can take place, and since massive stars have higher temperatures, these reactions occur mostly in massive stars. Here are some examples.

At higher temperatures ($\sim 5 \times 10^8$ K), *carbon burning* can occur,

$$^{12}C + {}^{12}C \rightarrow {}^{20}Ne + {}^4He + \gamma \tag{5.33}$$

$$^{12}C + {}^{12}C \rightarrow {}^{23}Na + {}^1H + \gamma \tag{5.34}$$

$$^{12}C + {}^{12}C \rightarrow {}^{23}Mg + n + \gamma \tag{5.35}$$

where n is a neutron.

Near 10^9 K, *oxygen burning* can occur,

$$^{16}O + {}^{16}O \rightarrow {}^{28}Si + {}^4He + \gamma \tag{5.36}$$

$$^{16}O + {}^{16}O \rightarrow {}^{31}P + {}^1H + \gamma \tag{5.37}$$

$$^{16}O + {}^{16}O \rightarrow {}^{31}S + n + \gamma \tag{5.38}$$

$$^{16}O + {}^{16}O \rightarrow {}^{32}S + \gamma \tag{5.39}$$

At temperatures exceeding 10^9 K, heavier nuclei can also combine with an α particle: for example, the sequence

$$^{24}Mg + {}^{4}He \rightarrow {}^{28}Si + \gamma \tag{5.40}$$

$$^{28}Si + {}^{4}He \rightarrow {}^{32}S + \gamma \tag{5.41}$$

$$^{32}S + {}^{4}He \rightarrow {}^{36}Ar + \gamma \tag{5.42}$$

$$^{36}Ar + {}^{4}He \rightarrow {}^{40}Ca + \gamma \tag{5.43}$$

For stars hot enough to undergo oxygen burning, the core would consist mainly of ^{28}Si and ^{32}S. Silicon has too high a Coulomb barrier to be able to fuse with itself, however, so silicon burning proceeds differently. It is broken up by high-energy photons, a process called *photo-disintegration*, followed by captures of light particles again in a series of reactions that eventually build up the iron-group elements near the peak of the binding energy curve of Fig. 5.2 [90].

One can see that nuclear burning in massive stars is what *forms* heavy elements up to the elements near the peak of the binding energy curve, which occurs at ^{56}Fe. The common phrase 'we are stardust' can almost be taken literally, especially since all life is carbon-based and the atoms within us were forged on stellar anvils.

Box 5.3

Starring ... η and λ Cas

These are both *main sequence* stars in Cassiopeia. They are outshone by the brighter off-main-sequence stars that form the classic 'W' shape of this constellation, but their core H $\rightarrow$ He burning properties are quite different. Both are in binary systems.

η **Cas:** $d = 5.95$ pc; V Magnitude = 3.44; Spectral Type = G0 V; $T_{eff} = 5{,}973 \pm 8$ K; $R = 1.039 \pm 0.004$ $R_\odot$; M = 0.972 ± 0.01 $M_\odot$; $L = 1.232 \pm 0.004$ $L_\odot$

Similar to the Sun, energy generation in its core is by the *PP-chain* and is *radiative* (Fig. 4.6).

λ **Cas:** $d = 116$ pc; V Magnitude = 4.77; Spectral Type = B8 Vnn; $T_{eff} = 12{,}000 \pm 1{,}000$ K; $R = 3.5$ $R_\odot$; M = 2.9 ± 0.4 $M_\odot$; $L = 255$ $L_\odot$

This is a hot, luminous star that is much farther away than η and so appears fainter. Its 'nn' spectral type means it is a very fast rotator. Nuclear energy generation is by the *CNO cycle* and is *convective*.

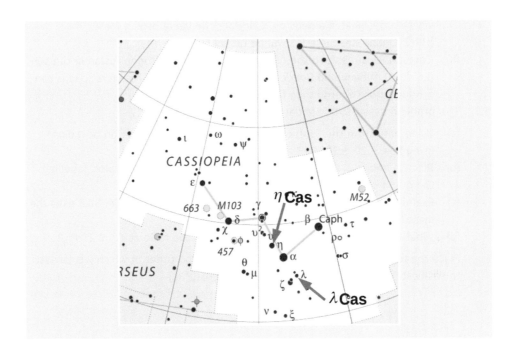

Online Resource

5.1 *Binding Energies per Nucleon*:
https://www.wolframalpha.com/widgets/view.jsp?id=22643e4a53683c92ff1b855a
0733b635

PROBLEMS

5.1. (a) Evaluate the mass decrement (g) for the PP-chain whose summary is given in Eq. 5.19. Evaluate the amount of energy released from this mass decrement in MeV. Express the results to three decimal places.

(b) When the two positrons that are released encounter free electrons, they annihilate and produce an additional amount of energy that must be added to the energy released from the mass decrement alone. Write a reaction equation for this energy (MeV), and evaluate it to three decimal places.

(c) What is the final energy produced by the PP-chain (MeV)? Compare your result to the value given in Eq. 5.19.

(d) Does all of this energy contribute to the solar luminosity? If not, estimate what fraction does (cf. Box 5.2 on page 121).

5.2. (a) An example of a fission reaction that takes place in reactors on Earth is

$$^{235}_{92}U + n \rightarrow \, ^{144}_{56}Ba + \, ^{90}_{36}Kr + 2n + \gamma \tag{5.44}$$

where n designates a neutron. Calculate the value of γ in MeV.
[HINT: Online Resource 5.1 will be useful.]

(b) Compare this result to the energy produced by hydrogen fusion in the sun (Eq. 5.19). Why would Box 5.1 on page 116 state that fusion reactions can produce more energy than fission reactions?

5.3. This problem uses GS98 SSM data from Online Resource 1.12.

(a) Using data from the SSM, calculate ϵ_{PP} and ϵ_{CNO} as a function of radius, using Eqs. 5.20, 5.21, 5.23 and 5.24.

(b) Plot ϵ_{PP} and ϵ_{CNO} as a function of solar radius on the same plot, labelling the graph clearly.

(c) Examine Fig. 5.4 and data from the SSM, and explain why Eq. 5.23 uses the mass fraction X_{CN} rather than X_{CNO}.

(d) What is $\epsilon_{CNO}/\epsilon_{PP}$ at the Sun's center? What is the ratio at $r = 0.25\ R_{\odot}$?

5.4. Place the reactions shown in Eqs. 5.40 to 5.43 in the order in which γ is largest to smallest. [HINT: Online Resource 5.1 will be useful.]

Chapter 6
The Modelled Star

But the sight of the stars makes me dream in as simple a way as the black spots on the map, representing towns and villages, make me dream. Why, I say to myself, should the spots of light in the firmament be less accessible to us than the black spots on the map of France?

Vincent Van Gogh [311]

We can indeed access the firmament, as Van Gogh desired, if not by visitation, at least by laying out the physical principles that are needed to *model* the stars. By so doing, we pass beyond the limitations imposed on us by observations of surface light alone. We can, in fact, delve into the very depths of these pinpricks of light in the sky and examine their properties from the surface to the core.

6.1 THE EQUATIONS OF STELLAR STRUCTURE

Consider a *stable*, spherically symmetric star. In fact, stars are *quasistatic*, which means internal structural changes are indeed occurring, but only at a slow rate that does not affect our requirement of stability here. Similarly, energy generation is taken to be stable in the same sense. Figure 6.1 sets out the geometry.

We can now make some fairly straightforward physical arguments that will lead to the *equations of stellar structure*. Our goal is to discover how different physical parameters vary with position r within the star. These parameters are M_r, $\rho(r)$, $P(r)$, L_r and $T(r)$: that is, mass, density, pressure, luminosity and temperature, respectively. Notice

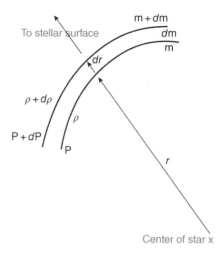

Figure 6.1 Slice showing a thin shell at some position *r* with respect to the stellar center. The pressure on the inside of the shell is P, and the density is ρ. The mass interior to the shell is M_r. The shell has width *dr* over which the pressure, mass and density change by *dP*, *dm* and *dρ*, respectively. Note that *r* is positive outwards and both *dP* and *dρ* are negative.

that we have written M_r rather than $M(r)$ since the relevant mass is the mass *interior to r*, whereas the pressure (for example) will be a value *at r*. (For small incremental mass, however, we will use a small m, as in *dm*.) Similarly, for L_r, the luminosity at any *r* includes whatever has been generated interior to it (an 'accumulation' of energy per second), so it is often written with the subscript.[1] We could as easily have *defined* $M(r)$ and $L(r)$ to mean the cumulative mass and luminosity, respectively, but in this section, it is helpful to use slightly different nomenclature for clarity.

In the next four subsections, we outline the *four principal equations of stellar structure*. We use *Eulerian coordinates*, which means the equations are written with respect to radius, *r*. An alternative form uses *Lagrangian coordinates*, which are the equations written with respect to mass, m. When there is motion, a Langrangian approach allows one to examine changes while 'travelling along' with the material. The transformation between the two forms for stationary material is explored in Prob. 6.1.

6.1.1 Conservation of Mass

Conservation of mass is a simple mathematical statement as to how the mass in a shell at radius *r* is related to the density, ρ, within the shell for a stable configuration. From the geometry of Fig. 6.1, the mass in a shell is

$$dm = 4\pi r^2 \rho\, dr \qquad (6.1)$$

where *dr* is the width of the shell.[2]

Equation 6.1 can be integrated quite simply between the limits m_1, r_1 to m_2, r_2, but only if the density, ρ, is a constant. Since, in general, ρ also varies with *r*, we

---

[1] Students of thermodynamics may wish to recall *intensive* and *extensive* quantities. T, P and ρ are intensive, whereas M_r and L_r are extensive.
[2] We have already seen this equation as Eq. I.14.

will remind ourselves of this fact by using a functional form for ρ to write the first equation of stellar structure,

$$\frac{dm}{dr} = 4\pi r^2 \rho(r)$$

(6.2)

6.1.2 Hydrostatic Equilibrium

The equation of hydrostatic equilibrium is really a re-statement of our stability criterion, i.e. any expansion or contraction of the star, if it exists, is on a very long timescale so that, at any given time, the star is mechanically stable. For stability, we require that the pressure within a shell, dP, balance the inwards gravitational pressure acting on the shell, i.e.

$$dP = -\frac{GM_r\,dm}{r^2}\frac{1}{4\pi r^2}$$

(6.3)

where we have expressed the gravitational pressure on the right-hand side as the gravitational force ($dF_r = GM_r\,dm/r^2$) per unit area ($4\pi r^2$). Each term on the right-hand side is positive, so the negative sign on the right-hand side ensures that dP is negative (decreasing outwards). Substituting Eq. 6.2 for dm gives

$$\frac{dP}{dr} = -\frac{GM_r\rho(r)}{r^2}$$

(6.4)

Again, dP/dr is a negative quantity.

We can simplify Eq. 6.4 by noting that at any r within the star, the local gravitational acceleration is

$$g(r) = \frac{GM_r}{r^2}$$

(6.5)

and therefore

$$\frac{dP}{dr} = -g(r)\rho(r)$$

(6.6)

Equations 6.4 and 6.6 are equivalent and represent the second equation of stellar structure.

6.1.3 Energy Conservation

For a star that is constant in luminosity, the radiative energy loss rate must balance the energy generation rate. In any given shell, then,

$$dL_r = \epsilon\,dm = \epsilon\rho(r)4\pi r^2\,dr$$

(6.7)

where we have used Eq. 6.1. Here, ϵ is the energy production rate (erg s^{-1} g^{-1}), introduced in Sect. 5.2. It is important to note that ϵ is not a constant but rather a function that is dependent on T, ρ and the chemical composition (μ, or X, Y, Z). The choice of ϵ

depends on which nuclear reactions are occurring (e.g. Eqs. 5.20, 5.23 and 5.28). Rearranging Eq. 6.7,

$$\boxed{\frac{dL_r}{dr} = \epsilon \rho(r) 4\pi r^2}$$

(6.8)

Like Eqs. 6.2 and 6.4, this equation applies quite generally to any stable stellar shell. However, once $r > r_{gen}$, where r_{gen} is the outermost radius at which energy is generated, $\epsilon = 0$ and $L_r = const$. For example, within the inner 25% of the Sun, 99% of the solar luminosity has been generated and $L_r \sim L_\odot$ at any larger radius (see Fig. C.3).

6.1.4 Energy Transport

We now need an equation that describes how energy is transported through the star. The possible equations were introduced in Chapter 4 of which the most important mechanisms for normal stars are radiative transport and convective transport.[3] Here, we rewrite Eqs. 4.21 and 4.29 for the sake of completeness:

$$\frac{dT}{dr} = -\left(\frac{3}{16\pi a c}\right)\left[\frac{\kappa_R \rho(r) L_r}{r^2 T^3}\right] \qquad \text{(Radiation)} \qquad (6.9)$$

$$= \left(\frac{\gamma_a - 1}{\gamma_a}\right)\frac{T}{P}\frac{dP}{dr} \qquad \text{(Convection)} \qquad (6.10)$$

Recall that γ_a is the adiabatic index (Eq. 2.31), and κ_R is the Rosseland mean opacity (Sects. 4.2.1, 4.2.2). For convection, note that the *stellar* temperature gradient is set equal to the *adiabatic temperature gradient* because there is negligible difference between the two, as outlined in Sect. 4.3.2. Convection involves *cyclic mass motion* but does not generally violate our assumption of stellar stability since there is *no net mass motion* within the star. Either of Eqs. 6.9 or 6.10 represents the fourth equation of stellar structure, depending on which one is dominant in a given region in a star.

6.1.5 Constitutive Relations

The four equations of stellar structure outlined in the previous sections, namely, conservation of mass (Sect. 6.1.1), hydrostatic equilibrium (Sect. 6.1.2), energy conservation (Sect. 6.1.3) and energy transport (Sect. 6.1.4), describe the star globally and show how stellar properties (M_r, $\rho(r)$, $P(r)$, L_r and $T(r)$) vary throughout the star.[4] It

---

[3] Electron conduction (Sect. 4.4) is most important in stellar remnants such as white dwarfs and in the degenerate cores of stars in certain post-main-sequence stages of evolution.
[4] These equations could also be expressed as functions of mass instead of radius (Prob. 6.1).

is evident, though, that we do not yet have a sufficient number of equations to solve for the unknowns.

We do, however, have relationships that describe specific properties of matter *at any location r*. Rather than describing how a property varies with radius, they describe the physical state of matter at some position. Thus, they are called *constitutive relations*, i.e. how is the material 'constituted'? Each constitutive equation is a function of density, pressure, temperature and chemical composition. We have already seen examples of these relations, summarized here as

$$P = P(T, \rho, [X, Y, Z]) \tag{6.11}$$

$$\kappa_R = \kappa_R(T, \rho, [X, Y, Z]) \tag{6.12}$$

$$\epsilon = \epsilon(T, \rho, [X, Y, Z]) \tag{6.13}$$

Equation 6.11 specifies the *equation of state*. We have seen a good example almost universally applicable to normal stars, namely the ideal gas law (Sect. 2.1.1). However, other equations of state are possible (e.g. Sect. 6.2.2.2).

Equation 6.12 specifies the *Rosseland mean opacity* as defined in Eqs. 4.1 and 4.10. Detailed numerical work is required to determine a value of κ_R for a given density, temperature and composition, and Rosseland mean opacities are computed as grids (tables) that can be referenced when modelling a star (e.g. Prob. 4.2). However, several algebraic expressions are known, such as Kramers' laws, and these are useful for understanding the Rosseland mean for specific types of opacity, as outlined in Sect. 4.2.2.

Equation 6.13 was considered in Sects. 5.2 and 5.4. It specifies the way in which energy is generated and is needed for Eq. 6.8. Examples for the PP-chain and CNO cycle are given in Eq. 5.20 and 5.23, respectively.

6.1.6 Boundary Conditions

In order to solve the equations of stellar structure, boundary conditions are also required. The simplest approach is to specify the following:

$$\left.\begin{array}{l} M_r \to 0 \\ L_r \to 0 \end{array}\right\} \quad as \quad r \to 0 \tag{6.14}$$

$$\left.\begin{array}{l} T \to 0 \\ P \to 0 \\ \rho \to 0 \\ M_r \to M_\star \\ L_r \to L_\star \end{array}\right\} \quad as \quad r \to R_\star \tag{6.15}$$

Here, $M_\star$ and $L_\star$ represent the total mass and total luminosity of the star, respectively. Some of these conditions are clearly first approximations. For example, at the surface of a star, the temperature is not zero but rather $T \to T_{eff}$, where T_{eff} is the star's effective temperature (Eq. 2.55). The choice of T_{eff} depends on a definition of the photosphere and actually varies quite strongly with frequency through the outer parts of a star and stellar atmosphere. Nevertheless, $T_{eff} \ll T_c$, where T_c is the central temperature. Compare, for example, $T_c = 1.5 \times 10^7$ K to $T_{eff} = 5780$ K for the Sun. Adopting $T_{eff} \approx 0$ changes the temperature gradient, ∇T, by only 0.04%! Similarly, the Sun's pressure in the photosphere is some 13 orders of magnitude lower than the central pressure, so approximating the surface pressure as zero is completely justified as a starting point for a solar model.

6.2 SOLVING THE EQUATIONS OF STELLAR STRUCTURE

As can be imagined from the previous sections, it is not a simple matter to integrate the equations of stellar structure, together with the constitutive relations and boundary conditions, to find the complete interior structure of a star. Recall that our goal is to find M_r, $\rho(r)$, $P(r)$, L_r and $T(r)$ for every position within a star and, if possible, to go even further by evolving the star over time. Although the full details of this process will not be pursued here, we can get a feel for the problem by outlining several approaches.

6.2.1 Numerical Solutions

The modern and most accurate approach is to tackle the problem numerically using complex computational codes. The standard solar model (SSM) introduced in Sect. 1.3.1 is one example of output from such a code. Many different approaches have been adopted and improved over the years (see e.g. [115]) in an attempt to ensure that the output is unique, stable, precise and convergent to physically reasonable values. For example, if small adjustments are made at the center of the star, large offsets at the surface could result, or vice versa. One practical starting point is to adopt a set of trial solutions for the entire star and then improve upon them in an interactive fashion. This is called the *Henyey method* after its inventor.

In this approach, the star is divided up into many shells of different radius, r, and thickness, Δr. For each shell, the differential equations of stellar structure are written in terms of mass, dm (Lagrangian form, Sect. 6.1), rather than dr. Because shells are designed to be narrow, the differential equations of each parameter, r, P, L and T, can be written as *linear equations* for each shell, effectively fitting straight lines centered at the midpoint of each shell. For pressure, for instance,

$$\frac{dP}{dm} = \frac{P_{[r+(\Delta r)/2]} - P_{[r-(\Delta r)/2]}}{m_{[r+(\Delta r)/2]} - m_{[r-(\Delta r)/2]}} = f(r, P_r, T_r, L_r) \qquad (6.16)$$

$$===> \frac{P_{[r+(\Delta r)/2]} - P_{[r-(\Delta r)/2]}}{m_{[r+(\Delta r)/2]} - m_{[r-(\Delta r)/2]}} - f(r, P_r, T_r, L_r) = 0 \qquad (6.17)$$

Similar equations are written for the other variables, and the set of equations is then written for every shell. The function $f(r, P_r, T_r, L_r)$ would be the relevant stellar structure equation that relates the unknown quantity to the other quantities at the mid-point of the shell. For pressure, for instance, this would be the hydrostatic equilibrium equation. Notice that the density, ρ, does not appear in the Lagrangian form of this equation (cf. Prob. 6.1).

Of course, values at the shell boundaries must match, and global boundary conditions must also be satisfied. Expressing differentials as linear slopes will not produce exact results initially. However, if an iterative approach is adopted, small corrections can be made to each solution at each iteration step. For example, Eq. 6.17 can be iteratively improved by forming the differential of f and applying appropriate mathematical techniques (e.g. the Newton-Raphson method, Online Resource 6.2) to improve the solution. Once a solution is found, the rates of the nuclear processes can be calculated as well as the changing chemical composition in each layer to develop solutions that follow the evolution of a star with time.

Once a stable equilibrium solution is found at some time, however, can we be sure that it is unique and that there are no other possible solutions? Considerable effort has focussed on this question. The general result is that models do tend to be locally unique except at points at which the stability of the star changes significantly. As the star evolves, moreover, its history is also a constraint that helps to direct its path [115].

More details about this and other numerical solutions can be found, for example, in [166, 186]. For a modern version of a numerical solution that includes effects such as convective overshoot, gravitational settling, diffusion and many other effects, see Online Resource 2.2 or the other models listed at the end of Chapter 8.

6.2.2 Conceptual and Analytical Approaches

While numerical models provide the best solutions to the equations of stellar structure, they can be rather opaque from a pedagogical point of view. Are there more conceptual approaches or simplifications that may only be approximate but still provide an understanding of the behaviours of the variables? In general, it is not possible to find a full analytical solution to the equations, but a variety of circumstances can lead to the understanding that we seek. For example, how do we even know that pressure is a declining function of radius from the center to the surface of a star? This question was tackled by Chandrasekhar as early as 1939, and a modified version of his argument is presented in Appendix E. We provide several other examples in this section.

6.2.2.1 At or Near the Centers of Stars

A good starting point is to explore the values and behaviour of stellar parameters at or near the center of a star. It is fairly easy to place boundary conditions on values at $r = R$. One need only measure $T = T_{eff}$, $L = L_\star$, and $M = M_\star$. However, historically, the center has been inaccessible. Nevertheless, there is much we can learn with some manipulation of the equations of stellar structure.

For example, suppose we let a star have constant density. We already made such an approximation in the Introduction when we examined a star's potential energy (Eqs. I.4, I.6). If we now combine the equation of continuity (Eq. 6.2) with the equation of hydrostatic equilibrium (Eq. 6.4) and adopt appropriate boundary conditions (e.g. Eqs. 6.14 and 6.15), we find (Prob. 6.2) a central pressure of

$$P_c = \frac{3G}{8\pi} \frac{M^2}{R^4} \tag{6.18}$$

Clearly, stars do not have uniform density, but one could generalize the result, as we did for potential energy in Eq. I.7, by including a multiplicative constant that takes into account a density distribution that declines, in some fashion, with radius. For example, when the central density, ρ_c, is higher than the star's average density, the central pressure can be written [251]

$$P_c \approx C\, GM^{2/3} \rho_c^{4/3} \tag{6.19}$$

where C is a unitless constant that ranges between 0.3 and 0.5, depending on the adopted density distribution, and ρ_c has now been explicitly included. For the Sun, using values of $\rho_c = 151$ g cm^{-3} and $C = 0.3$, the central pressure using Eq. 6.19 is within 11% of the SSM value. Such a result can be used as an initial boundary condition on the equation of hydrostatic equilibrium.

Our result of Eq. 6.18 shows that $P \propto M^2/R^4$. That is, if a star's radius increased by a factor of 2 (no change to the shape of its mass or density distribution), then its internal pressure (at the same r/R location) should decrease by a factor of 16. If *all* stars had the same density distribution, a situation called *homology* [255], then we could use the proportionalities of Eq. 6.18 to compare the behaviours of different stars. For example, if two homologous stars have the same size but one is a factor of 2 more massive than the other, then the more massive star will have an internal pressure that is a factor of 4 higher. Again, these results are approximations, but they are useful *scaling relations* that help to provide a physical understanding of the properties of stars. Scaling relations, with an emphasis on dimensional analysis, for a variety of stellar parameters are summarized in Box 6.1 on page 137, and we will pursue these concepts further in Sect. 6.3.

What about the behaviour of stellar parameters *near* the center? It is fairly clear that the *slopes* of the functions, e.g. dP/dr, etc. must vanish at the center in order to avoid discontinuities. An examination of the plots shown in Appendix C indeed shows that the slopes flatten as $r \to 0$. But analytical functions can also be explicitly found near the centers of stars by carrying out Taylor expansions of the variables.

These are outlined in Appendix D. Equations D.6, D.10, D.11 and D.12 give expressions for the mass, pressure, luminosity and temperature, respectively, of a star near its center.

6.2.2.2 Polytropes

Probably the best-known and exploited example of a simplification is when a star or stellar region is a *polytrope*. The first two equations of stellar structure, conservation of mass and hydrostatic equilibrium (Eqs. 6.1 and 6.4), describe the *structure* of a star. An examination of these two equations shows that there are three variables, M_r, P and ρ as a function of r. If we try to use these equations to solve for a star's structure, we need to find a relation between ρ and P so the two equations have only two unknowns. However, the connection between ρ and P is normally via the ideal gas law, $P \propto \rho T$, which adds another variable, T.

Fortunately, there are some circumstances for which pressure and density are related by the form

$$P = K\rho^{\gamma} = K\rho^{\frac{n+1}{n}} \tag{6.20}$$

where K is a constant, γ is the *polytropic exponent* and n is called the *polytropic index*. An object that has such an equation of state is called a *polytrope*. In such a case, the structure equations can be solved[5] without introducing the temperature variable. On the other hand, no information about temperature can immediately result from such a solution. The pressure *profile*, i.e. the form of the radial distribution, becomes steeper (more heavily weighted towards the center) as n increases.

What are relevant examples of polytropes?

The simplest polytrope is an isothermal sphere. With T = constant, the ideal gas law gives just $P \propto \rho$, as pointed out in Eq. 2.38. For an isothermal sphere, $\gamma = 1$ and $n \Rightarrow \infty$. A star, as a whole, is clearly not an isothermal sphere. However, there are instances in which the core of a star can be approximated as isothermal, similar to white dwarfs (Sect. 10.2).

An adiabatic ideal gas is also a polytrope. Recall that an adiabatic gas is one in which changes in temperature only occur as a result of expansion or compression, and not as a result of heat being exchanged with the surroundings (Sect. 2.1.8). As we showed in Eq. 2.42, $P \propto \rho^{\gamma_a}$ for an adiabatic gas, in which case the polytropic exponent is just the adiabatic index, γ_a. A typical gas with three degrees of freedom has $\gamma_a = 5/3$ (Eq. 2.35), and therefore the polytropic index is n = 1.5. Note that because the ideal gas law must also be obeyed in this case, once the structure (i.e. the pressure and density) is determined, the temperature profile is also fixed (Prob. 6.5).

---

[5] The equation to be solved is called the *Lane-Emden equation* (e.g. [44]).

Since convection, as outlined in Sect. 4.3.1, involves adiabatically rising and falling pockets of gas, such regions are also polytropes. In convective regions, the appropriate temperature gradient is given by Eq. 4.29, which can be rewritten as

$$\frac{dT}{dP} = \left(\frac{\gamma_a - 1}{\gamma_a}\right)\frac{T}{P} = \frac{2}{5}\frac{T}{P} \tag{6.21}$$

assuming a typical value of $\gamma_a = 5/3$. An integration of this equation gives $T \propto P^{2/5}$ and, from the ideal gas law, $T \propto P/\rho$, so

$$P \propto \rho^{5/3} \Rightarrow n = 1.5 \tag{6.22}$$

As can be seen, this is the same result given in the previous paragraph, as it should be, since both describe adiabatic ideal gases with the same adiabatic index. Here, we have simply started explicitly from Eq. 4.29, which was introduced in our discussion of convection.

Our final examples of polytropes are objects consisting of matter that is *degenerate*: that is, the matter has such a high density that particles are boosted to very high velocities according to the Heisenberg uncertainty principle (cf. Sect. 4.4). Stellar *remnants* such as white dwarfs and neutron stars, or regions of degenerate matter in the cores of some evolved stars, are examples. It is the degeneracy pressure from high-velocity particles that is holding the star up. In such cases, the temperature profile of the object is completely decoupled from its pressure and density profiles. In a white dwarf, electron degeneracy pressure is the dominant pressure, and in a neutron star, it is neutron degeneracy pressure. It is remarkable, for example, that the motions of tiny electrons can produce sufficient pressure to keep a white dwarf from collapsing in on itself; the temperature profile is irrelevant.

Examining two separate regimes in which particles are non-relativistic and relativistic, the polytropic equation of state can be written,

$$P = K_1 \rho^{5/3} \qquad \text{(non-relativistic)} \tag{6.23}$$

$$P = K_2 \rho^{4/3} \qquad \text{(relativistic)} \tag{6.24}$$

where K_1 and K_2 are constants. The polytropic indices are then 1.5 and 3 for the non-relativistic and relativistic cases, respectively. This means that a relativistic degenerate object has a pressure profile that is steeper than a non-relativistic one. We will discuss these objects in greater detail in Sects. 10.2 and 10.4 and also show that Eqs. 6.23 and 6.24 lead to important and non-intuitive properties for white dwarfs and neutron stars.

Polytropes can be simply calculated using Online Resource 6.1.

Box 6.1
Stellar Parameters, by Dimensions

Equations 6.25, 6.27, 6.29 and 6.33 show how a star's parameters can be expressed by using a combination of other quantities. Those equations apply to any star that can be described by an ideal gas and in which radiative diffusion is the energy transport mechanism. Let us summarize these expressions and check whether the cgs units are dimensionally correct:

$$\rho \propto \frac{M_r}{r^3} \rightarrow \frac{[g]}{[cm]^3}$$

$$T \propto \frac{m_H G M_r}{kr} \rightarrow \frac{[g][cm^3 g^{-1}s^{-2}][g]}{[erg\,K^{-1}][cm]} = \frac{[g][cm^2 s^{-2}]}{[erg\,K^{-1}]} = \frac{[erg]}{[erg\,K^{-1}]} = [K]$$

$$P \propto \frac{G M_r^2}{r^4} \rightarrow \frac{[cm^3 g^{-1}s^{-2}][g^2]}{[cm^4]} = \frac{[g\,cm\,s^{-2}]}{[cm^2]} = \frac{[dyn]}{[cm^2]}$$

$$L_r \propto \frac{acrT^4}{\kappa_R \rho} \rightarrow \frac{[erg\,cm^{-3}K^{-4}][cm\,s^{-1}][cm][K^4]}{[cm^2 g^{-1}][g\,cm^{-3}]} = \frac{[erg]}{[s]}$$

These relations apply to interior values (where r is taken depends on the specifics of the radial distributions and the problem at hand). They are also *scaling relations* because they help us to see how the parameter of interest scales when one of the other quantities changes. For example, considering the entire star, if two stars have the same mass and chemical composition but the radius of star 2 is 20% greater than the radius of star 1 ($R_2 = 1.2R_1$), then $\rho_2 = 0.58\rho_1$, $T_2 = 0.83T_1$, $P_2 = 0.48P_1$ and $L_{r_2} = L_{r_1}$. The final equality for luminosity is found by using the previously obtained proportionalities. Notice that the luminosity is equal for two stars that have the same mass and chemical composition, consistent with the mass-luminosity relation discussed in the text (e.g. Eq. 6.35).

6.3 THE VOGT-RUSSELL THEOREM, MASS-LUMINOSITY RELATION AND MASS-RADIUS RELATION

'If the pressure, P, the opacity, κ, and the rate of generation of energy, ϵ, are functions of the local values of ρ, T, and the chemical composition only, then the structure of a star is uniquely determined by the mass and the chemical composition.'

Such is a statement of the Vogt-Russell theorem. The *condition* in the statement is satisfied by the constitutive relations (Eqs. 6.11 through 6.13). Consider what a strong statement this is. With the number of equations that have been introduced previously, the Vogt-Russell theorem states that only the total mass and chemical composition are required to uniquely determine all properties of a star. How can this be, and is it true?

As it turns out, there are examples of conditions that violate this statement (see e.g. [114, 115]), so calling it a theorem rather loosely stretches what is normally accepted from a mathematical standpoint. For example, it is possible to 'construct' two different stable stars of the same mass and global chemical composition whose radii are very different, an example being a one-solar-mass main sequence star like the Sun and a one-solar-mass evolved subgiant star. *Nevertheless*, the argument that follows is not bad for main sequence stars, and, in fact, computer codes such as the one given in Online Resource 2.2 require only the mass and chemical composition (and time for stopping the calculations) as minimal inputs. It is therefore instructive to go through the steps.

In the following, we are interested in how a property scales with other properties (scaling relations, as described in Sect. 6.2.2), so we will consider the star 'as a whole' and take gradients to apply to the entire star, i.e. $dP/dr \approx P/R$, etc. This means we are ignoring details related to the radial profile of the property.

We start by writing the simple relation between mass and density,

$$\rho \sim \frac{M}{\frac{4\pi}{3}R^3} \propto \frac{M}{R^3} \tag{6.25}$$

and then insist that the star be in hydrostatic equilibrium. From Eqs. 6.4 and 6.25,

$$\frac{dP}{dr} = -G\frac{M_r}{r^2}\rho \Rightarrow \frac{P}{R} \sim -G\frac{M}{R^2}\frac{M}{\frac{4\pi}{3}R^3} \Rightarrow P \sim G\frac{3}{4\pi}\frac{M^2}{R^4} \tag{6.26}$$

$$P \propto [G]\frac{M^2}{R^4} \tag{6.27}$$

We have seen this relation several times before, e.g. Eqs. 6.18 and E.7, although with adjusted multiplicative constants, depending on the specifics of the radial density profile. The pressure refers to an interior pressure, and indeed it must, because the pressure at the surface goes to zero. It can be thought of as the central pressure but could be a pressure at some other interior point, depending on the density profile and what constants are used. Notice that the gravitational constant, G, has been retained in the proportionality of Eq. 6.27 because it has dimensions (cf. Box 6.1 on page 137). We will continue to put constants that have dimensions into square brackets.

The equation of state for normal stellar material is the ideal gas law, for which we adopt the form $P = \rho kT/(\mu m_H)$ (Eq. 2.4). Rearranging this equation for T and using

Eqs. 6.25 and 6.26 gives

$$T = \frac{\mu m_H}{\rho k} P \sim \frac{\mu m_H}{\frac{M}{\frac{4\pi}{3}R^3}k} G \frac{3}{4\pi} \frac{M^2}{R^4} \sim \frac{\mu m_H}{k} \frac{G}{R} \frac{M}{R} \tag{6.28}$$

$$T \propto \left[\frac{m_H G}{k}\right] \frac{M}{R} \tag{6.29}$$

which, again, would refer to an interior temperature. The constants that have dimensions are Boltzmann's constant, k, the gravitational constant, G, and the mass of a hydrogen atom, m_H. The constant, μ, is unitless and of order one, so it might have been ignored in Eq. 6.28, but we retain it because it contains information about the chemical composition.

We need to consider the flow of energy through the star, which we take to be via radiative transport. Let us re-write Eq. 4.21 in the following way (Prob. 6.7):

$$\frac{d}{dr}\left(\frac{1}{3}aT^4\right) = -\frac{\kappa_R \rho}{c} \frac{L_r}{4\pi r^2} \tag{6.30}$$

Rearranging this equation to solve for L_r,

$$L_r = -\frac{4\pi a c}{3} \frac{1}{\kappa_R \rho} r^2 \frac{d\left(T^4\right)}{dr} \tag{6.31}$$

and using star-wide gradients, as before,

$$L_{rad} \sim \frac{4\pi a c}{3} \frac{1}{\kappa_R \rho} R^2 \frac{T^4}{R} \sim \frac{4\pi a c}{3} \frac{RT^4}{\kappa_R \rho} \tag{6.32}$$

$$L_{rad} \propto \left[\frac{a c}{\kappa_R}\right] \frac{RT^4}{\rho} \tag{6.33}$$

where we have subscripted the luminosity as L_{rad} to indicate that the radiative luminosity is due to radiative transport rather than convective or other form of energy transport. The constants that have dimensions are the radiation constant, a, the speed of light, c, and the Rosseland mean opacity, κ_R. Note that κ_R could be a function (discussed shortly).

We can now insert Eq. 6.25 for ρ and Eq. 6.28 for T into Eq. 6.32 to find

$$L_{rad} \sim \frac{16\pi^2 a c}{9\kappa_R}\left(\frac{m_H G \mu}{k}\right)^4 M^3 \tag{6.34}$$

and therefore,

$$L_{rad} \propto \left[\frac{a c}{\kappa_R}\left(\frac{m_H G}{k}\right)^4\right] \mu^4 M^3 \tag{6.35}$$

Now suppose that the opacity follows a Kramers' law (e.g. Eq. 4.11 or 4.12), with κ_0 the appropriate constant that is dependent on the composition. Then,

$$\kappa_R = \kappa_0 \rho T^{-3.5} \tag{6.36}$$

$$\sim \kappa_0 \frac{M}{\frac{4\pi}{3} R^3} \left(\frac{\mu m_H G}{k} \frac{M}{R} \right)^{-3.5} \tag{6.37}$$

$$\sim \frac{3\kappa_0}{4\pi} \left(\frac{k}{m_H G} \right)^{3.5} \frac{R^{0.5}}{\mu^{3.5} M^{2.5}} \tag{6.38}$$

where, again, we have used Eqs. 6.25 and 6.28. Let us now insert Eq. 6.38 into Eq. 6.34 to find

$$L_{rad} \sim \frac{64\pi^3 ac}{27\kappa_0} \left(\frac{m_H G}{k} \right)^{7.5} \mu^{7.5} \frac{M^{5.5}}{R^{0.5}} \tag{6.39}$$

That is,

$$L_{rad} \propto \left[\frac{ac}{\kappa_0} \left(\frac{m_H G}{k} \right)^{7.5} \right] \mu^{7.5} \frac{M^{5.5}}{R^{0.5}} \tag{6.40}$$

The relations of Eqs. 6.35 and 6.40 represent two possible versions of the *mass-luminosity (M-L) relation*. The first equation ($L \propto M^3$) shows the functional dependence if κ_R is constant. This would be the case, for example, if electron scattering were the dominant source of opacity (cf. Eq. 4.13), such as in very massive hot stars. The second equation shows the functional dependence if κ_R follows a Kramers' law, as is more typical for lower-mass stars like the Sun. In this case, the dependence of luminosity on mass is much steeper ($L \propto M^{5.5}$), and there is also a dependence on R, although a much weaker one.

As can be seen from these results, the power in the M-L relation varies depending on the star's mass, and the opacity is an extremely important parameter in determining that power. Real stars, of course, cannot be treated as simply as has been done in this development. We can, however, simplify and generalize the M-L relation as

$$L \propto M^\gamma \tag{6.41}$$

where γ is the relevant power for the mass range being considered.[6]

All in all, one would expect a M-L relation for stars of different mass to show a varying slope that is steeper for low-mass stars like the Sun and shallower for

---

[6] The use of the Greek γ is standard in this relation, but do not confuse it with the adiabatic index, γ_a, or the polytropic exponent, γ.

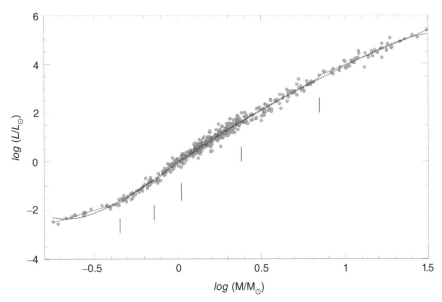

Figure 6.2 The observational mass-luminosity relation for main sequence stars. The small vertical lines delineate 'break-points' that separate of the different mass regimes given in Table 6.1. Red and blue curves represent different types of mathematical fits to the data. Credit: Z. Eker and collaborators. [89]/With permission of Oxford University Press.

high-mass stars. Observationally, this is indeed the case, as illustrated in Fig. 6.2 for main sequence stars. In this log-log plot, a single power law for all masses would be a straight line, but instead we see curvature. The corresponding numerical relations can be found in Table 6.1. For example, our Sun would lie in the regime in which $L \propto M^{5.743}$ (cf. $L \propto M^{5.5}$, Eq. 6.40) whereas a star of mass $M = 8 M_\odot$ would lie in the regime in which $L \propto M^{2.865}$ (cf. $L \propto M^3$, Eq. 6.35). Given the simplicity of our theoretical development, this agreement is surprisingly good.

However, it is far from the whole story. Some of the explanations for different slopes in different mass regimes (Table 6.1) appear to be related to the efficiency and types of nuclear energy generation. For example, the 'break-point' at 0.72 $M_\odot$ could be related to the increasing importance of the PPII and PPIII branches of the PP-chain of nuclear reactions (Fig. 5.3). And the break-point at 1.05 $M_\odot$ is likely related to the increasing importance of the CNO cycle compared to the PP-chain. The *shallower* slopes for the ultra-low- and low-mass stars are related to the fact that such stars can be fully convective (see Fig. 4.6), which was not considered in the previous development. Convection tends to mix material (Sect. 4.3.3), bringing cooler material from high regions into the core and lowering the nuclear burning efficiency. The net result is a shallower slope. More discussion and the uncertainties inherent in these relations can be found in [89].

Table 6.1 Main sequence mass-luminosity relations for six mass domains

Domain	Mass Range	Equation
Ultra Low Mass	$0.179 < M/M_\odot \leq 0.45$	$log(L) = 2.028\,log(M) - 0.976$
Very Low Mass	$0.45 < M/M_\odot \leq 0.72$	$log(L) = 4.572\,log(M) - 0.102$
Low Mass	$0.72 < M/M_\odot \leq 1.05$	$log(L) = 5.743\,log(M) - 0.007$
Intermediate Mass	$1.05 < M/M_\odot \leq 2.40$	$log(L) = 4.329\,log(M) + 0.010$
High Mass	$2.4 < M/M_\odot \leq 7$	$log(L) = 3.967\,log(M) + 0.093$
Very High Mass	$7 < M/M_\odot < 31$	$log(L) = 2.865\,log(M) + 1.105$

Credit: [89]. The mass, M, is in units of $M_\odot$, and luminosity, L, is in units of $L_\odot$.

At masses that are *higher* than the limit of Table 6.1 ($M \gtrsim 30\,M_\odot$), the slope, γ, is difficult to determine because of the paucity of extremely high-mass stars. This is because of the low number of high-mass stars formed in any star-forming event and also because of the relatively brief lifetimes of such stars (e.g. Fig. 7.1). According to [175], $\gamma \sim 3$ until masses exceed about 100 $M_\odot$, and then $\gamma \sim 1$ thereafter. A shallower slope is eventually expected because of the increasing importance of radiation pressure in high-mass stars (cf. Sect. 7.4.2) plus convection in the core (Fig. 4.6) [186].

We are not quite finished, nor have we demonstrated the Vogt-Russell theorem. We have, however, reduced our equations of stellar structure to equations (Eqs. 6.35 and 6.40) involving just three parameters: luminosity, L; mass, M; and chemical composition, represented via the mean molecular weight, μ, and the opacity, κ_0 or κ_R. (The second equation also has a weak dependence on radius, R.) These equations have made *no assumption about the source of energy*. They simply describe the energy loss. We must now, therefore, look at the energy generation rate.

The luminosity due to nuclear energy *generation*, L_{gen}, is given by Eq. 6.8, again crudely approximated over the whole star, as we did before.

$$\frac{dL_r}{dr} = \epsilon\,\rho(r)\,4\pi\,r^2 \Rightarrow \frac{L_{gen}}{R} \sim \epsilon\,\frac{M}{\frac{4\pi}{3}R^3}\,4\pi\,R^2 \tag{6.42}$$

$$L_{gen} \sim 3\,\epsilon\,M \tag{6.43}$$

where ϵ (erg s^{-1} g^{-1}) is the energy generation rate per unit mass and we have used Eq. 6.25. As indicated in Sect. 6.1.3, ϵ is not a constant; it is a function of both ρ and T, and the functional forms vary depending on which nuclear reactions are taking place. The strongest dependence, though, is on temperature (Sec. 5.4). For low-mass stars like the Sun, on the main sequence, hydrogen is burned via the *PP-chain*, for which we will write the dependence, following Eq. 5.22, as

$$\epsilon = \epsilon_0\,\rho\,T^4 \tag{6.44}$$

Here ϵ_0 *also* depends on the chemical composition. For higher-mass stars that burn according to the *CNO cycle*, the power on the temperature is much higher (cf. Eq. 5.25). However, here we will proceed with Eq. 6.44.

We now insert Eq. 6.44 into Eq. 6.43, making use of Eqs. 6.25 and 6.28, to find

$$L_{gen} \sim \frac{1}{4\pi} \left[\frac{m_H G}{k} \right]^4 (\epsilon_0 \mu^4) \frac{M^6}{R^7} \tag{6.45}$$

For a star in radiative equilibrium, we can equate Eq. 6.39 with Eq. 6.45 (i.e. $L_{rad} = L_{gen}$) so that we can solve for the radius, R. For simplicity, let us drop all constants, including those with dimensions, and write only the dependencies:

$$R \propto \left(\frac{\kappa_0 \epsilon_0}{\mu^{3.5}} \right)^{0.15} M^{0.08} \tag{6.46}$$

Each term in the parentheses depends on the chemical composition, so we can now see that *the stellar radius can be expressed in terms of only two parameters: the stellar mass and chemical composition!* This is the essence of the *mass-radius (M-R) relation*, demonstrated here for stars in which the opacity is dominated by a Kramers' law, which are undergoing hydrogen burning via the PP-chain and whose energy transport is radiative only. Such assumptions are rather specific and not really realistic since the full range of stellar parameters includes convection, other reactions such as the CNO cycle and other contributors to the opacity. However, the relation does argue for the existence of a M-R relation and suggests that a wide range of masses result in a smaller range of radii.

The observational M-*R* relation is shown in Fig. 6.3. Main sequence stars are between the two curves that delineate the zero age main sequence (ZAMS, blue) – the point at which stars begin their nuclear burning lifetimes – and the terminal age main sequence (TAMS, red) – the point at which stars have exhausted their hydrogen nuclear fuel and move off the main sequence. The most obvious change in slope occurs roughly at the transition between PP-chain dominance and CNO-cycle dominance. As can be seen from both this figure and Eq. 6.46 (and noted earlier), the dependence of radius on mass is weaker than it is for the mass-luminosity relation. For the same range of mass, the luminosity varies by nine orders of magnitude (Fig. 6.2), whereas the radius varies by less than two orders of magnitude. In other words, stars of different mass have very different luminosities but not so very different radii. The power derived in Eq. 6.46 is flatter than the observational result, so it is a poorer match to the observations.[7] An extension of the M-*R* relation to lower mass can be found in Fig. 7.3.

---

[7] The authors of [89] have fit the M-*R* relation with a polynomial, rather than a single power law, finding $R = 0.438 M^2 + 0.479 M + 0.075$ for the mass range $0.179 \leq M/M_\odot \leq 1.5$. For the mass range $1.5 \leq M/M_\odot \leq 31$, they use a relation $T_{eff} = -0.170 (\log M)^2 + 0.888 \log M + 3.671$, where T_{eff} is the effective temperature. Once T_{eff} is found, the radius can be determined from the Stefan-Boltzmann law (Eq. 2.56) and the M-*L* relations given in Table 6.1.

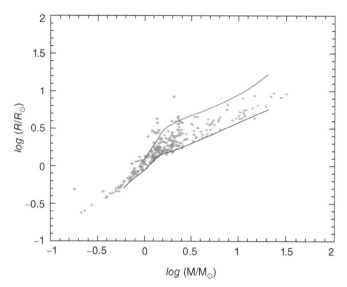

Figure 6.3 The observational mass-radius relation for main sequence stars. The blue curve represents the ZAMS (zero age main sequence) and the red curve represents the TAMS (terminal age main sequence). Grey circles represent stars on the main sequence, and green diamonds represent stars that are probable non-main-sequence stars. Credit: Z. Eker and collaborators [88]/With permission of IOP Publishing.

Finally (!) we have demonstrated (though far from proved) the Vogt-Russell theorem, 'the structure of a star is uniquely determined by the mass and chemical composition'. The essence of the argument is contained in the equations that we developed in this section, although the powers can vary for real stars due to assumptions that have been made and the crude way in which the derivatives were handled. Working backwards, given the chemical composition (contained in μ, ϵ_0 and κ_0) and mass, one can determine:

1. The radius, demonstrated via Eq. 6.46
2. The luminosity, demonstrated via Eq. 6.35 or 6.40
3. The density, demonstrated via Eq. 6.25
4. The pressure, demonstrated via Eq. 6.27
5. The temperature, demonstrated via Eq. 6.29

Because the equations in this section were derived from gross stellar properties with specific assumptions that do not apply to every star or every region within stars, they should not be used to do numerical calculations. They are meant only to illustrate functional dependencies of stellar properties, where possible. However, this development nicely illustrates how the Vogt-Russell theorum can be understood physically, using stellar equations that we already know.

Box 6.2
Starring ... Dubhe (α UMa)

Distance: $d = 37.7$ pc; Magnitude: V = 1.87; Spectral Type: K0II-K0III Effective Temperature: $T_{eff} = 4650$ K; Mass: M = 3.7 $M_\odot$ [125].

Van Gogh's love of the starry firmament led him to paint a number of scenes showing the beauty of the night sky. None, though, showed recognizable patterns as well as his *Starry Night Over the Rhone*, painted in the year 1888. Here, we see the Big Dipper, an asterism in the constellation of Ursa Major (the Big Bear). Two 'pointer stars' in the bowl of the Big Dipper are *Dubhe*, marked by the red arrow in the picture, and *Merak*, which is directly below it. Together, they form a line that points upwards to the North Star. Like other bright stars, Dubhe has been called by various names in different cultures throughout history. The Chinese called it Beiji Gouchen [4], and the Inuit of Igloolik, Nvt, saw it as part of a caribou [199].

Dubhe is a red giant star and has a companion that is a main sequence star of spectral type either A or F. The orbital period is 44.5 years. For such a bright star, it is surprising that there is uncertainty about some of Dubhe's parameters. Masses range from 3.7 to 4.5 $M_\odot$, and the secondary could be between 1.8 and 2.5 $M_\odot$. Newer spectroscopic measurements of Dubhe result in a rotation speed of 2.7 km s^{-1}, which, for a radius of $R = 26 R_\odot$, gives a rotation rate of less than 1.35 years [125].

Online Resources

6.1   *R. Townsend's mad star EZ-Web Poly-Web code for calculating polytropes*: http://www.astro.wisc.edu/~townsend/static.php?ref=poly-web

6.2   *The Newton-Raphson method for finding the root of a function*: https://mathworld.wolfram.com/NewtonsMethod.html

PROBLEMS

6.1. (a) Express each of the four main equations of stellar structure in *Lagrangian* form (differentials with respect to mass, dm) rather than Eulerian form (differentials with respect to radius, dr). Simplify the results where possible. For example, the equation of continuity (Eq. 6.2) can be rewritten simply by inverting dm and dr to find how the radius varies with mass,

$$\frac{dr}{dm} = \frac{1}{4\pi r^2 \rho(r)} \tag{6.47}$$

   (b) Compare the Lagrangian equations to the Eulerian equations. For each equation, indicate which version is simpler, i.e. has fewer variables in the expression.

6.2. For a uniform-density star, using the first two equations of stellar structure, derive Eq. 6.18.

6.3. Consider the energy transport equation for convection (Eq. 6.10). Using the appropriate substitutions, such as the equation of hydrostatic equilibrium (Eq. 6.4), find a simpler expression for dT/dr that involves only M_r, r and some constants.

6.4. For a uniform-density star:

   (a) Derive a simple expression for the normalized pressure, $P(r)/P_c$, as a function of normalized radius, r/R, where P_c is the central pressure (Eq. 6.18) and $P(r)$ is given by the Taylor expansion of Eq. D.10. Justify the use of the Taylor expansion equation.

   (b) Plot the result from $0 \le r/R \le 1$.

   (c) In a real star, how would you expect the curve to change?

6.5. Consider a convective region in which all pockets of gas are moving adiabatically with an adiabatic index of $\gamma_a = 5/3$. Suppose this region has a density distribution that can be described by $\rho(r) \propto (1 - r/R)$, where R is the radius of the star.

   (a) Find the proportionalities for $P(r)$ and $T(r)$ for this polytrope.

   (b) Of the three distributions, which one shows the steepest decline with r? Which is the shallowest?

   (c) Examine numerical data from the SSM (Online Resource 1.12) for a region of convection. Take a density ratio in this region, and determine whether $P(r)$ and $T(r)$ behave as predicted from part (a).

6.6. Consider a massive star within which electron scattering is the dominant opacity mechanism and the energy generation is by the CNO cycle. Starting with Eq. 6.35, derive a mass-radius proportionality relation for this star, as was done in Eq. 6.46. Retain only the constants that relate to the chemical composition. Compare your result to the plot of Fig. 6.3, and comment.

6.7. Show that Eq. 4.21 is equivalent to Eq. 6.30.

6.8. Let us briefly explore the online numerical code madstar (see Online Resource 2.2). We will compare two main sequence stars: a 3 $M_\odot$ star whose age is $t = 10^8$ years, and the Sun at its current age. Request a 'summary' output for these two stars, using a metallicity of $Z = 0.02$ and cgs units.

(a) What are the luminosity (in units of $L_\odot$), radius ($R_\odot$), surface temperature (K), central temperature (K), central density (g cm^{-3}) and central pressure (dyn cm^{-2}) of both stars? For the Sun, verify that the output agrees with the SSM.

(b) From the numerical results for the two stars, find the ratio of the central densities, the central pressures and the central temperatures. (Do these results surprise you?)

(c) Compare the numerical ratios to what you would expect from the rough scaling relations given in Eqs. 6.25 (density), 6.27 (pressure) and 6.29 (temperature). Comment on your results.

Chapter 7

The Quasistatic Star – Energies, Timescales and Limits

Nature does not hurry, yet everything is accomplished.

Lao Tzu [305]

When we place a dot, representing a star, on the H-R diagram (Fig. I.2), it would take a very long time until the dot would have to move. Stars on the main sequence, for example, spend the longest time of their nuclear burning lifetimes 'sitting' on the main sequence. But changes are indeed occurring in such a star. Like a child growing, from day to day, the growth is imperceptible; but when a year passes, the growth spurt becomes obvious. For a star like the Sun, billions of years

Astrophysics: Decoding the Stars, First Edition. Judith Irwin.
© 2023 John Wiley & Sons Ltd. Published 2023 by John Wiley & Sons Ltd.

could elapse before it moves perceptibly from position A to position B on the H-R diagram. In astronomy, we call 'A to B' changes *secular* (as opposed to *periodic*). With such long time periods, we generally do not see secular changes in any given star and instead have to piece together an evolutionary path from the known properties of many stars combined with theory. Stars are therefore *quasistatic* (or *quasistable*), which means the timescale for secular changes in a star is longer than the timescale required to establish stability, as governed by the equations of stellar structure (Sect. 6.1; see also Sect. 7.3). Except where evolutionary changes are occurring quickly, even stars that have evolved off of the main sequence can usually be considered quasistatic. In this chapter, we explore some of the issues related to the quasistatic star.

7.1 THE VIRIAL THEOREM

The virial theorem is widely used in astrophysics and reveals the relations between a star's potential energy its thermal energy and its total energy. If other types of energy are present (e.g. magnetic energy, turbulent energy), they can be included as well, although we do not include them here. There are a variety of ways to approach this theorem, but we will begin with the equation of hydrostatic equilibrium (Eq. 6.4), which is a statement that the internal pressure of a star has to balance the gravitational pressure. This equation says nothing about the source of internal pressure, only that there must be a balance in a stable star. We start by changing the variable in order to express the pressure in terms of mass rather than radius (Prob. 6.1):

$$dP = -\frac{G\,M_r}{4\,\pi\,r^4}\,dm \qquad (7.1)$$

Notice that the density, ρ, that was in the original equation of hydrostatic equilibrium is no longer explicitly present.

We will now multiply Eq. 7.1 by the volume $V = (4/3)\,\pi\,r^3$ at some radius r and integrate over the star from the center to the surface. At the center of the star, the radius is $r = 0$, the pressure is $P(r) = P_c$ and the mass is $M_r = 0$. At the surface, the radius is $r = R$, the pressure is $P(R) = 0$ and the mass is $M_R = M$.

$$\int_{P_c}^{P(R)=0} V \, dP = -\int \frac{G M_r}{4 \pi r^4} \left(\frac{4 \pi r^3}{3} \right) dm \tag{7.2}$$

$$= \frac{1}{3} \left[-\int_0^M \frac{G M_r}{r} \, dm \right] \tag{7.3}$$

The quantity in the square brackets is the *gravitational potential energy*, U_G, of the star (recall Eq. I.13), i.e.

$$U_G = -\int_0^M \frac{G M_r}{r} \, dm \tag{7.4}$$

By convention, the potential energy at infinity is set to zero, so gravitational potential energy is always negative, and the potential energy of a more-compressed object is more negative.[1] If a star were to expand (r larger), then U_G would become less negative (less tightly bound); if a star were to contract, then U_G would become more negative (more tightly bound). From Eqs. 7.3 and 7.4,

$$\int_{P_c}^{P(R)=0} V \, dP = \frac{1}{3} U_G \tag{7.5}$$

Let us now integrate the left-hand side of Eq. 7.5 by parts:

$$\int_{P_c}^{P(R)=0} V \, dP = P V \big|_{P_c}^0 - \int_0^{V(R)} P \, dV = -\int_0^{V(R)} P \, dV \tag{7.6}$$

The term $P V \big|_{P_c}^0$ disappears because at the surface of the star, $P(R) = 0$, and at the center of the star, $V = 0$. Finally, we equate the right-hand side of Eq. 7.5 with the right-hand side of Eq. 7.6 to find

$$\frac{1}{3} U_G = -\int_0^{V(R)} P \, dV \tag{7.7}$$

Students of thermodynamics will recognize the $-PdV$ term in Eq. 7.7 as expansive or compressive *work*.

We can change the variable from volume to mass using $dm = \rho \, dV$ to find

$$U_G = -3 \int_0^M \frac{P}{\rho} \, dm \tag{7.8}$$

Equation 7.8 is the *general virial theorem* expressed *globally*. By 'globally', we just mean the limits have been chosen to be from $r = 0$ to $r = R$. The expression could also be applied out to any mass, M_r, in a star by changing the upper limit.

---

[1] Recall that the work done by gravity is $W_G = -\Delta U_G = -\left(U_{Gf} - U_{Gi} \right) = -U_{Gf}$, where f and i represent the final and initial values, respectively. U_{Gf} is negative, so the work done by gravity in bringing an object from infinity to its current configuration is positive (force in the same direction as the displacement). The potential energy is negative.

A more common representation of the virial theorem relates the potential energy to the *kinetic energy*, which, for a gas, refers to the mean random motions of particles, i.e. the *internal thermal* energy. Let us now assume an ideal gas, which is the most universal equation of state since it applies to all stars with the exception of regions of degeneracy (e.g. Sect. 4.4). By standard thermodynamics, the average thermal energy (see Eq. 2.17) per unit mass is

$$u_{th} = \left(\frac{3}{2} kT\right) \frac{1}{\mu m_H} \tag{7.9}$$

$$= \frac{3}{2} \frac{P}{\rho} \tag{7.10}$$

where μ is the mean molecular weight, m_H is the mass of the hydrogen atom, and we have made use of the ideal gas law (Eq. 2.4) in going from Eq. 7.9 to Eq. 7.10.

Now the *total internal thermal energy*, U_{th}, requires an integration of u_{th} over all mass from the center to the surface of the star. We will apply this integration to Eq. 7.9 first:

$$U_{th} = \frac{3}{2} \int_0^M \frac{kT}{\mu m_H} \, dm \tag{7.11}$$

Eq. 7.11 relates the total internal thermal energy to stellar temperature. We now repeat the integration, but for Eq. 7.10:

$$U_{th} = \frac{3}{2} \int_0^M \frac{P}{\rho} \, dm \tag{7.12}$$

We can now compare Eq. 7.12 to Eq. 7.8 and, in so doing, find that

$$\boxed{U_{th} = -\frac{1}{2} U_G \quad \Rightarrow \quad 2 U_{th} + U_G = 0} \tag{7.13}$$

Equation 7.13 is one representation of the virial theorem, in this case relating thermal to gravitational potential energy. This simple equation presents a very important result with far-reaching consequences. It relates structure via gravitational energy (right-hand side) to temperature via internal energy (left-hand side). If we know the potential energy (recall that this is a negative quantity, Eq. 7.4), then the internal thermal energy (a positive quantity, Eq. 7.11) can be calculated, or vice versa, *without any other knowledge*. It is not necessary to know the source of the internal heating. The only assumptions are that the star is stable and that (in this example) the ideal gas law holds.

Equation 7.13 says that a more negative potential energy (higher mass and/or smaller radius) corresponds to a higher internal thermal energy. Immediately, we see that if two stars have the same radii but different masses, the more massive star will be hotter. If two stars have the same mass but different radii, the smaller star will be hotter. Similarly, if a star of fixed mass contracts quasistatically, U_G becomes more negative, so U_{th} increases and the star becomes hotter. And if a star expands, it cools – a result that has important consequences for the evolution of stars (Sect. 8.2). This behaviour is reminiscent of a contracting or expanding adiabatic gas that heats or cools, accordingly (Fig. 2.4). However, there is an important difference – stellar contraction is neither adiabatic nor isothermal, but something in between. The star heats during contraction but it is also radiating, and the way energy is shared between internal heating and radiation can easily be quantified.

For example, suppose that a star is contracting quasistatically. By Eq. 7.4, U_G becomes more negative. But by Eq. 7.13, only *one-half* of the original potential energy is converted into the internal thermal energy. Therefore, *the other half must be radiated away during the process of contraction*. In other words, the released gravitational potential energy is shared equally between internal thermal energy and radiative energy. We can write this as

$$U_{rad} = \frac{1}{2} U_G \qquad (7.14)$$

where U_{rad} is taken to be negative to specify that energy is *leaving* the system.

Since the *total* internal energy of a star, U, is the sum of the internal thermal and gravitational potential energies (essentially a sum of kinetic and potential energies), then (using Eq. 7.13)

$$U = U_G + U_{th} = U_G + -\frac{1}{2} U_G = \frac{1}{2} U_G \qquad (7.15)$$

$$= U_G + U_{th} = -2\, U_{th} + U_{th} = -U_{th} \qquad (7.16)$$

The total energy is negative, U_G is negative, and U_{th} is positive.

We can explicitly address the issue of the radiated energy by taking the time derivatives of these quantities. Using Eqs. 7.15, and 7.14,

$$\dot{U} = \dot{U}_G + \dot{U}_{th} = \frac{1}{2} \dot{U}_G = -L_{rad} \qquad (7.17)$$

where the star's luminosity, $L_{rad} = -\dot{U}_{rad}$, is taken to be positive, by convention.

The virial theorem is an important statement as to how the various energies relate to each other in a stable or quasistable system. This includes orbital systems such as our solar system[2] and systems containing many stars such as globular clusters (Sect. 7.5) or even some clusters of galaxies. Any system that has had sufficient time for its kinetic and potential energies to be shared in this way is said to be *virialized*. Systems that are losing or gaining significant mass or are being dynamically affected by neighbouring objects are examples of objects that are not virialized. The virial theorem can also be generalized to other equations of state besides the ideal gas law that we applied in this section. It is therefore one of the most useful and revealing theorems in astrophysics.

---

[2] A simple example is to consider a low-mass object of mass m in a circular orbit about a high-mass object of mass M (like the Earth's almost circular orbit about the Sun). By balancing the gravitational force with the centripetal force, we find $v^2 = GM/r$. The kinetic energy of the orbiting object is $E_K = (1/2)\, m\, v^2$, so substituting for v^2, $E_K = (1/2)\, m\, GM/r$. But the gravitational potential energy of the object in orbit is $E_p = -Gm\, M/r$, so $E_K = -(1/2)\, E_p$, consistent with Eq. 7.13.

7.2 TIMESCALES

In order to discuss what drives the relevant processes in star formation and evolution, it is important to consider what characteristic *timescales* are involved and what determines those timescales. In general, a timescale, τ, is expressed as

$$\tau = \frac{x}{\dot{x}} \tag{7.18}$$

where x is some physical property and $\dot{x}$ is the rate of change of that property. There are a variety of timescales one could consider, including buoyancy timescales in regions of convection, diffusion timescales, seismic timescales and others. However, the important timescales that are most relevant for stars as a whole are the *dynamical* timescale, the *thermal* timescale and the *nuclear* timescale.

7.2.1 The Dynamical Timescale

The dynamical timescale governs *structural* changes in stars. How fast could such changes occur? As we have done before (e.g. Sect. 6.3), let us consider a gross estimate. Suppose we take the physical property to be the radius of a star, R, and the change in radius with time (a velocity) is $\dot{R}$. As we argued in Sect. 1.2, the force of gravity dominates globally for stars, so we need an expression for $\dot{R}$ involving gravity that is applicable to the entire star. A reasonable choice is the escape velocity, $v_{esc} = \sqrt{2\,GM/R} = \dot{R}$, so that

$$\tau_{dyn} \sim \frac{R}{\dot{R}} = \frac{R}{\sqrt{2GM/R}} = \sqrt{\frac{R^3}{2\,GM}} = \sqrt{\frac{R}{2\,g}} \tag{7.19}$$

$$\sim 19 \sqrt{\left(\frac{R}{R_\odot}\right)^3 \left(\frac{M_\odot}{M}\right)} \ \text{min} \tag{7.20}$$

where g is the gravitational acceleration, as given in Eq. 6.5. Equation 7.20 expresses the result in easily understood units and shows that the approximate dynamical timescale for the Sun is about 19 minutes! This is *much* less than the age of the star. Equation 7.19 can be expressed as a function of density instead of R and M to find

$$\tau_{dyn} \sim \frac{1}{\sqrt{\frac{8\pi}{3}G\bar{\rho}}} \sim \frac{1}{\sqrt{G\bar{\rho}}} \tag{7.21}$$

where $\bar{\rho} = M/\left(\frac{4}{3}\pi R^3\right)$ is the average density. Here we are concerned with order-of-magnitude estimates, so the last term has dropped some constants. One could derive the dynamical timescale much more rigorously, but the result would agree with Eq. 7.21 to within factors of a few.

What does the dynamical timescale mean physically? An extreme interpretation is if all of the internal pressure suddenly disappeared, how long would it take for the star to collapse? This is the *freefall timescale*, and it is $\mathcal{O}(\tau_{dyn})$ (see Eq. 8.7). Alternatively, if an entire star were to explode, the timescale for the explosion would also be $\mathcal{O}(\tau_{dyn})$. When a star rotates, the rotation period cannot exceed τ_{dyn}, or the star would tear itself apart (e.g. Sect. 10.4.2). If there is a pressure perturbation, τ_{dyn} is how long it will take for the wave to propagate through the star (Prob. 7.1). If a star is stable to perturbations (Sect. 7.3.1), the return to hydrostatic equilibrium will occur on a dynamical timescale. Some stars oscillate radially,[3] usually involving specific stellar layers (e.g. the ionization zones). Again, the oscillations will be $\mathcal{O}(\tau_{dyn})$, though only the density of the oscillating layer will be relevant. That being the case, a measurement of τ_{dyn} could inform us of the density in the oscillating layer using Eq. 7.21. We will discuss stellar oscillations further in Chapter 9. In short, the dynamical timescale is a timescale related to motion.

7.2.2 The Thermal (Kelvin-Helmholtz) Timescale

Our second important timescale is the thermal timescale. The thermal timescale is related to the transfer of thermal energy. From Eqs. 7.15 and 7.16, $U = 1/2\, U_G = -U_{th}$, so neglecting factors of 2,

$$\tau_{th} \sim \frac{|U|}{L} \sim \frac{|U_G|}{L} \sim \frac{GM^2}{R\,L} \tag{7.22}$$

where L is the (positive) luminosity of the star. The use of the virial theorem indicates that our star is stable or quasistatic. Inserting numbers,

$$\tau_{th} \sim 32 \left(\frac{M}{M_\odot}\right)^2 \left(\frac{R_\odot}{R}\right)\left(\frac{L_\odot}{L}\right) \text{ Myrs} \tag{7.23}$$

So the thermal timescale of the Sun is about 32 million years. This is still much shorter than the lifetime of the Sun but considerably longer than the dynamical timescale.

The thermal timescale can be thought of as the time it would take for the star to radiate away all of its current thermal energy at constant luminosity. Put another way, if nuclear reactions suddenly turned off in the core, τ_{th} is the length of time that would pass before we would even notice (cf. Prob. 7.2).

The thermal timescale is very important when considering the quasistatic contraction of a star because gravitational potential energy is converted into thermal energy, and luminosity results (Eq. 7.17) from such a contraction. That is, a star could shine simply because of a slow contraction, even if there were no other source of energy in the interior. Historically, this is how Kelvin and Helmholtz surmised that the Sun's luminosity could be produced before it was understood that nuclear reactions were the source of power. The thermal timescale, however, is much too short compared to the lifetime of the solar system as measured from other sources, such as the ages of meteorites and the oldest rocks on Earth.

---

[3] Other types of oscillations also occur other than radial, as we shall see in Sect. 9.2.

Quasistatic contraction of a forming star before it reaches the main sequence occurs on the thermal timescale, as does expansion in an evolved star after the main sequence phase. Contractions and expansions are more likely to occur for specific regions within a star, however, as we shall see in future sections (e.g. Sect. 7.3.1).

7.2.3 The Nuclear Timescale

For the nuclear timescale, we have

$$\tau_{nuc} = \frac{E_{nuc}}{L} \tag{7.24}$$

where E_{nuc} is the total energy produced from all nuclear reactions that can occur and L is the luminosity produced by those reactions (i.e. the luminosity of the star).

The energy released during nuclear reactions was discussed in Sect. 5.3 and involves the mass decrement, Δm, which is the difference in mass between nucleons when they are free and when they are bound in a nucleus (Eq. 5.10). Let us express the mass decrement for a particular reaction, Δm_{nuc}, as a fraction of the mass of free particles that take part in the reaction, $m_{unbound}$, i.e.

$$f_{nuc} = \frac{\Delta m_{nuc}}{m_{unbound}} \tag{7.25}$$

$$f_{He} = \frac{\Delta m_{He}}{2m_p + 2m_n} = \frac{0.03035\,u}{4.03188\,u} = 0.75\% \tag{7.26}$$

where u is the atomic mass unit, and m_p and m_n are the masses of the proton and neutron, respectively. The second equation is explicitly written for the formation of helium (cf. Eqs. 5.12 and 5.13). Then the total energy available from such nuclear reactions is

$$E_{nuc} = f_{nuc}\, f_{avail}\, M\, c^2 \tag{7.27}$$

$$E_{He} = f_{He}\, f_{avail}\, M\, c^2 \tag{7.28}$$

where M is the mass of the star and f_{avail} is the fraction of the star's mass that is available for nuclear reactions. This will be less than one because nuclear reactions do not occur throughout the entire star. The second equation is again written explicitly for the formation of helium. We then have, from Eqs. 7.24 and 7.27,

$$\tau_{nuc} = \frac{f_{nuc}\, f_{avail}\, M\, c^2}{L} \tag{7.29}$$

For stars on the main sequence (ms) in which hydrogen is being converted into helium, we can rewrite Eq. 7.29 in more convenient units using f_{He} from Eq. 7.26:

$$\tau_{ms} \sim 10^{11}\, f_{avail} \left(\frac{M}{M_\odot}\right)\left(\frac{L_\odot}{L}\right) \text{ yrs} \tag{7.30}$$

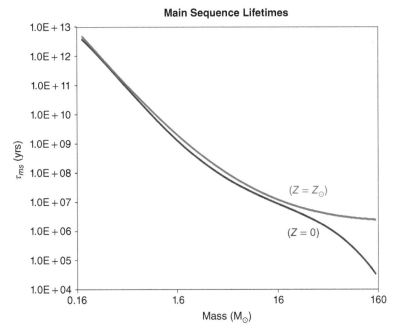

Figure 7.1 Main sequence lifetimes for stars with solar metallicity (Sect. 2.1.2), $Z = Z_\odot$ (green curve) and zero metallicity $Z = 0$ (magenta curve). Analytical expressions for these curves are from [329], as follows. Solar metallicity: $\log \tau = 9.96 - 3.32 \log M + 0.63 (\log M)^2 + 0.19 (\log M)^3 - 0.057 (\log M)^4$. Zero metallicity: $\log \tau = 9.81 - 3.55 \log M + 0.74 (\log M)^2 + 0.51 (\log M)^3 - 0.29 (\log M)^4 + 0.041 (\log M)^2$.

For the Sun, $f_{avail} \sim 0.1$ (e.g. [186]) implying that about 10% of the Sun's mass is actually available for nuclear reactions. Thus, for the Sun, $\tau_{nuc} \sim 10^{10}$ years. Since the age of the Sun is $t_\odot = 4.6 \times 10^9$ years (Eq. 1.5), the Sun has achieved about half of its total main sequence lifetime.

Because we know how the luminosity and mass are related for main sequence stars (Table 6.1), we can also find the mass-dependence of τ_{ms} (Prob. 7.4). A relation is shown in Fig. 7.1.

7.3 STABILITY

Our quasistatic star is undergoing slow changes, but at any given time, it can be considered in stable equilibrium. As we have seen, the equations of stellar structure (Sect. 6.1) and the virial theorem (Sect. 7.1) are essentially statements of equilibrium. The question, then, is whether the timescale for the change is greater than the timescale required for the star to adjust to those changes. If it is, then the star is stable.

Some stars *pulsate*, which is a *periodic* instability. However, the pulsation mainly applies to a fraction of a star's radius (or mass) and does not imply a gross stellar instability. For example, a pulsating star continues to oscillate back and forth through similar physical states. We defer discussion of pulsation to Sect. 9.1.

7.3.1 Stability against Perturbations

There are always perturbations in a star. Stars rotate, they have regions of convection, and surface activity can be dominated by magnetic field-related phenomena such as starspots, flares and winds. Nearby planets or other stars, if present, can also exert tidal forces on a star. The related pressure perturbations should then propagate at the sound speed (Eq. 2.43). We know, moreover, that sound waves travel through a star on a timescale $\mathcal{O}(\tau_{dyn})$ (cf. Prob. 7.1), which is the shortest of the timescales described in Sect. 7.2. Therefore, a star, globally, is able to adjust quickly to such perturbations. But how do we know that the adjustment is in the right direction? A stable star should damp out the perturbations with time, not amplify them.

Here, we do not probe such issues in detail but do offer one argument. Suppose, for example, that the luminosity generated in the core by nuclear reactions, L_{gen}, *increases* by some amount (we will write L_{gen} ↑). We will restrict the example to the stellar core since this is where nuclear reactions are occurring, but similar arguments could be extended to the whole star. What happens?

With an increase in the energy generation, this means the total internal energy, U, must have increased (U ↑; recall that U is negative). By the virial theorem (Eq. 7.15), U_G ↑ as well (recall that U_G is also negative). Since we have not claimed that there is any change in mass, the core must expand, (r ↑, Eq. 7.4), which decreases the density, ρ ↓ (Eq. 6.25). By the virial theorem again (Eq. 7.13), U_{th} ↓, which means the temperature must decrease, i.e. T ↓ (Eq. 7.11). Now we have a situation in which *both* ρ and T have decreased, but since the nuclear energy generation, ϵ, depends on both of these quantities (Eq. 5.9), ϵ ↓. This means L_{gen} ↓ (Eq. 6.43). We finally have a situation in which L_{gen} ↑ led to L_{gen} ↓. In other words, the perturbation corrected itself in the right direction for stability.

7.3.2 Secular Evolution of the Sun along the Main Sequence

Figure 7.2 shows the slow quasistatic secular changes that occur in the Sun as it gradually progresses along the main sequence. It indicates how solar properties change from the zero age main sequence (ZAMS) to the terminal age main sequence (TAMS), at which point the hydrogen in the core is exhausted. As can be seen from the figure, the hydrogen mass fraction in the center of the Sun, X_c (light blue curve), decreases with time as hydrogen is converted into helium. It is left as an exercise (Prob. 7.5) to explain changes in the remaining properties as the Sun moves along the main sequence.

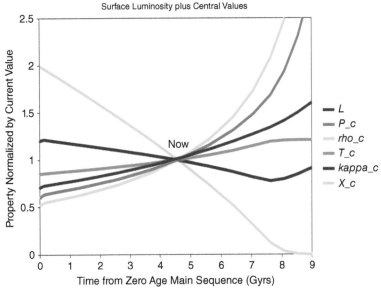

Figure 7.2 Figure showing the slow evolution of properties of the Sun on the main sequence. Data are from Online Resource 2.2, assuming a metallicity of $Z = 0.02$. All values have been normalized to current solar values, hence all values are 1 at the current age of the Sun ('Now', $t_\odot = 4.6$ Gyrs, Eq. 1.5). Dark blue: surface luminosity $L_\odot$. Red: central pressure P_c. Yellow: central density ρ_c. Green: central temperature T_c. Maroon: central Rosseland mean opacity κ_c. Light blue: central mass density of hydrogen X_c.

7.4 THE MINIMUM AND MAXIMUM STABLE STARS

7.4.1 The Lowest-Mass Stars

In Sect. I.1, we adopted the definition of a star as a self-gravitating object that was undergoing sustained nuclear fusion of hydrogen in its interior. At what mass is this no longer possible? The minimum temperature required for PP-chain hydrogen burning is $T_{min} \approx 4 \times 10^6$ K. This corresponds to a minimum hydrogen-burning stellar mass in the range $0.070\,M_\odot < M_{min} < 0.077\,M_\odot$. The value $M_{min} \sim 0.08\,M_\odot$ is frequently quoted. This is about 80 times the mass of Jupiter. At minimum mass, estimates of the effective temperature, luminosity and radius are $T_{eff} \sim 2075$ K, $L \sim 1.26 \times 10^{-4}\,L_\odot$ and $R \sim 0.086\,R_\odot$ [79]. This is the 'end of the main sequence'.

Figure 7.3 shows a low-mass extension to the mass-radius (M-R) relation presented in Fig. 6.3. On the right (blue triangles) is the M-R relation for low-mass stars.

Figure 7.3 The mass-radius relation for very-low-mass objects, showing the transition from stars (blue triangles on the **right**) to planets (magenta squares on the **left**). The black solid and dashed curves show results from models with two different ages. The blue dashed curve shows the expected result for objects with low metallicity (10% $Z_\odot$). The brown dwarfs Hat-P-2b and Corot-3b are labelled, as is the planet Jupiter, which is marked with a green dot. Credit: G. Chabrier, [47] / With permission of AIP Publishing.

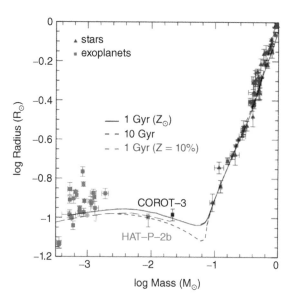

But then there is a marked change in slope with a shift to planet-mass objects on the left (magenta squares). The minimum in the curve corresponds approximately to the minimum stellar mass and radius quoted earlier. The separation in slope is quite striking, illustrating that different physics is involved in the two regimes. As the mass drops below ~0.08 $M_\odot$ (*log* M = −1.1), we enter the regime of *brown dwarfs* whose interior pressure is sustained by electron degeneracy (e.g. Sect. 4.4) rather than nuclear reactions, although some burning of deuterium and lithium is also possible. At lower masses, we see more 'normal' planets such as Jupiter (green dot). The study of substellar mass objects is an active area of current research, especially given the drive to understand *exoplanets* – planets in systems other than our own solar system. A challenge in observing these objects is that they are very faint, and it is usually difficult to separate their light from the light of the parent star.

7.4.2 The Highest-Mass Stars

The upper limit to the mass of a star is dictated by its luminosity. As we have seen, the luminosity is a strong function of mass, i.e. $L \propto M^\gamma$ (Eq. 6.41), where γ ranges from 2.0 to 5.7 on the main sequence (Table 6.1). More massive stars are also hotter, and since radiation pressure $P_{rad} \propto T^4$ (Eq. 2.58), whereas gas pressure $P \propto T$ (e.g. Eq. 2.4), radiation pressure starts to dominate in hot, high-mass stars. When we wrote down the equation of hydrostatic equilibrium (Eq. 6.4), we did not identify the source of the interior pressure, only that it must exist for a star to be stable. Now, if we wish to interrogate the stability of high-mass stars, we identify that pressure as radiation pressure and ask: what is the stellar *luminosity* above which the radiation pressure becomes so strong that the star is no longer in hydrostatic equilibrium? That luminosity is called the *Eddington luminosity* or the *Eddington limit*, and above this limit, a star begins to 'blow itself apart'. Eddington alluded to this issue in the quote on page xxiii. A derivation of this limit can be found in Example 7.7 of [154], the result being

$$L_{Edd} = \frac{4 \pi c G M}{\kappa} \tag{7.31}$$

where M is the star's mass, c is the speed of light, G is the gravitational constant and κ is the opacity. The Eddington luminosity depends only on a star's mass and opacity. At higher masses, it takes a greater radiative luminosity to destabilize the star. As for opacity, a high opacity means radiation is more easily blocked, like photons hitting a wall, so the radiation pressure is greater. If two stars have the same mass but different opacities, the higher-opacity star will destabilize at a lower Eddington luminosity.

We saw in Fig. 4.1 that at high temperatures, the gas is fully ionized and opacity is dominated by electron scattering, κ_{es} (units of cm^2 g^{-1}, Eq. 4.13), which is independent of frequency, density and temperature. To write Eq. 7.31 more simply in familiar units, we also introduce the ratio κ_{es}/κ, where κ is the true opacity, to take into account any departures from pure electron scattering. The result is

$$L_{Edd} = 3.2 \times 10^4 \left(\frac{M}{M_\odot} \right) \left(\frac{\kappa_{es}}{\kappa} \right) L_\odot \tag{7.32}$$

Values of the Eddington luminosity are now seen more easily. For hot, massive stars, the ratio κ_{es}/κ should be $\mathcal{O}(1)$, so a 30 M$_\odot$ star, for example, should have $L_{Edd} \sim 10^6 L_\odot$. Following the example at the beginning of Appendix F, the actual luminosity of a 30 M$_\odot$ main sequence star of spectral type O5 is $L \approx 4 \times 10^5 L_\odot$, so a 30 M$_\odot$ star falls below the Eddington limit. However, it is close enough that it should be experiencing strong radiatively driven stellar winds (discussed shortly). The Sun, by contrast, is highly stable. Having an opacity that is higher than κ_{es} (cf. Fig. 4.2), the Eddington luminosity will be $L_{Edd} > 10^4 L_\odot$ – much higher than the Sun's actual luminosity of 1 $L_\odot$. We conclude that the Sun is nowhere near a luminosity that would make it radiatively unstable.

Equation 7.32 implies that there should be an upper limit to the main sequence mass of a star. From Sect. 6.3, we saw that the approximate mass-luminosity (M-L) relation for high-mass stars is $L \propto M^3$ (Eq. 6.35). Let us extrapolate this relation to very high masses until it meets the Eddington luminosity. We set the opacity ratio to 1 to find

$$\frac{L_{max}}{L_\odot} = 3.2 \times 10^4 \left(\frac{M_{max}}{M_\odot} \right) = \left(\frac{M_{max}}{M_\odot} \right)^3 \tag{7.33}$$

$$\implies M_{max} \approx 180 \, M_\odot \tag{7.34}$$

Thus, we do not expect to see stable stars more massive than about 180 M$_\odot$ according to this development. As we saw in Sect. 6.3, however, it is difficult to measure the (M-L) relation for very-high-mass stars. Modifying the power of the (M-L) relation, according to different references, results in values of M_{max} that differ from Eq. 7.34 by factors of a few. The value $M_{max} = 150\,M_\odot$ is often quoted [e.g. 100, 329]. As stellar masses increase and their luminosities approach the Eddington limit, it is clear that the (M-L) relation should flatten until $L \propto M$, following Eq. 7.32. There should also be no stable stars with $L > L_{Edd}$.

What do we see observationally? The outer layers of a star will be affected first as a star's luminosity approaches the Eddington limit. An observable outcome is that high-mass stars, typically of O-type spectral class, experience very strong stellar winds. Compare the typical mass loss rate of an O-type star ($\approx 10^{-5}\,M_\odot\,yr^{-1}$) to that of our Sun ($10^{-14}\,M_\odot\,yr^{-1}$). Hot, massive stars with strong winds are called *Wolf-Rayet stars*. An example is shown in Fig. 7.4. As the figure suggests, the winds are not simple – they can be turbulent and episodic [222]. Another example of a massive star with outflow is given in Box 7.1 on Page 163.

It is actually rather difficult to find 'the most massive star'. Even though such a star would be very luminous, there are very few high-mass stars. This is because when stars are formed, many more low-mass stars are formed than high-mass stars (Sect. 8.1.5). These stars also do not stay on the main sequence for long (Fig. 7.1), and the mass loss itself complicates mass measurements. A currently known 'heavy-weight' star is R136a1, which is in a massive stellar cluster that is only 1.6 Myr old [68], located in the Large Magellanic Cloud. R136a1 is estimated to have a current mass of

Figure 7.4 The Wolf-Rayet star, WR 124. This star has a mass of $M = 21\,M_\odot$ and is losing mass at a rate $\dot{M} = 5 \times 10^{-5}\,M_\odot\,yr^{-1}$ [134]. Credit: Hubble Legacy Archive / NASA / Public domain.

215 M$_\odot$ and had an initial mass of 250 M$_\odot$ (uncertainties of $\approx$20%) prior to its mass loss [28]. When such high-mass stars are observed, there is some uncertainty as to whether the stellar mass refers to a single star or binary system, or whether the star has resulted from a merger [68]. If so, Eq. 7.32 would not apply as is.

Box 7.1

Starring ... η Carinae

Distance: d = 2.3 kpc; Magnitude: V = 4.0 (January, 2022);
Variable Type: LBV; Effective Temperature: T$_{eff}$ = 9400 K;
Mass: M = 100 M$_\odot$ [62]

A bright southern hemisphere star, η Carinae, in the constellation of Carina (the keel or hull of a ship), is one of the most massive stars known. A *luminous blue variable* (LBV), η Carinae is famous for a so-called 'great eruption' in the 1840s (see light curve), when it became the second-brightest star in the night sky. The eruption ejected 10 $\rightarrow$ 40 M$_\odot$ of material that we see today as a bipolar nebula called the *homunculus* ('little man', *Inset Picture*). A 'lesser eruption' followed in the 1890s. This star is in a binary system with a 5.54-year period, and its companion is also massive ($\sim$40 M$_\odot$) [131]. Both components of the binary have very strong radiation-driven stellar winds [62]. The system is currently brightening as the circumstellar material dissipates with time [131].

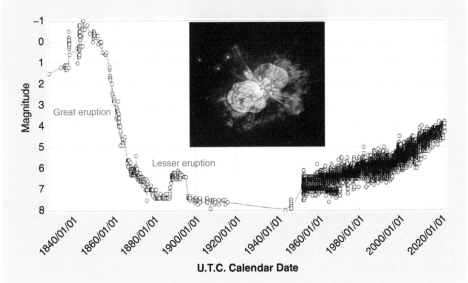

Visual magnitudes from 1827 to 2022. The inset picture shows the homunculus, and the location of η Carinae is marked with a yellow star. Credit: Adapted from Online Resource 7.2

7.5 A MAIN SEQUENCE PRIMER

In the remaining sections of this text, we will embark on a journey to understand stars that are *off* the main sequence. So far, the physical principles that we have been presented apply to stars that are on or off the main sequence: for example, the equations of stellar structure, radiative and convective transport, nuclear energy generation, the virial theorem, and so on. However, stars *on* the main sequence are more amenable to study because there are so many of them and they are relatively stable compared to their off-main-sequence counterparts. Let us then summarize what we know about main sequence stars.

As pointed out in Sect. 5.4, the main sequence is the locus of points on the H-R diagram (Fig. I.2) in which stars are converting hydrogen into helium in their cores. The timescale for this activity is governed by the nuclear timescale (Sect. 7.2.3), which is the longest timescale relevant to stars. Off-main-sequence reactions (Sect. 5.5) have shorter nuclear timescales than H → He. This means all stars spend the largest fraction of their 'lives' (i.e. the time during which interior nuclear reactions are taking place) on the main sequence. These lifetimes are shown in Fig. 7.1. Even a high-mass star, which may spend 'only' 10^6 yrs on the main sequence, spends an even shorter time off it. Consequently, the main sequence of an H-R diagram always shows the highest number of stars, simply because of the timescales involved.

As pointed out in Sect. 3.2, main sequence stars are also called *dwarfs* (luminosity class V) because these stars are the *smallest* that they will be at any point in their lives. A stellar remnant, such as a white dwarf, is smaller but is not a star by our original definition (Sect. I.1).

What about scaling relations? That is, as the mass, M, increases, how do other parameters vary?

The luminosity, $L \propto M^\gamma$ (Eq. 6.41) with γ given in Table 6.1, varies strongly with mass, and this M-L relation is shown in Fig. 6.2. As mass increases, so does the radius. However, the radius does not increase as strongly as the luminosity, as pointed out in Sect. 6.3. Equation 6.46 illustrates this point, though it is only a rough relation, given the approximations that went into its development. The observed M-R relation is shown in Figs. 6.3 and 7.3. A glance at Table F.1 also emphasizes the weak dependence of radius on mass. Between spectral types M3 and O8, for example, the absolute magnitude changes by 16.1. That magnitude change corresponds to a luminosity increase by a factor of almost 3 million, whereas the radius increases only by a factor of 26.3. As for surface temperature (the effective temperature), this is related to luminosity and radius via $L \propto R^2 T_{eff}^4$ (Eq. 2.56). We know that luminosity is a very strong function of mass, whereas radius is only a weak function of mass, so as mass increases, so does T_{eff}.

We now have the basics for understanding the main sequence as it appears on the H-R diagram. The theoretical H-R diagram would plot L as a function of T_{eff}, but the observational H-R diagram is a colour-magnitude (CM) diagram (cf. Sect. I.4 and Fig. I.2) in which an absolute magnitude is plotted against colour index. From Eq. 2.56,

$$log(L) = 2\,log(R) + 4\,log(T_{eff}) + const \qquad (7.35)$$

But $log(L)$ is directly related to the absolute magnitude (Eq. I.12), and $log(T_{eff})$ is directly related to the colour index (e.g. Fig. 3.3). There is a mapping, then, between theoretical and observational H-R diagrams, with the exact mapping depending on the observational filter bands used. Recall that temperature and luminosity are found without the need for any assumption of stellar radius (Sects. 3.1 and 3.2). If all stars were the same size, a plot of $log(L)$ versus $log(T_{eff})$ should result in a main sequence with a slope of exactly 4 (recall that temperature increases to the left in the H-R diagram). However, we see departures from such a slope. This alone leads us to conclude that *all stars do not have the same size*. In other words, if we had no measurement of radius at all, we could still infer the existence of a M-R relation just from a properly mapped H-R diagram.

What about the remaining stellar parameters? The *interior* temperature $T \propto M/R$ (Eq. 6.29), and again, the radius is a weaker function of mass, so the interior temperature also increases with mass. However, the density $\rho \propto M/R^3$ (Eq. 6.25) and interior pressure $P \propto M^2/R^4$ (Eq. 6.27). In these cases, the stronger dependence on radius has a greater effect, so ρ and P decrease with increasing mass (cf. Prob. 6.8).

We finally wish to consider the metallicity. The importance of metallicity has been stressed at various times, and we have seen how higher metallicities, Z, increase stellar opacities, κ_R (e.g. Eq. 4.11). For stars in which nuclear reactions occur via the PP-chain (Sect. 5.4.1) and the Rosseland mean opacity (Sect. 4.2.1) is dominated by a Kramers law, it can be shown (e.g. [255]) that the luminosity and temperature of a star depend on metallicity as

$$L \propto Z^{-1.15} \tag{7.36}$$

$$T \propto Z^{-0.31} \tag{7.37}$$

This means for two different stars of the same mass but different metallicity, the *lower-metallicity* star will burn hotter and brighter. Consequently, the lower-metallicity star will also have a shorter lifetime (Eq. 7.24). As can be seen by the powers in the previous equations, the effect on luminosity is stronger than on temperature. Thus, the main sequence for low-metallicity stars shifts to the left and up in the H-R diagram. This shift is illustrated by the bent arrow in Fig. 7.5, which shows a zoom-in of a small section of the H-R diagram. The solid black curve shows the main sequence for stars of solar metallicity, and the colours show the shifting main sequence for stars of progressively lower metallicity.

Figure 7.5 also shows how the main sequence position will shift between *Population I* and *Population II* stars. The stars that occupy the disk of our Milky Way galaxy are called Population I stars; these stars have metallicities similar to the Sun. Population II stars, on the other hand, are objects in the halo of the Milky Way and have lower metallicities. Many halo stars are in *globular clusters*, which are roughly spherical groups, typically containing $\sim 10^5$ stars that are virialized (Sect. 7.1). Halo stars also behave differently from disk stars in their dynamics. Disk stars show roughly circular rotational motion in a plane about the center of the Milky Way. Halo stars also orbit the center of the Milky Way but in more random orientations instead of

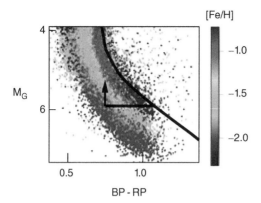

Figure 7.5 Zoom-in of a colour-magnitude diagram from Gaia data showing the main sequence for different metallicities. More negative values indicate more metal-poor stars. The solid black curve represents the main sequence of the Hyades cluster, which has a metallicity that is close to solar at [Fe/H] ~ +0.1 to +0.2 (Eq. 3.1). The arrow shows how a star of given mass shifts its position (hotter: to the left; brighter: up) when the metallicity decreases. Wavelengths of the Gaia BP and RP filters are given in Table 3.1. Credits: Adapted from ESA/Gaia/DPAC, Carine Babusiaux and co-authors of the paper [106].

a plane. These are all clues as to the origin of these two populations. Presumably the halo stars formed very early and rapidly during the formation of the Milky Way galaxy, using up the available gas in this region so that star formation then ceased. By contrast, in the disk, with its spiral arms and copious gas clouds, stars have continuously formed up to the present day. Different generations of stars eventually eject metal-rich material into the interstellar medium from which new stars are formed (Chapter 8). Hence, metallicity increases with time in the disk but not in the halo. Thus metallicity is an important clue as to the history of star formation in our Galaxy.

Online Resources

7.1   *American Association of Variable Star Observers (AAVSO)*: https://www.aavso.org
7.2   *AAVSO light curve plotter*: https://www.aavso.org/LCGv2

PROBLEMS

7.1.   Re-derive the dynamical timescale, τ_{dyn}, using the sound speed, c_s, as the variable $\dot{R}$ instead of the escape velocity. Compare your result to Eq. 7.19, and comment.

7.2. (a) Calculate the time it would take a photon to travel in a straight line from the center of the Sun to its surface if there were no interactions en route.

(b) For radiative diffusion, a photon takes a 'random walk' in which the number of steps needed to go a total distance d is $N = \left(d/\bar{l}\right)^2$, where $\bar{l}$ is the mean free path (Box 1.1 on page 7). Adopt a mean free path of $\bar{l} = 0.3$ cm, and determine the time (yrs) it takes for a photon to diffuse from the center to the surface of the Sun. Assume that interaction times are negligible.

(c) Write an expression for the diffusion time of a photon, assuming that each interaction takes 10^{-8} s, and re-evaluate the diffusion time for the Sun.

(d) Compare your result from (c) with the *thermal timescale* of the Sun. Do they agree to order of magnitude?

7.3. Suppose that the Sun's luminosity were generated by a slow contraction instead of nuclear reactions. For simplicity, let the density of the Sun be constant.

(a) Write an equation for the change in gravitational potential energy, $\Delta U_G = U_{G_f} - U_{G_i}$, where U_{G_f} is the final potential energy and U_{G_i} is the initial potential energy. The equation will include mass and radius. [HINT: Eq. I.15 will be useful.]

(b) Now suppose that the initial radius is much larger than the final radius. Calculate the age of the Sun if it has been shining at a constant rate of $L_{\odot}$ due to gravitational contraction.

(c) Compare this thermal timescale to a measurement of solar system age from meteorites.

7.4. (a) Equation 7.30 provides an estimate of the main sequence (ms) lifetime, τ_{ms}, of a star. Use the M-L relations from Table 6.1 to find the relation $\tau_{ms}(M)$ for each mass range. Assume that $f_{avail} = 10\%$ for all stars and that L for any star does not change with time over the ms lifetime.

(b) Plot a graph of τ_{ms} (yr) for the mass range $0.179 < M/M_{\odot} \leq 31$. Use a logarithmic scale. Compare your results with Fig. 7.1.

(c) Compare τ_{ms} for a 0.3 $M_{\odot}$ star to the age of the universe. Should we see any 0.3 $M_{\odot}$ stars that have evolved off the ms? Explain. Compare τ_{ms} for an 11 $M_{\odot}$ star to the age of a typical globular cluster. Should we see any ms stars of this mass in globular clusters? Explain.

7.5. Refer to Fig. 7.2 and explain why each of the properties $L_{\odot}$, P_c, ρ_c, T_c, κ_c and X_c increases or decreases from the zero age main sequence until now. Try to refer to a variety of equations to back up your claims. Use the virial theorem at some point in your arguments, and *predict* the behaviour of the energy generation rate at the center, ϵ_c.

Chapter 8

The Forming and Ageing Star – Evolution to and from the Main Sequence

Look up at the stars, Jean-Luc. Look up.

Star Trek: *Picard*, season 2 episode 1, 'The Star Gazer'

8.1 THE FORMING STAR

In the Introduction to this text, the physics of stars was introduced as a mature discipline. But a significant gap in our knowledge involves *star formation*. Much effort has been expended on this problem, both theoretically and observationally, and considerable progress has been made. We know, for example, that stars form out of dense regions within the molecular clouds (predominantly H_2) that permeate the interstellar medium (ISM) of galaxies. Yet we still cannot confidently provide the kind of details that we would like. Why is the issue so problematic? It is mainly because

Astrophysics: Decoding the Stars, First Edition. Judith Irwin.
© 2023 John Wiley & Sons Ltd. Published 2023 by John Wiley & Sons Ltd.

the so-called 'details' cannot be neglected. Reiterating the quote from page xviii in the Introduction, physicists can solve the problem, but only for 'spherical chickens in a vacuum'. Dense regions in molecular clouds are neither spherical nor in a vacuum. Nevertheless, it is always helpful to start with such simplifying assumptions and see what we can learn. We will look at the complicating details after.

8.1.1 The Jeans Criterion and Free-Fall Timescale

Recall that the virial equilibrium implies stability. It follows that the equations describing virial equilibrium (Eqs. 7.13, 7.15 and 7.16) should break down for a collapsing cloud. From Eq. 7.13, $2U_{th} + U_G = 0$, so for *instability*, we require

$$|U_G| > 2U_{th} \tag{8.1}$$

That is, gravity 'wins' over the internal thermal energy, and the cloud collapses. As indicated in Sect. 7.1, if there were other energy sources such as magnetic energy or turbulence, they would have to be included, but here we are including only the energies that are certain to be present and significant in order to have a tractable problem.

Using Eq. I.15 for U_g and Eq. 7.11 (integrated) for U_{th}, we can rewrite the previous inequality as

$$\frac{3}{5}\frac{G M^2}{R^2} > \frac{3 M k T}{\mu m_H} \tag{8.2}$$

where M, R and T refer to the mass, radius and temperature of the cloud, respectively. The constant μ is the mean molecular weight, and G and k are the gravitational constant and Boltzmann's constant, respectively. Notice that for the potential energy relation, we have used the constant 3/5 applicable to a *uniform-density* cloud. In reality, molecular clouds may be 'clumpy', but any given clump can be approximated as uniform density to first order. For the thermal energy equation, we have also assumed that the cloud is *initially* isothermal. If we now introduce the initial density $\rho = M/[(4/3)\pi R^3]$ and rearrange the previous equation, we find the collapse criterion,

$$M > \left(\frac{5 k T}{G \mu m_H}\right)^{3/2} \left(\frac{3}{4 \pi \rho}\right)^{1/2} \tag{8.3}$$

$$R > \left(\frac{15 k T}{4 \pi G \mu m_H \rho}\right)^{1/2} \tag{8.4}$$

These two relations make the same statement, but the equations are simply reorganized to represent mass or radius.

As we have done before (cf. the Schwarzschild criterion introduced in Sect. 4.3.1), it is useful to identify exactly where stability breaks down. Equations 8.3 and 8.4

can be rewritten as equalities, identified as the *Jeans mass*, M_J, and *Jeans radius*, R_J, respectively:

$$M_J = \left(\frac{5\,kT}{G\,\mu\,m_H}\right)^{3/2}\left(\frac{3}{4\,\pi\,\rho}\right)^{1/2} \propto \left(\frac{T^3}{\rho}\right)^{1/2} \tag{8.5}$$

$$R_J = \left(\frac{15\,kT}{4\,\pi\,G\,\mu\,m_H\,\rho}\right)^{1/2} \propto \left(\frac{T}{\rho}\right)^{1/2} \tag{8.6}$$

A cloud (typically a cold, $T \sim 10-20$ K, molecular cloud) of fixed mass M_c that is greater than M_J will begin to collapse, and the cloud of radius R_c that is greater than R_J will begin to collapse. We should therefore not see any stable clouds whose masses are greater than M_J, under the assumption of thermal gas pressure support only.[1]

Equations 8.5 and 8.6 show the dependences of the Jeans quantities on temperature and density. If a cloud of fixed mass begins to collapse, its density, ρ, will increase. So whether the collapse continues or halts depends on the behaviour of the temperature, T. In other words, the future of the collapse depends on whether the cloud can rid itself of heat. Continuing with our simple example of a spherical cloud with no other complicating processes, let us consider two cases: the cloud is *isothermal* and the cloud is *adiabatic*. A helpful illustration is Fig. 2.4, which shows these two cases for an ideal gas under compression in a piston.

In the isothermal case, the cloud is efficiently radiating away its heat, and T is constant in Eq. 8.5. Then $M_J \propto \rho^{-1/2}$, so M_J *decreases* as the cloud density increases during collapse. Therefore, if $M_c > M_J$ initially, then $M_c > M_J$ continually throughout the collapse.

In the adiabatic case, by contrast, the core becomes opaque, and no heat can escape. For an adiabatic ideal gas, Eq. 2.41 tells us that $T \propto \rho^{\gamma_a-1}$, where γ_a is the adiabatic index. A typical value of γ_a is 5/3, corresponding to three translational degrees of freedom (Sect. 2.1.7). Thus the temperature increases during collapse, as it must when heat cannot escape. Then Eq. 8.5 leads to $M_J \propto \rho^{1/2}$. We now see that the Jeans mass *increases* during collapse.[2] If $M_c > M_J$ initially, then at some point, $M_c < M_J$, and the collapse will halt. In this simple development, we continue to see how important opacity is to stellar astrophysics. It is the opacity that will decide whether radiation can escape (isothermal) or be trapped (adiabatic).

How long will it take for a cloud to collapse? This will be the dynamical timescale, which was given (to within factors of a few) in Eq. 7.21 and described in Sect. 7.2.1.

---

[1] Slightly different versions of this derivation exist, including, for example, a comparison of the free-fall time to the sound-crossing time. However, the results agree to within a factor of two.
[2] H_2 can have more degrees of freedom, should they be active. For temperatures above $\sim$300 K, two rotational motions about two axes can also be excited [273], for a total of $f = 5$. This gives $\gamma_a = 7/5$ and a weaker, but still positive, $M_J \propto \rho^{1/10}$. Additional vibrational degrees of freedom require temperatures above $\sim$1000 K.

Table 8.1 Properties of dark clouds, clumps and cores

Property	Clouds	Clumps	Cores
Mass ($M_\odot$)	$10^3 - 10^4$	$50 - 500$	$0.5 - 5$
Size (pc)	$2 - 15$	$0.3 - 3$	$0.03 - 0.2$
Mean density (cm^{-3})	$50 - 500$	$10^3 - 10^4$	$10^4 - 10^6$
Velocity extent[a] (km s^{-1})	$2 - 5$	$0.3 - 3$	$0.1 - 0.3$
Gas temperature (K)	$10 - 20$	$10 - 20$	$8 - 12$

[a] Measurable via spectral line widths.
Credit: Adapted from [25].

The *free-fall timescale*, t_{ff}, is the dynamical timescale specific to the gravitational collapse of a spherically symmetric object in the absence of other forces [54]:

$$\tau_{ff} = \sqrt{\frac{3\pi}{32\,G\,\rho}} \tag{8.7}$$

As collapse proceeds, the density of a cloud increases and the free-fall timescale shortens. Therefore, the *initial* cloud density, ρ_0, ultimately determines the timescale. Notice that the free-fall timescale is independent of the initial cloud *radius*. Assuming that the initial density is constant, τ_{ff} will be the same for all radii within the collapsing cloud. This means such a cloud continues to be of uniform (but increasing) density during the collapse, a situation called *homologous collapse* [44].

Characterizing the basic properties of molecular clouds is somewhat challenging, depending on which size scale is observed, but Table 8.1 provides some guidance. For example, consider a molecular cloud core with parameters $M_c = 3M_\odot$ and $T = 10$ K, and an initial number density of $n_{H_2} = 10^5$ cm^{-3}. Its mass density is then $\rho = \mu\,n_{H_2}\,m_H$ (cf. Eqs. 2.3 and 2.4), where μ is the mean molecular weight and m_H is the mass of a hydrogen atom. For a pure H_2 cloud, $\mu = 2$, but this must be increased to $\mu = 2.8$ to take into account heavier elements, so $\rho = 4.68 \times 10^{-19}$ g cm^{-3}. The Jeans mass (Eq. 8.5) is then $M_J = 2.35 \times 10^{33}$ g $= 1.2\,M_\odot$. The cloud's mass (3 $M_\odot$) is greater than this, so the cloud will collapse on a free-fall timescale (Eq. 8.7) of $\tau_{ff} = 3.1 \times 10^{12}$ s $\sim 10^5$ years. The end result, a *protostar* (Sect. 8.1.3), will have a typical density of $n_{ps} \sim 2 \times 10^{20}$ cm^{-3} (for solar metallicity) [200], an increase of 15 orders of magnitude!

8.1.2 Real Star Formation

Gravitational collapse is not as simple as the previous scenario portrays. The ISM is a dynamic place. Stellar winds (e.g. Fig. 7.4) and supernovae (Sect. 10.3) can disrupt interstellar gas, along with associated shock waves.[3] Supersonic and subsonic

---

[3] Shock waves occur when matter is travelling faster than the speed of sound.

turbulence are also pervasive, possibly driven by supernovae and/or galactic differential rotation [96]. A distribution of turbulent density fluctuations results. This kind of activity can disrupt clouds but can also ensure that some regions are sufficiently compressed beyond the Jeans mass density that collapse will start. In any collapsing region, additional sources of internal pressure (e.g. turbulence or magnetic pressure) must also be present because the effective free-fall time scale appears to be longer than predicted by Eq. 8.7, which ignored these effects [99].

Generally, clouds fragment, and *hierarchical*, rather than homologous, collapse ensues. The inner parts of clouds are denser than the outer parts, leading to smaller values of M_J (Eq. 8.5) and shorter collapse timescales (Eq. 8.7). Moreover, denser regions are more efficiently shielded from external heating sources, and that shielding tends to keep these parts colder, limiting their thermal support. Thus higher-density regions collapse faster, but they are still surrounded by molecular gas, so the collapsing region is not in a vacuum and will experience some external pressure. Similarly, other pockets within the parent cloud may also begin to collapse, and these pockets can interact with each other gravitationally or via magnetic torques.

Thus stars do not form in isolation but tend to form in groups or clusters (see also Sect. 8.1.5). Stellar clusters that are formed within the disk of the Milky Way are called *open clusters*[4] and will eventually disperse as they experience the shear of galactic rotation and the influence of the gravitational effects of other stars.

With these complications in mind, let us continue and examine what happens after free-fall stops.

8.1.3 Protostars

A cloud core becomes opaque and free-fall ceases once the density reaches approximately 10^{-13} g cm^{-3} [84], corresponding to the onset of the adiabatic phase. This stage, in which hydrostatic equilibrium is achieved, is called the *first hydrostatic core* (FHSC). The FHSC contracts quasistatically but is enshrouded by its parent molecular cloud and continues to experience infall from above. Infalling material represents an effective decrease in radius, so gravitational potential energy is released (Eq. 7.4), which is balanced by an increase in internal temperature (recall the virial theorem, Eq. 7.13). Once the central temperature reaches 2000 K, however, this is hot enough to dissociate H_2. The available energy that is released by infall then contributes to H_2 dissociation rather than exclusively to an increase in internal temperature. Without a sufficient increase in temperature, there is not enough pressure to hold up the FHSC, and a second collapse occurs. The process then repeats so that the collapse again stops once dissociation is complete, and the collapsed object is again opaque.

The end result is a *protostar* that is gravitationally bound and quasistable, but whose mass continues to increase because of continuing infall [84]. Mass

---

[4] These are distinguished from *globular clusters* found in the low-metallicity environment of the Milky Way halo (Sect. 7.5).

accretion rates can be estimated from $\dot{M} \sim (c_s{}^3/G)$ g s^{-1}, where c_s is a characteristic sound speed in the cloud. For a temperature of 30 K, this yields (Eq. 2.44) about 10^{-5} M$_\odot$ yr^{-1}. However, accretion is largely dominated by the angular momentum of the collapsing gas and the formation of a flattened *accretion disk*. Thus, the accretion is not spherical symmetrical. It is also not uniform with time, but *episodic*. The result is 'bursts' of accretion over time [179].

The evolution of a protostar proceeds on the thermal (Kelvin-Helmholtz) timescale (Sect. 7.2.2) although the lifetime in this enshrouded, accreting state is uncertain. For a solar-mass star, ~0.5 Myr has been quoted [84], but a few Myr could also be possible. The protostar has a size that varies depending on its mass and also on its metallicity. A 1 M$_\odot$ protostar with solar metallicity has a radius of $R_{ps} \approx 3 R_\odot$ [200]. However, a high-mass star can be considerably larger. Massive stars have high radiation pressure (Sect. 7.4.2) but also higher accretion rates. For example, a 10 M$_\odot$ protostar with an accretion rate of $\dot{M} = 10^{-3}$ M$_\odot$ yr^{-1} could have a radius as large as $R_{ps} \sim 100 R_\odot$ [149].

Can we 'see' the infall? The most direct evidence comes from the velocity profiles of protostars, as shown in Fig. 8.1 *Left*. Here we see an *inverse P Cygni profile*. Recall that a P Cygni profile shows *blueshifted absorption lines*, indicating outflow (Sect. 3.6). Absorption lines are seen when the gas is right in front of a background continuum source. An inverse P Cygni profile shows *redshifted absorption lines*, indicating that foreground gas is moving away from us towards the central object. This characteristic profile leaves no doubt that gas is infalling. In the example shown, the infall is likely onto the accretion disk, rather than directly onto the protostar.

But now the complications become more evident. A collapsed object that had a small initial rotation will have a very large final rotation by conservation of angular momentum. Material that rains down from above increases its rotational velocity and contributes to the flattened accretion disk closer to the protostar. The expected final rotational speed from an object that has collapsed from a cloud core to a protostar is extremely large, to the point that the star could tear itself apart by its rotation (Prob. 8.4). Yet we see newly formed stable stars. This means there must have been some mechanism to rid the forming star of angular momentum.

The most widely observed phenomenon is the formation of bipolar outflows or jets. Such outflows were first detected as small ionized regions in dense, dusty molecular clouds. These are called *Herbig-Haro objects*, and we now know that they trace supersonic shock waves from outflows. An example of an outflowing jet is shown in Fig. 8.1 *Right*. The details as to how material flows through an accretion disk and into perpendicular outflows are currently the subject of much scientific inquiry, but the presence of magnetic fields is likely critical to this process [20, 257]. Bipolar outflows carry away angular momentum and appear to be a ubiquitous phenomenon in the star formation process.[5] A video of an evolving protostellar jet is shown in Online Resource 8.1. The disks that surround protostars can still be present once

---

[5] *Magnetic braking* also plays a role in reducing angular momentum. See Box 8.5 on page 202.

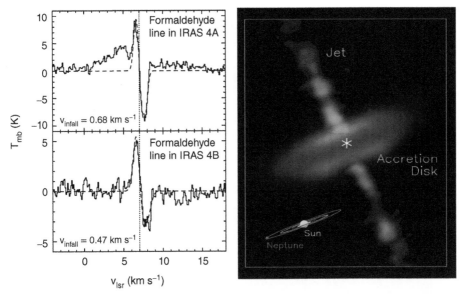

Figure 8.1 Evidence of infall *and* outflow associated with protostars. **Left**: An 'inverse P Cygni' profile in a line of the formaldehyde molecule in the source NGC 1333 IRAS 4; source 4A is at **Top** and source 4B is at **Bottom**. T_{mb} is a measure of the brightness of the line, and V_{lsr} is a measure of the velocity with respect to a Galactic standard. The infall velocity is marked. Credit: James Di Francesco, [77]/With permission of AIP Publishing. **Right**: Molecular jet (green) and accretion disk (orange) in the HH 212 protostellar system. The dark lane across the disk equator is due to dust. The size of our solar system is shown at lower left for size comparison. Credit: Lee et al. (2017), National Astronomical Observatory of Japan (NAOJ).

the surrounding envelope has dissipated [84]. Called *protoplanetary disks*, these are regions in which planets can form (Box 8.1 on page 178).

It is clear that *both* inflows *and* outflows are present in the vicinity of young stars. Bipolar outflows associated with protostars, stellar winds and supernovae that result from massive short-lived stars are all present and are collectively referred to as *feedback*. Such outflows are contributors to driving away surrounding material and eventually exposing the young pre-main-sequence stars.

While Table 8.1 provides a simple overview of dense molecular clouds, some clouds are even larger. Giant molecular clouds (GMCs), which are undergoing hierarchical star formation, can be up to 100 pc across and have masses of order $\sim 10^{6}$ M$_{\odot}$. This is sufficient for the shear of differential galactic rotation to 'stretch out' clouds and lengthen the collapse time [54]. Figure 8.2 is an illustration. Here we see the large Orion star-forming complex with two star-forming filaments, Orion A and Orion B (labelled). Orion A alone is 90 pc long and tilted with respect to the plane of the sky [129]. Such a filamentary structure is common in star-forming regions. The *Top* picture

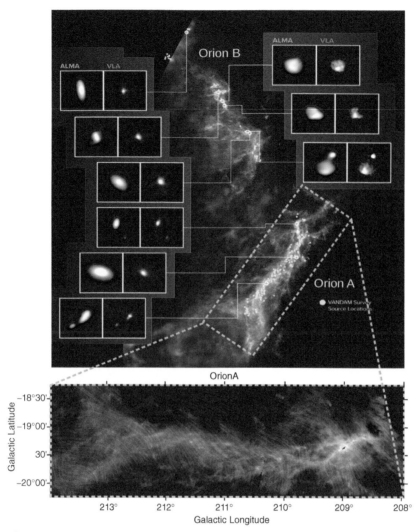

Figure 8.2 The Orion molecular cloud star-forming region. **Top**: The background blue image shows dusty regions imaged by the Herschel space telescope. Yellow dots are the positions observed in the VANDAM (VLA/ALMA Nascent Disk and Multiplicity) Survey of Orion protostars. Offset panels reveal the protostars as imaged by the (colour-coded) ALMA and VLA telescopes. Credit: Alma Observatory/Wikimedia/CC BY 4.0. **Bottom**: Blowup and reorientation of the Orion A region that is boxed in the top image. Blue (wavelength $\lambda = 160\ \mu\text{m}$), green ($\lambda = 250\ \mu\text{m}$) and red ($\lambda = 500\ \mu\text{m}$) represent emission from dust at progressively lower temperatures. The drapery pattern of lines represents the magnetic field component in the plane of the sky, inferred from 353 GHz polarization observations. The well-known Orion Nebula is roughly at the position of the blue emission on the right. Credit: [287] Soler et al. (2019)/EDP Sciences.

shows numerous protostars (*Inset Boxes*) embedded in the Orion A and B star-forming complexes. The *Bottom* picture shows Orion A with its magnetic field threaded like drapery across it. The role of magnetic fields is not completely understood in the star-forming process, but they appear to be a key ingredient. They provide an extra source of pressure that could inhibit infall [143], but the orientation of the field is important. These fields can restructure cloud geometry and actually aid in the star-forming process, depending on how they are oriented [287].

8.1.4 From Protostar to the ZAMS

Once the main mass accretion ends, our protostar can be considered a *pre-main-sequence star* and 'appears' on the H-R diagram. The starting position is debated [298] because of the complications outlined in the previous section, but a general result is that the pre-main-sequence star is in hydrostatic equilibrium, its mass is constant, and it is initially fully *convective*. At this stage, the dominant contribution to the opacity is the H$^-$ ion. The extra bound electron is taken from an abundance of free electrons that originate from the ionization of heavier metals. Even though the metal abundance is low compared to hydrogen and helium, at temperatures from about 3000 to 5000 K, these metals are the dominant source of free electrons because of the high number of electrons per atom as well as their generally lower ionization potentials [166]. The opacity law for the H$^-$ ion is *not* a Kramers law (Sect. 4.2.2) but rather has a strong positive dependence on temperature, $\kappa_{H^-} \propto \rho^{1/2} T^9$ (Eq. 4.14). High opacities favour steeper temperature gradients (Sect. 4.2.3), and steeper temperature gradients favour convection (Sect. 4.3).

Pre-main-sequence stars, at this stage, follow quasi-static contraction. Although 'star' is in their name, they are not yet stars by the definition given in Sect. I.1 because of the lack of sustained nuclear reactions in the interior. We will see, however, that a few nuclear reactions do take place prior to the star settling onto the zero age main sequence (ZAMS). By the virial theorem, the luminosity of the star is mainly supplied by gravitational (Kelvin-Helmholtz) contraction (Eq. 7.17).

Numerical modelling is needed to follow such evolution. However, some simplifying assumptions help us to understand the behaviour of the evolutionary tracks on the H-R diagram. For example, a fully convective star can be approximated by adopting a polytropic equation of state (Sect. 6.2.2.2) whose form is $P \propto \rho^{\gamma_a}$, where γ_a is the adiabatic index. Once the structure is determined, the temperature follows from the ideal gas law. The interior solution can then be linked to photospheric solutions to find the relationship between the surface effective temperature, the luminosity and the mass. It can be shown [166] that T_{eff} has only a very weak dependence on luminosity and a marginally higher dependence on mass when a fully convective star contracts, i.e.

$$T_{eff} \propto L^{0.05} M^{0.2} \tag{8.8}$$

Consequently, the effective temperature remains almost constant with changing luminosity, and a star of given mass 'slides down' a very steep vertical curve on the

H-R diagram. These initial vertical tracks are called *Hayashi tracks* after Chushiro Hayashi, who presented a consistent formulation of evolving fully convective stars [139, 140]. Hayashi tracks are on the right-hand side of any given pre-main-sequence track in Fig. 8.3 (labelled for the $1 M_\odot$ star). Equation 8.8 also has a weak dependence on mass, so a more massive star has a Hayashi track that is shifted to the left (higher temperature) compared to a less massive star. The result is that Hayashi tracks for different masses tend to be close together in the H-R diagram, ranging from $T_{eff} \sim 3000$ K ($log\,T = 3.5$) to $T_{eff} \sim 5000$ K ($log\,T = 3.7$) for the stellar masses plotted. To the right of any specific Hayashi track is a 'forbidden region' for the mass in question. This means temperatures are too low for the star to be in hydrostatic equilibrium in that area.

Box 8.1

T Tauri Stars

T Tauri stars, named after the prototype in the constellation Taurus, are low-mass ($\lesssim 3\,M_\odot$) pre-main-sequence stars on Hayashi tracks. They are highly variable in luminosity due to non-steady accretion from a surrounding disk. Modelled accretion rates are $\mathcal{O}\left(10^{-10}-10^{-9}\right)\,M_\odot\,yr^{-1}$ [63], very low compared to the protostar stage discussed in Sect. 8.1.3, but sufficient to excite high levels of stellar *coronal* activity, including variable X-ray and radio emission. The coronal activity is similar to what we see in the Sun today (Sect. 1.2.5) but much more powerful.

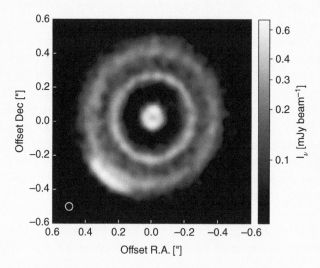

The figure shows a dusty protoplanetary disk around the young T Tauri star HD 143006, using the Atacama Large Millimeter/submillimeter Array

(ALMA) at a wavelength of λ 1.25 mm [248]. Bright rings and dark gaps trace overdensities and depletion of dust grains, respectively. Rings and gaps are known to be a consequence of dynamical interactions with planets. At a distance of $d = 165$ pc, the displayed region is ≈ 200 AU on a side. Credit: Courtesy of Laura Perez.

If temperature is not changing much on a Hayashi track, what is producing the change in luminosity? As indicated previously, the gravitational contraction itself produces a luminosity by the virial theorem (e.g. Prob. 7.3), and the Stefan-Boltzmann law (Eq. 2.56) tells us how luminosity and effective temperature on the H-R diagram relate to each other. Since $L \propto R^2 T_{eff}^4$, the culprit is the radius, R, and the steep reduction in luminosity along a Hayashi track results from a decreasing radius. For example, a low-mass star of $M = 0.2 \, M_\odot$ starts on its Hayashi track at $R = 1.68 \, R_\odot$ (point 1 in Fig. 8.3 *Bottom*), and by the time it reaches the ZAMS (point 2), its radius is only $R = 0.22 \, R_\odot$ (Table 8.2). Stars less massive than about $0.5 \, M_\odot$ never leave the Hayashi track, settling onto the ZAMS once their core temperatures are high enough to ignite sustained nuclear reactions. Like all low-mass main sequence stars, they convert hydrogen into helium by the PP-chain (Sect. 5.4.1).

Table 8.2 provides additional data for the points labelled in Fig. 8.3. For example, the last column indicates what fraction of the total luminosity L is supplied by contraction, L_G, i.e. L_G/L. For the 0.2 $M_\odot$ star, we see that contraction provides

Table 8.2 Pre-main-sequence data

$\frac{M}{M_\odot}$	Ref.[a]	$log\left(\frac{t}{yrs}\right)$[b]	$log\left(\frac{L}{L_\odot}\right)$[c]	$log\left(\frac{T_{eff}}{K}\right)$[d]	$\frac{R}{R_\odot}$[e]	$log\left(\frac{T_c}{K}\right)$[f]	$\frac{L_G}{L}$[g]
0.2	1	5.354	−0.562	3.509	1.67	5.991	0.054
	4	8.933	−2.305	3.516	0.22	6.821	0.0004
1.0	1	4.961	1.014	3.642	5.58	6.154	0.133
	2	6.855	−0.275	3.642	1.27	6.762	0.999
	3	7.476	0.001	3.765	0.99	7.139	0.237
	4	7.517	−0.073	3.762	0.92	7.144	0.019
6.0	1	4.258	2.396	3.686	22.4	6.299	0.774
	4	5.620	3.117	4.289	3.19	7.439	0.713

[a] Reference position shown in Fig. 8.3.
[b] Elapsed time from the initial model in which the radius and luminosity are large and the star starts its quasistatic contraction at constant mass. The ZAMS is taken to be the point at which the core hydrogen fraction drops to 0.998 of its initial value.
[c] Luminosity.
[d] Effective temperature.
[e] Radius.
[f] Central temperature.
[g] Fraction of the luminosity supplied by gravitational contraction.
Credit: Online Resource 8.5/University of Pisa.

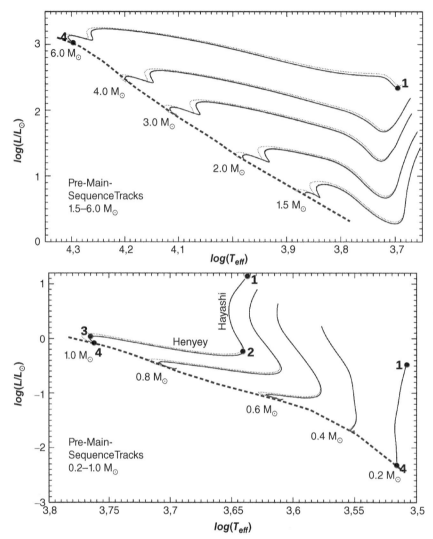

Figure 8.3 Pre-main-sequence evolutionary tracks for 'classical' models that do not include rotation, magnetic fields or accretion. These models use abundances of $X = 0.707$, $Y = 0.278$ and $Z = 0.015$, a deuterium abundance of $X_D = 2 \times 10^{-5}$ and a convective mixing length parameter (Sect. 4.3.2) of $\alpha_{ML} = 1.68$. Each solid curve is accompanied by a very faint dashed curve, superimposed in some cases, whose differences reflect varying mixtures of heavy elements for the same value of Z. The dashed blue line approximates the zero-age main sequence. Data corresponding to the numbered points are given in Table 8.2. A Hayashi track is the roughly vertical part at the right of any given track, and a Henyey track is the roughly horizontal part (both labelled for the $1\,M_\odot$ star). Credit: Emanuele Tognelli & Pier Giorgio Prada Moroni, adapted from [298].

only 5% at the beginning of its Hayashi track. This is because of some deuterium burning, which is also contributing to the luminosity initially. Deuterium is present in the ISM and is therefore present as an initial input to pre-main-sequence evolutionary models. A typical abundance for the disk of our Galaxy is $X_D = 2 \times 10^{-5}$. The specific deuterium burning reaction is $^2H + ^1H \rightarrow {}^3He + \gamma$, which is the second line in the PP-chain shown in Fig. 5.3 on page 118. This reaction destroys deuterium when the temperature reaches 10^6 K, but that temperature is too low to ignite the reaction that *forms* deuterium (first line of Fig. 5.3). Consequently, deuterium is quickly depleted at this stage. Deuterium burning is a characteristic of Hayashi tracks for most masses. It acts to slow down the contraction, shifting from the thermal timescale to the nuclear timescale, while activated. However, it does not produce enough internal pressure to reverse the general gravitational contraction during the Hayashi phase. For example, once the 0.2 $M_\odot$ star's luminosity decreases to $log(L) = -1.23$, deuterium has been used up, deuterium burning ceases, and the entire luminosity is produced by gravitational contraction alone.

Stars more massive than ~0.4 $M_\odot$ show a different behaviour in their pre-main-sequence evolution. Their tracks show a shift to the left (higher temperature) in the H-R diagram. This roughly horizontal section is called a *Henyey track* (labelled for the 1 $M_\odot$ star). The change in direction is due to the increasing importance of a growing radiative core. As the star continues to contract and the interior temperature is sufficient to ionize more hydrogen, the opacity starts to follow a Kramers law ($\kappa \propto \rho/T^{3.5}$), decreasing with increasing temperature (see Fig. 4.2). The lower opacity produces a radiative core, allowing more energy to flow into the convective envelope. A smattering of PP-chain reactions also occur, contributing to the luminosity. From points 2 to 3 for a 1 $M_\odot$ star, for example, the contribution of gravitational contraction to the total luminosity decreases from $\approx$100% to only $\approx$24% (Table 8.2), the remainder being supplied by nascent nuclear reactions. The radius still decreases, though not as much as on the Hayashi track, and the temperature and luminosity increase. Even higher-mass stars, such as the plotted 6 $M_\odot$ star, experience more PP-chain reactions right at the beginning, core temperatures are higher at equivalent stages of evolution, and the core remains radiative until the full set of CNO cycle reactions begin at the ZAMS.

The behaviour of the pre-main-sequence tracks continues to vary depending on mass and the combination of nuclear reactions that may be occurring. For a 1 $M_\odot$ star, there is a small 'hook' from point 3 to point 4 in Fig. 8.3. We know that the CNO cycle of nuclear reactions is minor in 1 $M_\odot$ stars on the main sequence (e.g. Prob. 5.3), but it is not zero. The first step in the reaction, as displayed in Fig. 5.4, activates at a lower temperature than the other steps in the cycle and begins to play a part near point 3. Because of the strong dependence on temperature (Eq. 5.25), the core becomes convective again, and this convection causes the core to expand somewhat. Expansion produces cooling, reducing the amount of energy pumped into the overlying layers, so there is a slight decrease in luminosity and surface temperature. Finally, at point 4, the star reaches the main sequence. The full range of

PP-chain reactions are activated, with only a minor contribution from the CNO cycle. The weaker temperature dependence of the PP-chain (Eq. 5.22) makes the core radiative again. Our $1\,M_\odot$ star on the ZAMS is radiative throughout, except for an outer convective envelope, similar to our Sun today. For higher-mass stars, there are more bumps and wiggles in the final pre-main-sequence phase. These are related to establishing equilibria of the secondary elements in the CNO cycle.

As we can see, these serpentine tracks on the H-R diagram are generally understood, but they vary with mass, and connecting them to observations is challenging. The strongly varying timescales provide an illustration. For example, a $6\,M_\odot$ star will take 4×10^5 yrs to reach the main sequence (Table 8.2) and then spend 6.8×10^7 yrs on the main sequence (Fig. 7.1). Recall that the longest part of a star's life is *on* the main sequence. This star will already be leaving the main sequence before the $0.2\,M_\odot$ star even reaches it. The contrast with higher-mass stars is even more striking. Stars greater than $8\,M_\odot$, whose tracks are not shown in Fig. 8.3, will be formed, live their lives on the main sequence, evolve off of the main sequence, and end their lives as supernovae, all in $\sim 10^7$ years or less. In the meantime, a $1\,M_\odot$ has yet to reach the main sequence. Star-forming regions are complex, with accretion and feedback occurring at the same time as the quasistatic evolution that results from internal changes. The classical pre-main-sequence evolutionary tracks presented in Fig. 8.3 can also be altered if rotation, accretion and magnetic fields are included [e.g. 179]. In the next section, we will look at the star-forming region in a more statistical way.

8.1.5 Number, Mass and Luminosity Functions

To look at star formation from a broader perspective, we need to know how many stars in a certain mass range are formed in any given star-forming episode. As molecular clouds fragment and collapse, is the end result more high-mass stars or more low-mass stars, for example? A description of this distribution of mass is called the *initial mass function* (IMF), introduced in Box 8.2 on page 183. The boxed figure reveals the basic form, as well as two different types of fit to the data. Note also the stellar/substellar boundary. Although objects with $M \lesssim 0.08 M_\odot$ do not form stars (Sect. 7.4.1), there is no reason to believe that the collapsing cloud 'knows' this. The physics of a collapsing cloud should be independent of the eventual central temperature of the final object, which determines whether nuclear reactions can be sustained in the core.

An IMF describes how many stars land on the ZAMS in some mass range or some logarithmic mass range, as shown in Box 8.2 and Fig. 8.4. In *non*-logarithmic form,[6] the IMF is

$$\xi(M) \equiv \frac{dN}{dM} \tag{8.9}$$

---

[6] Recall that $d\,\ln M = 1/M\,dM$. Therefore, $dN/d\,\ln M = M\,dN/dM$. If $dN/dM \propto M^\alpha$ (slope α), then $dN/d\,\ln M \propto dN/d\,\log M \propto M^{\alpha+1} \propto M^\Gamma$ (slope Γ). Thus, it is important to note whether an IMF plot is of $dN/d\,\log M$ or dN/dM because $\Gamma = \alpha + 1$.

where N is the number of stars found in the mass interval of interest. Finding the total number of stars over some broader mass range, say from M_{min} to M_{max}, requires an integration:

$$N = \int_{M_{min}}^{M_{max}} \xi(M)\, dM \tag{8.10}$$

The total luminosity or the total mass, respectively, in that mass range can be found from

$$L = \int_{M_{min}}^{M_{max}} \xi(M)\, L(M)\, dM \tag{8.11}$$

$$M = \int_{M_{min}}^{M_{max}} \xi(M)\, M\, dM \tag{8.12}$$

and a solution of Eq. 8.11 would require knowledge of the mass-luminosity (M-L) relation for the ZAMS. The relation for the main sequence is shown in Fig. 6.2 and given in numerical form in Table 6.1.

What, then, is a good description of the IMF? According to [177], a good fit for the disk of the Milky Way is a piecewise power law with a flatter slope for low-mass stars and a steeper slope for high-mass stars, i.e.

$$\xi(M) = k\left(\frac{M}{0.07}\right)^{-1.3\pm0.3} \qquad 0.07 < M \lesssim 0.5 \tag{8.13}$$

$$\xi(M) = \xi(0.5)\left(\frac{M}{0.5}\right)^{-2.3\pm0.36} \qquad 0.5 < M \lesssim 150 \tag{8.14}$$

$$\text{where} \qquad \xi(0.5) = k\left(\frac{0.5}{0.07}\right)^{-1.3\pm0.3} \tag{8.15}$$

Equation 8.15 ensures that both power laws give the same result at $M = 0.5\,M_\odot$. The constant k is a normalizing factor that allows the curve to shift up or down depending on the number of stars in the sample of interest. The curve flattens even more for sub-stellar objects such as brown dwarfs (Sect. 7.4.1).

Box 8.2
The Initial Mass Function (IMF)

The initial mass function is often expressed in logarithmic form as $\phi(M) = dN/d(log\, M)$, representing the number of stars per unit (logarithmic) mass interval as a function of mass. It represents the mass *distribution* of stars as they first 'land' on the ZAMS and is critical to understanding the mechanics of star formation. A star's initial mass determines its luminosity, its lifetime and the eventual return of enriched, higher-metallicity material to the interstellar

(continued)

(*continued*)

medium for future generations of stars. Galaxy evolution also depends on its star formation history. Hence, obtaining a robust IMF is an important goal of current active research. To what extent is the IMF 'universal', and to what extent can it vary with other properties, such as metallicity or local conditions [22], for example?

The IMF is difficult to measure because most young star clusters in the disk have already seen some evolution. The most massive stars may already have evolved off of the main sequence, and dynamical effects can preferentially eject low-mass stars and cause high-mass stars to migrate towards the center. Many stars are binaries or multiples, and for clusters that are still associated with their parent molecular cloud, dust extinction corrections can strongly vary with location. Since it is the star's luminosity that is initially measured, an accurate M-*L* relation (Fig. 6.2) is also needed [232].

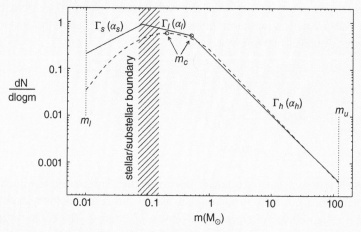

The figure shows how the IMF can be parameterized. A piece-wise power law (solid curve) is one possibility, with different slope, Γ or α, depending on whether or not the IMF is expressed in logarithmic form (footnote 6). Another possibility is a log-normal distribution (dashed curve). Different mass regimes (substellar, *s*, low mass, *l*, or high mass, *h*) are marked, along with the characteristic mass (subscript *c*), which is at the peak of the log-normal distribution or at M = 0.5 $M_\odot$ for the piece-wise distribution. Suggested lower and upper mass limits are subscripted *l* and *u*, respectively. Note the stellar/substellar boundary. Credit: Courtesy of A. Hopkins [146].

To understand the process of star formation, much effort has been undertaken to connect the mass function of dense cloud cores, called the *dense core mass function* (DCMF), with the IMF. An example is shown in Fig. 8.4 in which the DCMF (black histogram with error bars) is plotted along with various stellar IMFs. The stellar IMF

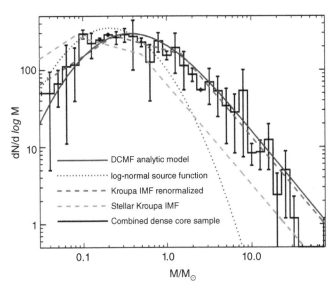

Figure 8.4 The dense core mass function (DCMF) is plotted as a histogram in black with error bars. Various stellar IMFs are plotted in other colours. Credit: Courtesy of A.A. Klishin and I. Chilingarian [170].

(green dashed curve) appears to follow the DCMF except for a mass offset. When the stellar IMF is bumped up in mass by a factor of about four (shifted to the right), it forms a renormalized IMF (the red dashed curve). The renormalized stellar IMF now matches the DCMF quite well. Such information can indicate the *efficiency* of star formation, which, in this particular example, indicates that approximately 25% of the mass of a dense core may actually turn into a star [170]. Values from about 20 to 50% have been quoted in other studies. Such an approach can provide a direct link between observed molecular cloud cores and the stars that are formed from them.

It is clear from Fig. 8.4 that, in any given star formation episode in the Milky Way, many more low mass stars are formed compared to high mass stars. There is some evidence, however, that modifications to Eqs. 8.13 to 8.15 may be required for different environments. Galaxies that experience strong *starbursts*, in which many stars are formed in a relatively short time, may form a greater number of high-mass stars than low-mass stars [177]. Such IMFs would be called *top-heavy*.

8.2 THE AGEING STAR

Of a list of the 288 brightest stars in the night sky, only about 19% are main sequence stars [87]. Given that stars spend the longest part of their lives on the main sequence, one might expect that any snapshot of the sky would be dominated by main sequence stars. Indeed, this is what the H-R diagram shows (Fig. I.2). But as we will see, ageing stars become brighter when they leave the main sequence, and this fact weights our naked-eye perception of the night sky towards evolved stars.

In this section, we seek to understand what happens once a star has used up the hydrogen in its core and finally leaves the main sequence. Some prior slow evolution occurs *on* the main sequence, as described in Sect. 7.3.2. But once core H → He burning has ceased, the star is now 'ageing'. By our definition in Sect. I.1, the ageing star is still a star because sustained nuclear reactions are occurring in its interior, although not always in the core. The main sequence lifetime is governed by the H → He nuclear burning timescale (Sect. 7.2.3), whereas the post-main-sequence lifetime is often on the thermal timescale (Sect. 7.2.2), which is much shorter. Consequently, post-main-sequence evolution occurs in the 'blink of an eye' by comparison to the main sequence. However, that blink contains changes that can include mass loss, variability, sharp turns in the H-R diagram and dramatic transformations in a star's size, temperature and luminosity. We would like to understand what happens in this ageing process.

Numerical modelling is required, so we will make use of the *Modules for Experiments in Stellar Astrophysics (MESA)* as implemented by *MESA Isochrones & Stellar Tracks (MIST, v. 1.2)* in Online Resource 8.9 [56, 80, 243–245]. These models adopt solar abundances of $X = 0.7154$, $Y = 0.2703$ and $Z = 0.0142$ and apply the Henyey method of solving the equations of stellar structure (Sect. 6.2.1). We use solar metallicity [Fe/H] = 0 (Eq. 3.1) and assume no rotation. Mass loss is included but not magnetic fields.

The way in which a star ages depends on its mass, so let's start with the mass that is most important (at least, to us!): a one-solar-mass star.

8.2.1 Post-Main-Sequence Evolution of a 1 $M_\odot$ Star

Figure 8.5 shows the post-main-sequence evolution of a 1 $M_\odot$ star with some key points labelled. The numbers are sequential in time and start at point #5 so as not to confuse them with any points on the pre-main-sequence tracks of Fig. 8.3. Notice how the tracks zig-zag up and down, overlapping each other in some regions. Table 8.3 provides some data for the numbered points.

Point #5 (●) is the terminal age main sequence (TAMS), defined as the point at which the hydrogen fractional mass in the core has reduced to the value $X = 10^{-12}$. In other words, the hydrogen in the core has been used up. The star's age in this model is 9.9 Gyrs. Recall that the current age of the Sun is 4.57 Gyrs (Eq. 1.5), so the Sun will not reach this point for another 5.3 Gyrs! At the TAMS, the stellar radius has already increased to $R = 1.6\,R_\odot$ and the luminosity to $L = 2.3\,L_\odot$. The core consists of inert helium and is surrounded by a hydrogen-burning shell followed by an inert hydrogen envelope.[7] It is interesting that by the TAMS, we are already seeing some

---

[7] The envelope has abundances that are essentially the same as when the star formed, but since hydrogen is clearly the dominant element initially, we will refer to the envelope as being an inert *hydrogen* envelope.

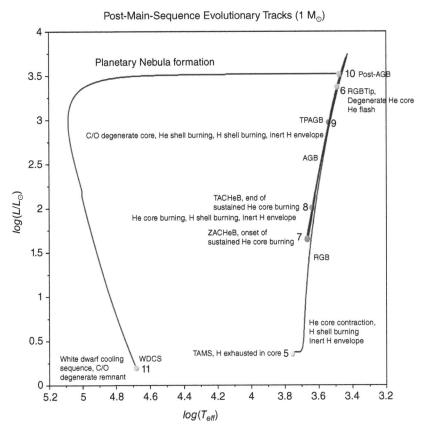

Post-Main-Sequence Evolutionary Tracks (1 M⊙)

Figure 8.5 Evolutionary tracks of a 1 M⊙ star. Numbered points are consecutive in time, and data for these are given in Table 8.3. TAMS (●): Terminal age main sequence. RGB: Red giant branch. RGBTip (●): Tip of the red giant branch. ZACHeB (●): Zero-age core helium burning. TACHeB (●): Terminal-age core helium burning. AGB: Asymptotic giant branch. TPAGB (●): Thermally pulsing asymptotic giant branch. Post-AGB (●): Post-asymptotic giant branch. WDCS (●): White dwarf cooling sequence.

differentiation of the elements within the star, i.e. helium in the core and hydrogen in the envelope.

With no more nuclear reactions in the core, there is insufficient pressure to hold it up, so the core contracts. The hydrogen-burning shell continues to produce helium, though, and that helium 'ash' adds to the mass of the inert helium core below it. Core contraction continues,[8] releasing gravitational potential energy (Eq. 7.4) and thereby

---

[8] If the core mass exceeds a value called the *Schönberg-Chandrasekhar limit*, it cannot hold up the weight of the layers above it, and contraction continues at an accelerated pace. See [44] for details.

Table 8.3 Post-main-sequence data

M^a	Ref.[b]	Phase[c]	$\log(t)^d$	$\log(L)^e$	$\log(T_{eff})^f$	$\log(R)^g$	$\log(T_c)^h$	$\log(\rho_c)^i$
0.6	5	TAMS	10.7736	−0.362	3.67	0.00374	7.27	3.64
	6	RGBTip	10.8033	3.27	3.48	2.188	7.84	5.94
1.0	5	TAMS	9.9965	0.36	3.75	0.193	7.31	3.26
	6	RGBTip	10.0545	3.38	3.49	2.237	7.87	5.96
	7	ZACHeB	10.0545	1.67	3.66	1.032	7.04	4.63
	8	TACHeB	10.0587	2.01	3.64	1.256	8.27	4.73
	9	TPAGB	10.0592	3.02	3.53	1.965	8.04	6.02
	10	post-AGB	10.0593	3.52	3.48	2.332	7.95	6.20
	11	WDCS	10.0594	0.20	4.68	−1.733	7.81	6.36
5.0	5	TAMS	8.0000	3.13	4.14	0.7995	7.62	2.46
	6	RGBTip	8.0036	3.36	3.61	1.989	8.11	4.02
	9	TPAGB	8.0627	4.18	3.52	2.567	8.36	6.55
	11	WDCS	8.0632	2.17	5.24	−1.865	8.05	7.14
24	5	TAMS	6.8571	5.22	4.47	1.187	7.92	1.66
	10	C-burn	6.9019	5.36	3.58	3.036	8.94	5.34

[a] Mass ($M_\odot$).
[b] Reference position shown in Fig. 8.5.
[c] Phase: TAMS (terminal age main sequence), RGBTip (tip of the red giant branch), ZACHeB (zero-age core helium burning), TACHeB (terminal-age core helium burning), TPAGB (thermally pulsating asymptotic giant branch), post-AGB (post-asymptotic giant branch), WDCS (white dwarf cooling sequence), C-burn (end of carbon burning).
[d] Log of the time (years) from the beginning of the pre-main-sequence track.
[e] Log of the luminosity ($L_\odot$).
[f] Log of the effective temperature (K).
[g] Log of the radius ($R_\odot$).
[h] Log of the central temperature (K).
[i] Log of the central density (g cm^{-3}).
Credit: Adapted from Online Resource 8.9.

creating a contraction-driven luminosity, L_{grav}. This L_{grav} increases the effectiveness of the hydrogen burning in the surrounding shell and adds to its luminosity. The net result is an increase in the luminosity of the star, dominated by the hydrogen-burning shell, and the star moves up in the H-R diagram along the *red giant branch* (RGB).

As the luminosity increases, the star's envelope absorbs some of the radiative energy that originates below it, so the outer layers expand. Expansion results in some cooling (cf. adiabatic expansion, Sect. 2.1.8), so as the star climbs the RGB track, it also moves to the right (cooler) in the H-R diagram. The star is now a *red giant*. Notice that what is happening to the surface (what is actually observed) is opposite to what is happening to the core: the expanding envelope becomes less dense and cooler, whereas the contracting core becomes denser and hotter. This is sometimes called *mirroring*, the mirror being at the hydrogen-burning shell.

As the stellar envelope cools, the H^- ion becomes an important source of opacity and, as we have seen before (e.g. Sect. 8.1.4), convection is favoured. A convective envelope that was already present at the TAMS becomes deeper as the star ascends the RGB, and eventually reaches down into regions in which the chemical composition has been modified by nuclear reactions. When that occurs, processed materials from deep within the star are convected upwards, altering the observable chemical composition at the surface. This is called the *first dredge-up*. (Stars that are more massive can experience additional dredge-ups at later phases in their post-main-sequence evolution.) An ascent up the RGB with a deepening convective envelope is essentially the opposite of the descent down the Hayashi track during star formation that we saw in Sect. 8.1.4.

In a time 1.4 Gyrs after the TAMS, the star reaches the *tip of the red giant branch* (RGBTip) at point #6 (●). The RBGTip identifies the point after leaving the main sequence in which the stellar luminosity is a maximum (or T_{eff} is a minimum) and before core He burning has progressed. (This end point is difficult to see in Fig. 8.5 because of the overlapping tracks that come later.) The change in the star along the RGB is impressive. The luminosity has increased to $L = 2400\,L_\odot$, the surface has cooled to ~3000 K and the radius has increased to $R = 173\,R_\odot = 0.80$ AU. This is larger than the orbit of Venus! It is interesting to imagine what Earth's daytime sky will look like when our Sun reaches this red giant phase.

At the RGBTip, the evolutionary track makes an abrupt reversal in direction, and to understand why, we need to take a closer look at the helium core. Along the RGB, the core mass continues to increase due to the addition of helium from the hydrogen-burning shell above it. At the same time, this core continues to contract, and its density increases dramatically. Between the TAMS and RGBTip, for example, the core mass increases from 0.04 $M_\odot$ to 0.47 $M_\odot$, and its central density increases from ~10^3 g cm^{-3} to ~10^6 g cm^{-3} (Table 8.3). As the higher density is approached, the core becomes degenerate (Sect. 4.4), so the pressure is supplied by electron degeneracy pressure (equation of state, $P \propto \rho^\gamma$, Eq. 6.20), and its structure is independent of its temperature. Thermal particle pressure ($P \propto \rho\,T$, Eq. 2.4) still exists but is negligible by comparison. Once the core temperature reaches ~10^8 K, helium burning is ignited via the triple-α process (Sect. 5.5.1). Triple-α burning is highly temperature sensitive (e.g. $\epsilon_{3\alpha} \propto T^{41}$, Eq. 5.28), so even a small temperature gradient can cause sudden ignition as soon as the temperature threshold is achieved, like an on/off switch. However, the pressure of a degenerate core adjusts to changing densities, not temperatures, so once helium burning begins, the core does not expand and cool as it would for normal matter. The initial helium burning contributes to a rise in temperature, producing even more burning, in a violent runaway process called the *helium flash*.

Helium *flashes* is actually a better representation of what occurs because of a peculiar property of the core. An important cooling mechanism is from neutrino losses, and the dominant neutrino production process in the core is via the decay of plasmons (quanta of electromagnetic fields from plasma oscillations) into neutrino/anti-neutrino pairs [30, 163]. Neutrinos readily escape from anywhere in the core because of their low interaction cross-section, but they are more readily produced at the denser center. Consequently, neutrino cooling creates a temperature inversion in the core such that the highest temperature is not at the center itself but rather at a larger radius off-center. Therefore, the helium flash begins off-center [e.g. 98] and propagates inwards! Neutrino cooling determines the depth at which helium burning begins but does not influence the subsequent process, which involves a series of flashes that propagate towards the center [113]. The first flash is the strongest, with a luminosity that can achieve $\sim 10^{10}$ $L_\odot$ in a short time (of order decades, [221]), comparable to the luminosity of an entire galaxy [113]! Yet that luminosity is absorbed within the star, so we do not externally 'see' the He flash itself. Instead, we see its consequence in the form of a sharp track reversal at point #6.

The helium flash phase takes $\lesssim 2$ Myrs [220] and ceases when the rising temperature raises the ideal gas pressure to rival the degeneracy pressure and finally lift the degeneracy. The core then expands and cools and does not pump so much energy into the outer layers, so the luminosity drops. The outer layers respond by contracting and heating up. Again, notice that the expanding, cooling core and contracting, heating outer layers are behaving oppositely. The star moves from point #6 to point #7 in Fig. 8.5, a process that takes only 1.7×10^6 years in the plotted model.

Point #7 (●) to point #8 (○) denotes zero-age core He burning (ZACHeB) and terminal-age core He burning (TACHeB), respectively. In this segment, The triple-α process now converts helium into carbon (Eq. 5.27) and then oxygen (Eq. 5.30), the two primary products. The star is on the 'helium burning main sequence', and evolution proceeds similarly to the hydrogen burning main sequence that we saw in Sect. 7.3.2. The timescale for this phase is governed by the nuclear timescale for helium burning and lasts $\sim 10^8$ yrs. This roughly vertical section becomes more horizontal as the ZAMS mass changes (cf. the small loop in the 5 $M_\odot$ star track in Fig. 8.6). Observationally, the helium burning section is referred to as the *horizontal branch*, but it is more obvious and horizontally stretched out when groups of low-metallicity stars, such as in globular clusters, are observed.

After point #8, there is insufficient helium in the core to continue sustained helium burning, and the star begins to climb the *asymptotic giant branch* (AGB). The AGB is so named because the track generally approaches the RGB asymptotically from the left, although it essentially overlaps the RGB for the 1 $M_\odot$ star shown in Fig. 8.5. What happens physically is similar to the previous climb of the RGB – a contraction and heating of the core, and expansion and cooling of the outer layers. But the internal structure is now more complex, and there is further differentiation of the elements. The core is inert carbon and oxygen (C/O), a helium-burning shell surrounds it, a hydrogen-burning shell is next outwards, and finally there is an inert hydrogen envelope. Along the AGB, the C/O core also becomes degenerate.

Point #9 (●) specifies the thermally pulsing AGB (TPAGB) and specifies the onset of *thermal pulsations*. Thermal pulsations derive from an instability between two shell-burning regions when there are mismatches between the temperature dependence, energy generation rates and rates of fuel exhaustion in the two shells. For example, a helium-burning shell can use up its fuel, become narrow and contribute little luminosity compared to the hydrogen-burning shell above it. However, the upper hydrogen-burning shell continues to provide helium to the lower level, eventually building up sufficient helium fuel at the right temperature and pressure to ignite the lower helium-burning shell again. The re-ignited helium-burning shell expands, and its density decreases, but the expansion of this thin shell is insufficient to change the pressure from the layers above, so (recall $P \propto \rho T$) the temperature increases, producing a brief runaway thermal pulse [186]. These pulses are thought to begin once the difference in mass between the H-burning and He-burning shells is less than 0.1 $M_\odot$, and it is this point that is identified as #9 in Fig. 8.5. Thermal pulses occur intermittently along the AGB and contribute to mass loss during this phase.

The AGB star now has an extended envelope and a small, degenerate C/O core. This extended envelope is highly susceptible to mass loss via stellar winds, pulsations and/or radiation pressure on dust grains. Evolution is then towards the post-AGB at point #10 (●), which is where the H-rich stellar envelope falls below 20% of the current stellar mass.[9] The stellar mass is no longer 1 $M_\odot$, though. By #10, it has lost 40% of its initial mass, most of which occurred post-TPAGB.

At point #10, thermal pulses have ended, and this point marks the beginning of the evolutionary track that shifts to the left on the H-R diagram. This phase is arguably the least understood of all post-main-sequence evolution [26]. With no new nuclear burning starting and only some hydrogen burning occurring in a depleted envelope, the luminosity remains approximately constant. The envelope contracts, the temperature increases (Eq. 2.56), and the star moves horizontally to the left on the H-R diagram. The speed of the horizontal crossing is set by the speed with which the H-rich envelope is consumed [26]. Once the temperature achieves ~30,000 K (log(T) = 4.5), a high-speed radiation-driven wind develops and can interact with previously ejected material [186]. This circumstellar expanding material can be ionized by the central star, forming a visible *planetary nebula* (Sect. 10.1) around it. At this phase, the star is known as a *central star of a planetary nebula* (CSPN).

By the time the temperature is at its maximum (farthest to the left in Fig. 8.5), the radius has decreased to $R = 0.07\,R_\odot$ and the final vestiges of hydrogen shell burning are fading away. Thereafter, the luminosity declines until it reaches the *white dwarf cooling sequence* (WDCS) at point #11 (●). The white dwarf is essentially the degenerate C/O core of the evolved 1 $M_\odot$ star. At this point, a mass of M = 0.54 $M_\odot$ has been packed into a radius of only $R = 0.018\,R_\odot = 2\,R_\oplus$ (see also Box 8.3 on page 192). This is no longer a star but a *stellar remnant* in which no new nuclear reactions will take place. White dwarfs will be discussed again in Sect. 10.2.

---

[9] Some models define this point as occurring when the mass of the envelope is 1% of the mass of the star [26].

Box 8.3

The Initial-Final Mass Relation (IFMR)

The IFMR is the relation between a star's initial mass and its final mass. Not to be confused with the IMF (Box 8.2 on page 183), the IFMR is a probe as to how much mass is lost from a star as its evolution progresses. Mass loss is a crucial factor in how stars age and is one of the key uncertainties in models of post-main-sequence evolution, especially for massive stars.

The figure shows the IFMR from [69]. The data points represent masses (M_f) of white dwarfs (WDs) in open and globular clusters. Spectroscopy of WDs provides a value of the gravitational acceleration, g (Sect. 3.2) and effective temperature, T_{eff}. Using $g \propto M/R^2$, M_f can be found from the mass-radius relation (Sect. 10.2.1). Cluster fitting (Sect. 8.2.3) provides the age of the cluster, including its white dwarfs, and their location on the WD cooling curve is a function of their initial mass, M_i, obtained from stellar evolution codes.

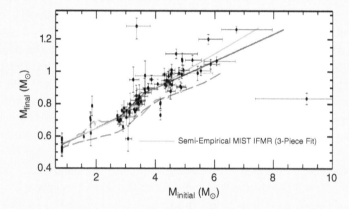

Other approaches to finding the IFMR have also been pursued, but most involve studies of WDs.

Credit: Courtesy of P.-E. Tremblay, from [69].

The data in the figure give (for masses in $M_\odot$):

$$M_f = (0.080 \pm 0.016)\,M_i + (0.489 \pm 0.030) \quad (0.83 < M_i < 2.85)$$
$$M_f = (0.187 \pm 0.061)\,M_i + (0.184 \pm 0.199) \quad (2.85 < M_i < 3.60) \qquad (8.16)$$
$$M_f = (0.107 \pm 0.016)\,M_i + (0.471 \pm 0.077) \quad (3.60 < M_i < 7.20)$$

Consequently, a 1 $M_\odot$ star produces a WD remnant of ~0.57 $M_\odot$, and a 7 $M_\odot$ star results in a ~1.2 $M_\odot$ WD.

8.2.2 Post-Main-Sequence Evolution of Stars of Different Mass

Stars of different mass age differently, and this is illustrated for a few masses in Fig. 8.6. Table 8.3 provides data for the points marked in the figure. Examination of Table 8.3 immediately shows how post-main-sequence timescales vary with mass.

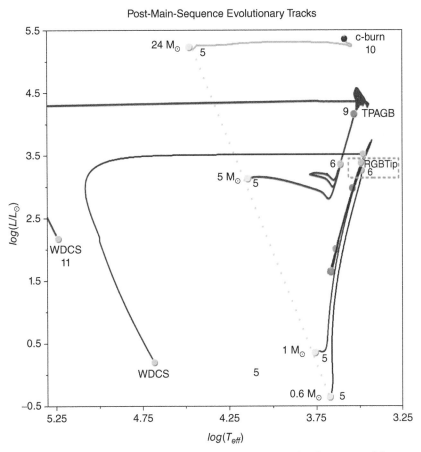

Figure 8.6 Post-main-sequence evolutionary tracks for stars of four different masses marked near the large yellow dots. Numbered points are consecutive in time, and data for these are given in Table 8.3. Numbering, colouring and acronyms are the same as in Fig. 8.5 and Table 8.3. Each track starts at the TAMS, which is denoted with a yellow dotted line. A dashed purple rectangle encloses the RGBTip for masses 0.6 $M_\odot$ and 1 $M_\odot$. The track for a 5 $M_\odot$ star (black) is cut off slightly at the left to save space.

Just as we saw for the main sequence (Fig. 7.1), more-massive stars age more quickly than less-massive stars. When measured from the TAMS, for example, a 0.6 $M_\odot$ star takes $\sim 4 \times 10^9$ yrs to reach the tip of the red giant branch, whereas a 5 $M_\odot$ star takes only $\sim 8 \times 10^5$ yrs – a rather striking difference. Higher-mass stars also achieve higher luminosities and higher central temperatures, just as they did on the main sequence.

The timescales and relations found for the main sequence cannot be used for post-main-sequence evolution, however. The main sequence represents a relatively uniform data set in which stars of all masses are converting hydrogen into helium in their cores with properties that are only slowly varying with time and do not change by large amounts (e.g. Fig. 7.2). By contrast, post-main-sequence evolution of different masses can involve many different types of nuclear reactions in shells or cores, degeneracy, significant mass loss, and changes of luminosity and radius that cover many orders of magnitude. General post-main-sequence relations like those that we found in Sect. 6.3 are therefore not easily found except for restricted stellar regions or evolutionary phases.

In this section, we will not look in detail at the post-main-sequence tracks of stars of different masses as we did for the 1 $M_\odot$ star in Sect. 8.2.1. Instead, we will focus on a number of mass ranges and point out some salient and interesting features in each. We will also indicate what stellar remnant should result, but remnants will be discussed in more detail in Chapter 10. Note that the boundaries as to what is considered a 'very-low-mass star' or a 'low-mass star', etc. can vary, depending on the details of stellar models.

8.2.2.1 Very-Low-Mass Stars

Stars whose masses range from the lowest mass possible for a star (M $\sim$ 0.08 $M_\odot$, Sect. 7.4.1) to $\sim$0.6 $M_\odot$ evolve very slowly. Evolution on the main sequence, for example, takes longer than 15 Gyrs for all stars of mass $\lesssim$1 $M_\odot$ [51]. Since the current age of the universe (the *Hubble time*) is 13.7 Gyrs, these stars have not yet evolved off of the main sequence. Given that no very-low-mass post-main-sequence stars exist, it hardly seems necessary to explore how they age. However, these are the most *numerous* stars, as a glance at Fig. 8.4 verifies. Consequently, the future of our Galaxy, and possibly the universe, actually depends on what happens to these stars.

Very-low-mass stars on the main sequence are generally referred to as *red dwarfs* because their low temperatures (of order 3500 K) lead to Planck curves that peak in the red to infrared part of the spectrum (Eq. 2.51). Their low masses and luminosities lead to the longer main sequence lifetimes (e.g. Eq. 7.30). As we have seen, low temperatures also favour higher opacities (Sect. 3.1), and higher opacities favour convection (Sects. 4.2.3, 4.3). Therefore, red dwarfs can be fully convective, and this mixing leads to a fairly homogenous composition within any given star. This mixing also contributes to their long lifetimes because star-wide convection makes fuel from the entire star accessible for burning, rather than just in the core, so hydrogen

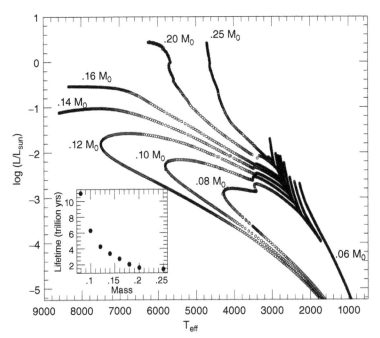

Figure 8.7 Evolutionary tracks of very-low-mass stars. The mass is marked on each track. *Inset Diagram*: The corresponding main sequence lifetimes for the different masses. Tracks generally start at lower right, move up and to the left, and finally move back down and to the right again for the lower masses. Note that only the 0.25 $M_\odot$ star follows a bona fide upwards red giant track. Credit: Courtesy of Peter Bodenheimer [188].

burning can be kept active for longer. We found, for example, that the Sun had only ~10% of its hydrogen fuel actually available for burning (f_{avail} in Eq. 7.30). A star with an initial mass of $M = 0.1\,M_\odot$ remains fully convective for 5.7 *trillion* (10^{12}) years! Consequently, its available fuel supply is about the same as the Sun [1].

Modern numerical codes generally do not evolve very-low-mass stars into the far future. However, we show one early example of numerically determined evolutionary tracks in Fig. 8.7 from [188]. The behaviour of these tracks is very peculiar, compared to the tracks of Figs. 8.5 and 8.6. Small red dwarfs remain physically small and grow hotter to become blue dwarfs, eventually running out of fuel and fading away as degenerate helium white dwarfs. Only the 0.25 $M_\odot$ star in the figure climbs its RGB, and this is presumably the least massive star that will expand to become a red giant [1].

Why is there a difference? As a fully convective star gradually turns hydrogen into helium, its opacity decreases because the probability of a photon interacting with helium is lower than that of hydrogen. This happens in the core first, eventually decreasing the opacity sufficiently that the core becomes radiative rather than convective. When the core becomes radiative, helium produced in the core is no longer mixed into the outer layers. There is then an opacity gradient in the star such that the outer layers have higher opacity (a higher fractional hydrogen content), whereas the core has a lower opacity (a higher fractional helium content). The higher-mass 0.25 $M_\odot$ star is evolving somewhat faster than stars at lower masses, so this opacity gradient occurs at an earlier evolutionary time than for the lower-mass $\lesssim 0.25\,M_\odot$ stars. Consequently, there is a greater envelope/core opacity difference for the 0.25 $M_\odot$ star, i.e. this star has relatively more hydrogen in its envelope compared to the lower-mass stars. Core photons in the 0.25 $M_\odot$ star hit an 'opacity wall', driving the radius outwards so the star becomes red giant. By contrast, the more slowly evolving lower-mass stars have more helium in their envelopes when the core becomes radiative, so photons can escape more easily. As the low-mass star's luminosity increases, as is normal for post-main-sequence stars, an increase in temperature, rather than radius, results (recall $L \propto R^2\, T^4$, Eq. 2.56), and the star becomes a hot blue dwarf rather than a cool red giant [188].

There is, of course, much more to this picture. For example, other important sources of opacity exist, such as from the H^- ion, hydrogen ionization, molecules and grains, depending on temperature (cf. the M5V stellar spectrum in Fig. 3.1). Nevertheless, the basic result is that to become a red giant, the opacity of the envelope needs to be relatively high so that outgoing photons can be absorbed in the envelope and cause expansion. The importance of opacity is, again, obvious.

Very-low-mass stars should ultimately lead to a remnant that is a helium white dwarf. The mass is insufficient to start helium burning, whether or not the star becomes a red giant. Given the length of time that this process takes, we should not observe any helium white dwarfs that result from normal stellar evolution. Nevertheless, some helium white dwarfs are indeed observed. Those that we *do* see likely result from *tidal stripping*, e.g. stripping of the outer layers of red giants by nearby companions [e.g. 290], or possibly have non-solar abundances originally [7].

8.2.2.2 Low-Mass Stars

Stars in the mass range $0.6\,M_\odot \lesssim M \lesssim 2.2\,M_\odot$ may be considered low-mass stars. For an entertaining and instructive animation of a low-mass star's evolution, see Online Resource 8.2.

In this mass range, the helium core along the RGB is degenerate, so helium ignition at the RGBTip occurs with helium flashes [221] as we saw for the 1 $M_\odot$ star in Sect. 8.2.1. This is a unifying trait for this mass range, so we will focus on the star's journey up the RGB.

Because the degenerate core is small (of order the size of the Earth), yet its mass is significant and growing, the gravity at the core surface is very large (Eq. 6.5).

The Lagrangian form of the equation of hydrostatic equilibrium (Prob. 6.1) is $|dP/dm| \propto m/r^4$. Therefore, with r small, the pressure drops precipitously as mass increases outwards in the hydrogen-burning shell. The extended envelope at higher r exerts negligible pressure on the shell region compared to the gravitational pressure of the core. In general, the properties of the degenerate core have a far greater effect than the overlying layers on the properties of the luminosity-producing shell. For example, it can be shown [166, 258] that *approximate* relations for temperature and luminosity in the *shell* are, respectively, $T \propto M_c$ and $L \propto M_c^8$, where M_c is the *core* mass. This latter relation is a M-L relation for low-mass stars on the RGB, although it relates the luminosity to the *core* mass rather than the total mass of the star as we saw for the main sequence (Sect. 6.3).

Although the structure of the degenerate core is independent of temperature, its temperature and that of the surrounding shell still increase as helium from the hydrogen-burning shell falls onto the core and increases its mass. Once the core mass reaches $M_c \approx 0.48\,M_\odot$, the temperature is high enough ($\sim 10^8$ K) to ignite helium burning with helium flashes, as described in Sect. 8.2.1. The key points are that even though the stellar mass may vary, the luminosity at the RGBTip depends on the *core mass*, and helium flashes start at a well-defined core mass/temperature. This means low-mass stars should all reach the RGBTip at around the same luminosity. This is illustrated, in a limited fashion, by the light purple dashed box in Fig. 8.6, which encloses the RGBTip for stars of mass $0.6\,M_\odot$ and $1\,M_\odot$.

We know that evolutionary tracks make a sharp reversal in the H-R diagram at the RGBTip, so the RGBTip for low mass stars is a well-defined region in the H-R diagram with a *well-defined luminosity* (although it varies somewhat with metallicity). There-fore, the RGBTip can be used as a *standard candle*, which is an object or objects with a known luminosity (cf. Box 2.1 on page 42). A measurement of the magnitude of the RGBtip (say, a V magnitude) together with the adopted luminosity (L) can provide the *distance* via Eq. I.11. This is a common method of obtaining the distances to galaxies for which an H-R diagram can be obtained.

Remnants of stars in this mass range are C/O white dwarfs.

8.2.2.3 Intermediate-Mass Stars

Stars in the mass range $2.2\,M_\odot \lesssim M \lesssim 8\,M_\odot$ progress similarly to low-mass stars, with a climb up the RGB, a reversal, helium core burning, a trip up the AGB and mass loss. However, different internal physical conditions in the intermediate-mass stars mani-fest as differently appearing twists and turns in the H-R diagram tracks. This is clear when a $5\,M_\odot$ star is compared to a $1\,M_\odot$ star in Fig. 8.6.

As with all stars, properties at the terminal-age main sequence (TAMS) starting point are different, and this affects later evolution. For example, main sequence hydrogen burning is via the CNO cycle, and its dependence on temperature is steeper than the PP-chain, so the core is convective (Fig. 4.6).[10] More massive main sequence stars have lower densities (Sect. 7.5), and this is also the case for the *central* density, as can be seen in the $log(\rho_c)$ entries at the TAMS in Table 8.3. The key difference for intermediate-mass stars compared to low-mass stars is that by the RGBTip, the helium core is *not* degenerate. For example, the 5 $M_\odot$ star has a central density of only $\sim 10^4$ g cm^{-3} (Table 8.3) compared to a density of order $\sim 10^6$ g cm^{-3} that would be needed for degeneracy. Therefore, when the temperature is high enough to ignite helium burning, there are no helium flashes, and the star remains in equilibrium as helium burning progresses.

The tracks still reverse at the RGBTip, and the star then follows a loop in which sustained core helium burning occurs. This is called the *blue loop*. For the 5 $M_\odot$ star, the loop is centered at $log(T/K) \sim 3.7$ and $log(L/L_\odot) \sim 3.1$. The star then climbs the AGB, where, similar to low-mass stars, the C/O core becomes degenerate. The outer layers, by contrast, are highly distended compared to the small, dense Earth-sized core. For example, when thermal pulses begin at point #9, the size of the 5 $M_\odot$ star is $R \sim 370 R_\odot = 1.7$ AU. This would engulf the orbit of Mars if such a high-mass star were at the center of our solar system! On the AGB, the star experiences instabilities and rapid mass loss (Box 8.3 on page 192), eventually exposing the degenerate C/O core. The remnant is therefore a C/O white dwarf.

8.2.2.4 High-Mass Stars

A high-mass star is any star with M $\gtrsim$ 8 $M_\odot$ up to the highest masses thought possible (Sect. 7.4.2). A dramatic change occurs at this mass because most high-mass stars are expected to explode as *supernovae* (SN)[11] [e.g. 212]. The 8 $M_\odot$ boundary is somewhat uncertain, and a range of about $\pm 1 M_\odot$ has been quoted [284], though sometimes a higher uncertainty is suggested. The consequences of the SN are dramatic not only for the star but also for the surrounding interstellar medium (ISM), which becomes turbulent and disrupted. High-mass stars are also the key to heavy element formation and dispersal into the ISM.

Even before leaving the main sequence, high-mass stars (O and B spectral types) have significant winds and may already have luminosities close to the Eddington limit (L_{Edd}, Sect. 7.4.2). Their subsequent evolution also will not exceed L_{Edd}, so post-main-sequence tracks tend to be horizontal in the H-R diagram. This is evident for the 24 $M_\odot$ star shown in Fig. 8.6. Rotation, which has not been taken into account in the models presented here, contributes to mass loss and makes a bigger difference in the behaviour of post-main-sequence tracks of high-mass stars than it does for lower-mass stars [56]. Timescales are very short. Only ~ 0.8 million years (Table 8.3) elapses between the TAMS and point #10, which is labelled C-burn and represents the end of carbon burning (see Eqs. 5.33, 5.34 and 5.35 for sample reactions). From

---

[10] Core convection actually starts at somewhat lower masses of M $\sim$ 1.3 $M_\odot$ (Sect. 4.3).

[11] Very-high-mass stars may skip the supernova stage if an outwards shock is inadequate (Sect. 10.3.1).

the end of carbon burning to the formation of an iron core is only a few thousand years! All told, following the post-main-sequence evolution of high-mass stars is challenging.

What happens en route to the iron core? As the star's luminosity remains about the same, the C/O core that was formed from earlier helium burning does *not* become degenerate by the time temperatures are high enough to ignite carbon burning ($\sim 5 \times 10^8$ K). This has important consequences because nuclear burning does not cease at this stage but carries on, with the star going through a rapid sequence of nuclear burning stages. Products of carbon burning such as neon, oxygen and magnesium form and, at the appropriate temperature, also start to burn. Oxygen burning ignites at T $\sim 10^9$ K, leading to silicon, magnesium and sulphur (e.g. see Table 5.1). After each fuel is exhausted in the core, the burning continues in a shell, the core contracts and heats up, and then the next species starts to burn in the core. As each burning stage begins in the core, previously formed elements are shifted outwards by comparison. Consequently, the lightest elements are in the outer envelope followed by increasingly heavier elements towards the center. The differentiation of elements, first introduced in Sect. 8.2.1, is now extreme. Figure 8.8 provides a sample illustration in which different elements are dominant in different shells, with nuclear burning shells in between. Convection occurs at various stages and can mix material in different shells.

In spite of the uncertainties involved with highly evolved high-mass stars, a general picture of the final stages in the life of a massive star can be obtained from these models. The final core consists of ^{56}Fe with some additional iron group elements such as chromium (Cr). Our earlier discussion of binding energy (Sect. 5.3) showed that fusion reactions are endothermic (they require energy) for elements with the mass of iron or higher (Fig. 5.2). This has the drastic consequence of leading to a supernova explosion, of which more will be said in Chapter 10. The remnant is a neutron star or black hole.

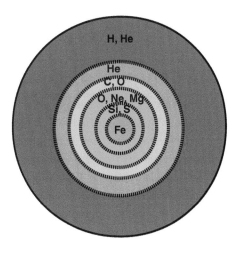

Figure 8.8 Example schematic illustrating the differentiation of elements in a high-mass star. Elements are labelled, and the changing colours represent increasing temperatures towards the core of iron. The hatched separations represent nuclear-burning shells between each. Sizes are not to scale.

Box 8.4

Starring … Albireo (β Cygni)

Distance: $d \sim 100 \rightarrow 120$ pc

β **Cygni A:** Spectral Type: K2II; Apparent Magnitude: V = 3.21; Effective Temperature: T_{eff} = 4,383 K; Mass: M = 5.2 $M_\odot$; Luminosity: L = 1,259 $L_\odot$

β **Cygni B:** Spectral Type: B8Ve; Apparent Magnitude: V = 5.21; Effective Temperature: T_{eff} = 13,200 K; Mass: M = 3.7 $M_\odot$; Luminosity: L = 230 $L_\odot$

Most easily seen in the northern summer sky is an asterism called the *summer triangle*. Three bright stars form the corners of the triangle. They are Vega in the constellation Lyra (the lyre), Altair in the constellation Aquila (the eagle) and Deneb in the constellation Cygnus (the swan). Cygnus looks like a cross in the sky; hence its familiar moniker, the *northern cross*. Right in the middle of the summer triangle or, equivalently, at the foot of the cross, is the star Albireo.

Albireo looks like a single star to the naked eye, but a small telescope reveals that it consists of two principle stars. This is one of the prettiest binaries in the sky because its two stars show their contrasting colours beautifully. The figure gives a telescopic view. Star *A* is a red giant, and star *B* is a hot blue dwarf. The actual sizes of the stars are not apparent in the picture, though. Star *A* just looks bigger because it is brighter and its light is spread out over a larger detectable area as a result of the telescope's resolution.

8.2.3 Connecting Theory with Observations

In this chapter, we have seen a rather complex story in which stars are formed over various mass ranges with different timescales. They remain on the main sequence for most of their lives and then follow different post-main-sequence pathways, again with different mass-dependent timescales and behaviours. Theoretical H-R diagrams, i.e. T_{eff}-L plots, were shown in Figs. 8.3, 8.5, 8.6, and 8.7. How can we put these together and relate them to observational H-R diagrams, i.e. colour-magnitude (CM) plots, such as shown in Fig. I.2 in the Introduction? That CM diagram included all stars of measured distance using the Gaia satellite (Box 3.1 on page 60) in its first data release. In that diagram, we have no problem seeing the broad swaths of the main sequence plus the red giant branch, but it is difficult to deduce much more.

An important theoretical ⇒ observational step is to produce *isochrones*. These are curves that connect stars of the same (iso) age (chrone), but different masses on an H-R diagram. Numerical codes such as the one introduced at the beginning of Sect. 8.2 (see also Online Resources 8.3 through 8.9) include isochrones in their outputs. Steps must then be taken to convert from theoretical values to observational quantities. That is, the temperature must be converted to a colour index in the filter band of interest (Sect. 3.1), and the luminosity must be converted to an absolute magnitude in the band of interest (Sect. 3.2). With these transformations made, we are well-positioned to compare data to theoretical expectations.

The usefulness of isochrones is clear when a group of stars can be identified that were all formed at approximately the same time, or at least in the same star-forming episode. As pointed out in Sect. 8.1.2, stars form in clusters. In the disks of galaxies, including the Milky Way, these open clusters continue to form to the present day. Any given open cluster is relatively young because, over time, cluster stars disperse from dynamical effects. By contrast, globular clusters formed very early in the history of the Milky Way. They have low metallicites, and star formation ceased long ago (Sect. 7.5). Although there is a distribution of ages for globular clusters, they are very old on average and are dynamically stable unless they are close to the disrupting effects of the Galaxy disk.

Figure 8.9 shows fitted isochrones for the open cluster the Pleiades *(Left)* and the globular cluster Ruprecht 106 *(Right)*. The filter bands are different for the two clusters, precluding the possibility of a quick numerical comparison between the two. Nevertheless, the shape of the fitted isochrones immediately indicates which cluster is older. As a cluster ages, more and more stars move off the main sequence and onto post-main-sequence phases, starting with the highest-mass stars and moving to lower-mass ones. Consequently, the point at which stars turn off of the main sequence, the *turnoff point*, shifts down the main sequence to lower and lower-mass stars as the cluster ages. The turnoff point alone gives an estimate of the age, but fitting the entire isochrone to the data is more accurate. Isochrones fit to these two clusters reveal ages of 100 Myr for the Pleiades and a much older 12.0 Gyr for Ruprecht 106. Because isochrones shift somewhat with different metallicity, such fits can provide estimates of the metallicity as well. A few stars appear to be hotter and more luminous than the current age of the cluster would suggest – as if they were recently formed. This is most obvious in the CM diagram of Ruprecht 106, in which a smattering of points can be seen to the upper left of the main sequence curve. These are called *blue stragglers* and may represent stars that have either collided and coalesced with another star or experienced mass transfer from a companion star in the densely populated globular cluster environment.

Once isochrone fitting has been carried out for a variety of clusters, we can search for other properties that might correlate with stellar age. If successful, observations

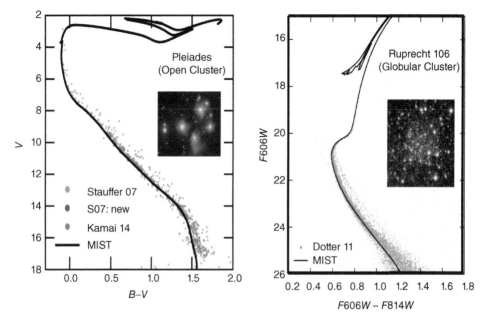

Figure 8.9 Colour-magnitude (CM) diagrams of the open cluster the Pleiades (**Left**) and the globular cluster Ruprecht 106 (**Right**) with fitted isochrones. Different filters are used for the colour index (x-axis) and magnitude (y-axis) between the two plots. Dots represent the data, and the fitted isochrones (100 Myr for the Pleiades and 12.0 Gyr for Ruprecht 106) are represented with solid curves. Credit: Courtesy of Jieun Choi, from [56]. *Insets*: Optical images of the clusters. Pleiades: NASA, ESA, AURA/Caltech, Palomar Observatory, Ruprecht 106: ESA/Hubble & NASA, R. Cohen.

of such properties for any given star could then tell us its age. Some examples for stars that have convective envelopes are given in Box 8.5 on page 202.

Box 8.5

Magnetic Braking and Chronology

We saw in Sect. 8.1.3 that jets can rid a star of angular momentum, causing the star's rotation to slow down. A related mechanism is called *magnetic braking*. The picture shows an artist's concept of a star whose magnetic field is connected, like curved wires, to a surrounding disk of material. The more slowly orbiting disk material 'brakes' the star due to this magnetic connection. Such braking can even occur without a surrounding disk, although not as strongly. This is because magnetic fields are coupled to charged particles that leave the star as winds, and these magnetized outflows also carry away angular momentum. The consequence is that both rotation and magnetism on the star diminish, and this continues as long as rotation and magnetism are connected via a global dynamo [217]. Dynamos (Sect. 1.4) are present in 'sun-like' stars, i.e. stars that have convective envelopes and radiative cores, or $0.4 \lesssim M/M_\odot \lesssim 1.3$ [70], so stars in this mass range should show decreasing magnetic activity and slower rotations with time.

Observations bear this out. With lower magnetic field strengths, there is less chromospheric heating, and this affects the strengths of spectral lines. A standard proxy for chromospheric magnetic activity is ionized calcium, specifically the easily measured Ca II H and K lines (Fig. 3.5), which are found to decay with stellar age [283]. Ages measured this way are called *chromospheric ages*, and uncertainties of order 40–60% have been quoted for this method [330]. Alternatively, dating stars by observing their rotation periods is called *gyrochronology* [21], and age uncertainties of ~15% [21] have been quoted for this method. Gyrochronology does not work as well for solar-like stars that are more than about half-way through their main sequence lifetimes, though. Once the rotation period has slowed to a value that is about the same as the global convective turnover time, the slow-down stalls. Rotation and magnetism become decoupled, and the star's rotation remains constant while the magnetic field continues to decline [217].

Figure 8.10, from [106], shows another H-R diagram from the Gaia satellite. This diagram represents over 4 million stars with low foreground dust extinction, representing a region of about 1.5 kpc around the Sun that we will call the 'solar neighbourhood'.[12] It goes to a much fainter magnitude than the first Gaia H-R diagram

---

[12] The distance to the center of the Milky Way is 8 kpc, by comparison.

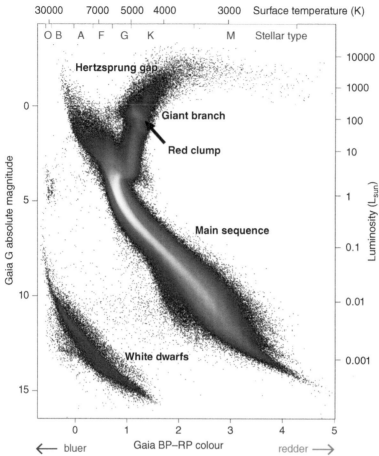

Figure 8.10 H-R diagram from the Gaia satellite representing over 4 million stars within ~1.5 kpc from the Sun. Certain regions are labelled (see text). The x-axis gives the Gaia colour index (**bottom**) and corresponding effective temperatures and spectral types (**top**). The y-axis is the absolute magnitude in the Gaia G-band (**left**) and corresponding stellar luminosity (**right**). The lowest number of stars in a region are shown as black, then red, orange, and finally yellow, representing the highest number of stars. The Sun is at position (0.82, 4.67) (Eqs. 1.14, 1.15). Credit: Gaia collaboration [106].

that was shown in Fig. I.2. The black to yellow colour sequence represents increasing numbers of stars in any region on the plot.

The main sequence is clear, having the highest number of stars, as expected, given that stars live the greater part of their lives on the main sequence. We can also see that the main sequence continues to the upper left beyond the giant branch, including many hot young stars. These stars are not a few blue stragglers as we saw for globular clusters. Rather, the presence of a strong main sequence towards the blue end indicates that some star-formation has been recent. Therefore, the region of the solar neighbourhood that is being probed must include star-forming regions. Indeed, we expect this to be the case for stars in the disk of our Galaxy. The number of stars trickles off at the far upper left, though, because there are fewer massive stars compared with low-mass stars (Fig. 8.4).

The giant branch (labelled) includes the RGB as well as the AGB, which overlap to some extent (cf. Fig. 8.2.1), although the AGB extends farther towards the cool side on the right. The *red clump* represents low-mass stars that burn helium in their cores. The way in which the stars pile up in this region is strongly dependent on their age and metallicity. For the red clump, the metallicities are high (i.e. solar), whereas the corresponding feature for low-metallicity stars would be more stretched out along the horizontal branch [106]. Notice where the giant branch starts near the main sequence. As indicated in Sect. 8.2.2.1 and elsewhere, very-low-mass stars have not yet left the main sequence, no matter how large a volume of our Galaxy is being probed.

Between the upper hot main sequence and the giant branch is the *Hertzsprung gap*. There are some hot evolving stars in this region (cf. Fig. I.2, for example), but not many, compared to other highly populated areas of the diagram. The fact that there are few hot massive stars to begin with, plus the short timescales for their post-main-sequence evolution, results in a low probability of finding stars in this 'gap'. Nevertheless, remarkably, historical records of naked-eye observations record detectable changes in the colours of quickly evolving stars in this region. For example, 2000 years ago, the star Betelgeuse was recorded as being yellow in colour, whereas now it is red [226].

Probably the most obvious new feature is the white dwarf cooling curve. The Gaia satellite has delineated this sequence beautifully, seen as the curve to the lower left of the main sequence. See Sect. 10.2 for more information about this curve.

It is now quite clear that the H-R diagram is an essential tool to understanding stars and their evolution, and comparing observational to theoretical plots is a necessary step. The basic features, such as the main sequence, are always present. However, there is some variation between H-R diagrams. For example, the region of space being probed highlights different features. If we are looking at a specific globular cluster, we do not see an extension of the main sequence towards the upper, blue end, whereas if we are looking at the solar neighbourhood that includes star-forming regions, we do. If our probed region includes both high- and low-metallicity stars – say, disk and halo stars, respectively – then the main sequence will be broader, as we saw in Fig. 7.5. The sensitivity of our instruments is also clearly important. The ability of the Gaia satellite to detect very faint objects led to the beautifully delineated white dwarf cooling curve as well as the faint cool main sequence in Fig. 8.10.

Online Resources

8.1 *ESA Video showing the evolving stellar jet HH 47*: https://esahubble.org/videos/heic1113b

8.2 *Jieun Choi's animation of a star moving through the H-R diagram*: https://jieunchoi.github.io/images/2M_movie.gif

8.3 *Padova database of stellar evolution tracks*: http://stev.oapd.inaf.it/cgi-bin/cmd

8.4 *Dartmouth Stellar Evolution Database*: http://stellar.dartmouth.edu/models

8.5 *Pisa Stellar Models*: http://astro.df.unipi.it/stellar-models

8.6 *Geneva Grids of Stellar Evolution Models*: https://www.unige.ch/sciences/astro/evolution/en/research/geneva-grids-stellar-evolution-models

8.7 *BASTI (Bag of Stellar Tracks and Isochrones)*: http://basti-iac.oa-abruzzo.inaf.it

8.8 *The Modules for Experiments in Stellar Astrophysics (MESA) Stellar Evolution Code*: http://mesa.sourceforge.net

8.9 *MESA Isochrones & Stellar Tracks (MIST)*: https://waps.cfa.harvard.edu/MIST/

PROBLEMS

8.1. (a) Show that the Jeans mass (Eq. 8.5) and Jeans radius (Eq. 8.6) can be written, respectively, as

$$M_J = 92.2 \left(\frac{T^{3/2}}{\mu^2 \, n^{1/2}} \right) \quad M_\odot \tag{8.17}$$

$$R_J = \frac{9.64}{\mu} \left(\frac{T}{n} \right)^{1/2} \quad \text{pc} \tag{8.18}$$

where T is the temperature, μ is the mean molecular weight, and n is the number density.

(b) Find the Jeans mass and Jeans length for the following spherical clouds, each of which are 5 pc in radius (assume pure hydrogen). Compare with the mass and radius of the cloud itself, and comment as to whether these clouds should collapse. Express your results in $M_\odot$ and pc.

   (i)   A spherical molecular cloud of density $n_{H_2} = 10^3 \text{ cm}^{-3}$ and temperature T = 30 K

   (ii) A diffuse HI cloud of density $n_{HI} = 1 \text{ cm}^{-3}$ and temperature T = 80 K

   (iii) An HII region of density $n_{HII} = 50 \text{ cm}^{-3}$ and temperature T = 10^4 K

8.2. Show that

$$t_{ff} \approx \left(\frac{R}{g} \right)^{1/2} \tag{8.19}$$

8.3. (a) For a typical temperature of a pre-collapse molecular cloud, calculate the sound speed (Eq. 2.44), justifying your choice of parameters.

(b) Compare the sound speed to typical random velocities of 10 to 20 km s^{-1} in the interstellar medium. Do you expect there to be shock waves in the cloud that could compress some gas and push it above the Jeans mass?

8.4. Suppose a spherical molecular cloud core has a mass of $M_c = 1\,M_\odot \approx M_J$ and starts to collapse. Its initial temperature and rotation velocity are $T = 10$ K and $v_{rot} = 0.3$ km s^{-1}, respectively (e.g. Table 8.1).

(a) Calculate the cloud's initial density, ρ_0, and radius, R_0.

(b) Use conservation of angular momentum to compute the rotation velocity of the resulting protostar whose radius is $R_{ps} = 3\,R_\odot$ (Sect. 8.1.3). State any assumptions.

(c) By balancing the centripetal acceleration with the gravitational acceleration in a spherical rotating object of mass M and radius R, we find that the rotational speed at the equator must not exceed

$$v_{max} = \sqrt{\frac{GM}{R}} \qquad (8.20)$$

or the star's fast rotation will tear it apart. Does the rotation velocity computed in part *(b)* exceed v_{max} or not?

(d) Comment on the need for such a collapsing object to remove angular momentum if we are to observe the stars that we see today.

8.5. (a) Estimate the thermal timescale (Eq. 7.23) of a main sequence star for the following masses: i) 1 $M_\odot$ and ii) 6 $M_\odot$. [HINT: Tables 6.1 and F.1 will be useful.]

(b) Compare this thermal timescale to those given for a star of the same mass undergoing contraction onto the main sequence as given in Table 8.2.
Do they agree to order of magnitude? Why might there be differences?

8.6. Let us arbitrarily designate two mass ranges as 'low mass' ($0.07 \leq M/M_\odot \leq 0.5$) and 'high mass' ($0.5 \leq M/M_\odot \leq 150$).

(a) State any assumptions and, using the IMF function given in Sect. 8.1.5, find:

(i) The ratio of the number of low-mass stars compared to the total number of stars, N_{low}/N_{total}, and high-mass stars compared to the total number of stars, N_{high}/N_{total}.

(ii) The ratios of luminosities, L_{low}/L_{total} and L_{high}/L_{total}

(iii) The ratios of masses, M_{low}/M_{total} and M_{high}/M_{total}

(b) Are most stars low mass or high mass? Is most of the mass in low-mass stars or high-mass stars? Is most of the light in low- or high-mass stars?

8.7. Consider the two stars that make up Albireo, shown in Box 8.4 on page 200.

(a) Plot the approximate locations of these two stars on Fig. 8.10.

(b) Determine the radii of both stars in $R_\odot$. Which star is larger?

(c)   Make the assumption that the given current mass is approximately the same as the initial mass, M_i. What will be the mass of the stellar remnant of each star?

8.8.   Consider the solar metallicity 24 $M_\odot$ star of Fig. 8.6 whose corresponding data are given in Table 8.3.

(a)   From the table, find the time it takes for this star to move from the TAMS to the point called C-burn. Compare this time to the time the star spends on the main sequence, and comment on the existence of the Hertzsprung gap. [HINT: Fig. 7.1 will be helpful.]

(b)   Such a star is off the chart shown in Fig. 8.10. Explain why this is the case.

Chapter 9
The Variable Star – Pulsation

Simplicio: A [celestial] body is inaugmentable, inalterable, invariant, and finally eternal, ... nothing is seen to be altered, either in the remotest heavens, or in any integral part of heaven. ... thus Aristotle proves the incorruptibility of heaven!

Salviati: Excellent astronomers have observed many comets generated and dissipated in places above the lunar orbit, besides the two new stars of 1572 and 1604, which were indisputably beyond all the planets.

Galileo Galilei, Dialogue Concerning the Two Chief World Systems, 1632 [108]

Since Tycho Brahe made famous the 'new star' that was observed in 1572, now known as Tycho's supernova, and later Kepler's supernova appeared in 1604, the world began to realize that the heavens are *not* unalterable. Aristotle's view that the heavens were perfect had held sway for 1000 years, but the sudden appearance of new stars in the sky led to a shift in philosophy and eventually put an end to that notion. Galileo used Tycho's and Kepler's supernovae as examples that there can be changes, or *variations* in the starry realm. He presented his

Astrophysics: Decoding the Stars, First Edition. Judith Irwin.
© 2023 John Wiley & Sons Ltd. Published 2023 by John Wiley & Sons Ltd.

arguments in the form of a debate between Simplicio (who followed Aristotle), Salviato (who held to Copernicus[1]) and a less-
partisan friend named Sagredo. Other changes in the heavens
were also being noticed. The star o Ceti (Mira) was first seen to
vary by the German astronomer Fabricius in 1596 [181], a change
that could also have contributed to the 'Copernican Revolution'.
That some stars vary in brightness, however, had been noted by
other cultures much earlier. There is some evidence, for example,
that the ancient Egyptians noticed changes in brightness of the
eclipsing binary (Sect. 3.5.1), Algol, as early as 1244–1163 B.C. [159]!

Our modern understanding that stars can indeed change has extended far beyond
wonderment at the appearance of a 'new star'. We have moved into an era of *tempo-
ral astronomy* in which changes in the sky not only are frequent but also provide us
with physical information about the changing object, as we will soon see. We will look
at supernovae in Sect. 10.3. In this chapter, we will focus on less destructive variations
in stars.

What do we actually mean by a *variable star*? The slow secular change in stellar
properties both on (Sect. 7.3.2) and off (Sect. 8.2) the main sequence has already
been examined. All stars evolve, yet we would not characterize all stars as variable.
Generally, we expect a variable star to show some change on a measurable timescale.
Changes could be manifest in a variety of ways, such as variable polarization, magnetic fields, rotation rates, spectral characteristics and others [e.g. 39, 224, 238]. More
classically, however, a variable star is usually one that shows detectable *photometric
variability*, i.e. variability of its magnitude with time. A plot of such variation is called
a *light curve*, as first introduced in Sect. 3.5.

We have already seen examples of light curves. Figure 3.9 showed the stylized light
curve that results from an eclipsing binary system. And the figure shown in Box 7.1
on page 163 showed the observed light curve of the luminous blue variable η Carinae.
It is clear that the reasons for the variability are quite different. In the former case,
one star passes in front of the other, causing a reduction in brightness. In the latter, a
massive star is undergoing strong eruptions that produce brightenings of many magnitudes when large quantities of mass are ejected. In fact, there are so many different
types of variability that it is challenging to categorize them all. Online Resource 9.2

---

[1] He held to the Copernican view that the Sun, rather than the Earth, was at the center of the
'world'.

provides a sample list.[2] Even our own Sun, stable as it is, would nevertheless be seen from a distance as a rotating, spotted variable star.

Figure 9.1 drives home this point with a visual attempt at categorizing variable stars, i.e. a 'variability tree'. Some non-stellar objects are also included, such as asteroids and active galactic nuclei (AGN). This is an evolving field, and the categories can shift as active research provides new insights on these fascinating stars (cf. [94] and [107]).

On the left-hand side of the figure are objects that vary for 'extrinsic' reasons. These objects show variations due to changes in geometry, such as rotation (shown in orange) or eclipses (in blue). An example of rotational variability is the star Spica (α Virginis), which has a close companion. Consequently, both stars are ellipsoidal due to their gravitational interaction (similar to Regulus, Box I.1 on page xxi) and present more (or less) light towards the observer as the system rotates. A list of eclipsing binaries can be found in Online Resource 9.5. Some stars fall into several categories. For example, Algol (β Persei) is both an eclipsing binary and ellipsoidal in shape.

On the right-hand side of the figure are objects that vary for 'intrinsic' reasons. This means some physical change is occurring in the star itself. Examples that are eruptive (shown in red) include luminous blue variables (LBVs) such as η Carinae and P Cygni (Box 3.3 on page 74) and Wolf-Rayet (WR) stars as shown in Fig. 7.4. Cataclysmic variables (magenta) include novae and supernovae, among others. *Novae* are produced when mass transfers from an evolved red giant star onto a companion white dwarf. Nuclear reactions start at the surface of the white dwarf under degenerate conditions. This leads to a *thermonuclear runaway* and rapid mass loss from the white dwarf [57] with associated brightening (see Box 10.1 on page 241). Supernovae will be discussed in Sect. 10.3.

A large number of variable stars, and perhaps the most 'classical' of all variables, are the pulsators[3] (green filled dots in Fig. 9.1), and these are the ones we focus on in the current chapter.

9.1 PULSATION

Table 9.1 gives a list of the most common pulsating stars, with brief descriptions in the table caption. There is a bewildering variety. We will look at pulsation types and modes in more detail in Sect. 9.2, but note that the simplest type of pulsation is when a star expands and contracts in a radial fashion, as if it were breathing. The star brightens and dims as it contracts and expands periodically, and these changes in brightness, as well as the period, can be measured.

The pulsating star types of Table 9.1 are plotted on the H-R diagram in Fig. 9.2. Data are from the Gaia satellite and include objects within 1 kpc of the Sun. There

---

[2] Well over 100 different variability types are listed as of July 1, 2022.
[3] Do not confuse 'pulsator' with 'pulsar' (Sect. 10.4).

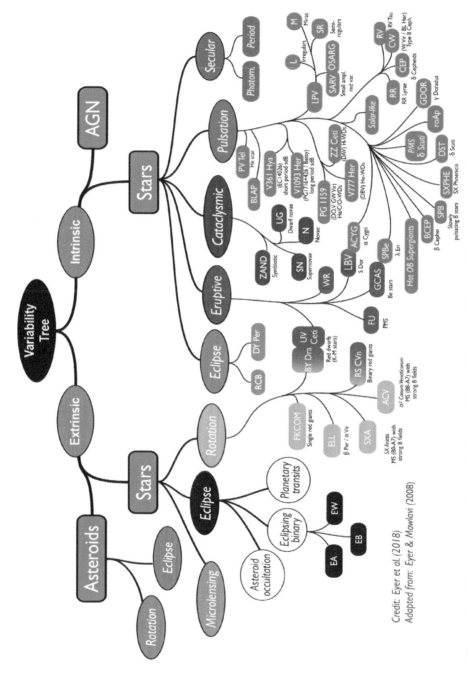

Figure 9.1 Chart showing the main types of variable stars, along with some non-stellar objects such as active galactic nuclei (AGNs) and asteroids. Credit: Courtesy of Laurent Eyer, from [107].

Table 9.1 Observational data for the most common pulsating variable stars.

Class name[a]	Period[b]	Change in brightness (mag)[c]
1. Long-period variables (LPVs)[d]	50–1000 d	Up to 8
2. α Cygni stars[e]	1–50 d	Few 0.1
3. δ Scuti stars[f]	30 min–6 h	Up to few 0.1
4. SX Phoenicis stars (Pop II)[g]	0.7–1.9 h	Up to 0.7
5. γ Doradus stars[h]	0.3–3 d	Up to few 0.01
6. RR Lyraes (Pop II)[i]	0.2–1 d	0.2–2
7. Slowly pulsating B (SPB) Stars[j]	0.4–5 d	Less than 0.1
8. β Cephei stars[k]	0.1–0.6 d	Up to 0.30
9. Classical Cepheids[l]	1–100 d	0.1–1.5
10. Type II Cepheids (Pop II)[m]	1–100 d	0.3–1.6
11. PV Telescopii stars[n]	0.1–1 d	~ 0.1
12. Rapidly oscillating Am and Ap stars[o]	5–20 min	~ 0.01
13. V361 Hydrae (or EC 14026) stars[p]	1–10 min	Up to 0.03
14. V1093 Her (or PG 1716) stars[q]	1–3 h	Up to ~ 0.01
15. ZZ Ceti stars[r]	0.5–25 min	Up to 0.2

[a] A class is usually named after the prototype, or first star that was discovered with the variability attributes. See also Figs. 9.1 and 9.2.
[b] Days (d), hours (h), minutes (min), seconds (s).
[c] Amplitude of the (visual) magnitude variation.
[d] LPVs are red giants, including Miras, semi-regular variables, slow irregular variables and small-amplitude red giants.
[e] Luminous supergiants that pulsate in non-radial modes.
[f] A and F stars, mainly with short pressure-mode (p-mode) periods.
[g] Analogous to δ Scuti stars but for Population II.
[h] A and F stars that pulsate mainly in high-order gravity modes (g-modes).
[i] Population II horizontal branch stars that pulsate in p-modes.
[j] B stars that pulsate non-radially in g-modes.
[k] O or B stars that vary in both p- and g-modes.
[l] Also called δ Cephei stars, that pulsate in radial p-modes.
[m] Population II stars pulsating in p-modes. Includes BL Herculis, W Virginis, RV Tauri subclasses and yellow semi-regulars.
[n] Rare H-deficient supergiants with complex light variations.
[o] Chemically peculiar A stars with non-radial p-modes.
[p] Subdwarf B stars on the hottest part of the horizontal branch that pulsate in p-modes.
[q] Subdwarf B stars on the hottest part of the horizontal branch that pulsate in g-modes.
[r] White dwarfs with non-radial g-mode pulsations.
Credit: Adapted from [94] and [107], supplemented with data from Online Resources 9.5 and 9.5 and some adjustments from Prof John Percy (private communication).

is considerable crowding in the figure, with different symbols overlapping in some regions. However, some types cluster in the figure very clearly, such as the classical Cepheids (# 9, green dots) and the long-period variables (#1, red dots), both of which are off the main sequence and therefore represent evolved stars. On the main sequence, most pulsators are high-mass stars that have low-density envelopes in comparison to lower main sequence stars (e.g. Sect. 7.5).

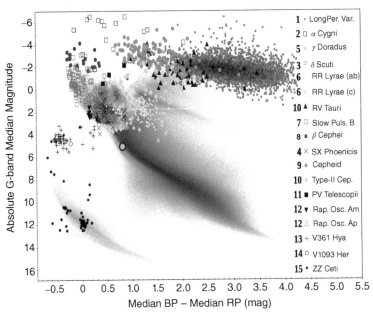

Figure 9.2 Variable stars in the H-R diagram from the Gaia satellite [107]. The x-axis is the colour index, and the y-axis is the absolute magnitude in the Gaia filter bands (Table 3.1). The data represent objects with parallax greater than 1 mas (closer than 1 kpc) along with a variety of other criteria. The colours represent pulsating variables, coded according to the legend at right. The numbers on the legend correspond to the numbers in Table 9.1. The Sun is shown at position (0.82, 4.67) (Eqs. 1.14, 1.15). The region of diagonal lines indicates *approximately* the instability strip that contains classical Cepheids (green-filled dots), RR Lyrae stars (yellow-filled dots and open turquoise circles) and δ Scuti stars (red open diamonds). Credit: Courtesy of Laurent Eyer, from [107].

Along the lower main sequence, there is a notable absence of pulsators. Since most stars are low-mass stars (Fig. 8.4), this suggests that most stars do not pulsate [181]. However, this does not mean there are no disturbances, variability or even low-level oscillations that might fall below the detectability level of the instruments. For example, the Sun experiences a variety of low-level oscillations [295]. The Sun's luminosity also varies periodically with the 11-year sunspot cycle (see Fig. 3.6 in [154]). But as we learned in Sect. 1.2, the Sun shows no evidence of pulsations that change its size.

Pulsation is an instability – but that instability is stable! That is, the star varies in a periodic fashion, but the timescale of the periodicity is shorter than the timescale over which any variability persists in the star. On the H-R diagram, for example, there are various *instability strips* through which a star may enter and then leave again during its slow evolution. The approximate location of one of these strips is shown with diagonal lines in Fig. 9.2.[4] Recall the zig-zag tracks that occur in post-main-sequence evolution (Fig. 8.6). As a star ages, it can become a pulsator for a while, and then its pulsation can cease as it slowly advances along its evolutionary track in the H-R diagram. It may even enter the same instability strip again if its track reverses.

Usually, we expect that instabilities should damp out with time. An example of stability against perturbations was given in Sect. 7.3.1. Yet observationally, we do see long-lived pulsations. An example is the long-period variable *o* Ceti (Mira). Mira's period is 332 days, so Mira has undergone more than 450 pulsations since it was first noticed to vary in the year 1596. Similar conclusions can be reached for shorter-period pulsators. The classical Cepheid δ Cephei was first observed to vary by John Goodricke in 1784 [135]. With a period of 5.4 days, this star has pulsated over 16,000 times since its discovery! Pulsation is *not* a transient phenomenon. Consequently, some *driving mechanism* is needed to keep the pulsation going. We will look at these mechanisms in Sect. 9.4. However, there can be slow, measurable changes in pulsation characteristics that can be compared with evolutionary predictions to help determine the physical properties of the star.

Mira varies by an astonishing ~7 magnitudes (see Online Resource 9.5 for data). This corresponds to the entire range of stellar magnitude that the human eye can detect in a dark night sky (e.g. Table 3.2 of [154]). Historically, a combination of larger changes in brightness, and timescales that are easily measured, has conspired to populate the observational database of pulsating stars. In Table 9.1, for example, these tend to be the long-period variables (#1), the RR Lyraes (# 6) and the Cepheids (#9 and #10). Each of these types has been well-studied, leading to important relations that are fundamental to our knowledge of the distance scale in the universe (Box 3.1 on page 60; see also Sect. 9.5).

9.2 ASTEROSEISMOLOGY

Stars are like musical instruments. Sound waves have tone (the frequency of the signal), intensity (the amplitude of the signal) and timbre (the combination of the fundamental frequency and its various harmonics). To generate the waves, some initial perturbation is required. For musical instruments, this could be a musician blowing into a flute or air forced into a column of a pipe organ. For stars, there may be a variety of perturbations from various dynamical processes, including rotation, convection, magnetic-related activity, tidal effects from companions and others. The result of such perturbations is the propagation of waves through the star. Like music, sound waves (also called *acoustic waves*) propagate at the sound speed (Sect. 1.2.1, Box 2.1 on page 42).

---

[4] This particular instability strip can look almost vertical in an H-R diagram, depending on how the x- and y-axes are scaled.

Figure 9.3 A solar flare (white filamentary feature) produces a sunquake resulting in waves, like ripples, propagating along the surface. A movie is available at Online Resource 9.7. The data were obtained from the Michelson Doppler Imager onboard the Solar and Heliospheric Observatory (SOHO) spacecraft immediately following a moderate-sized flare on July 9, 1996. SOHO is a joint project of the European Space Agency and NASA. Credit: NASA.

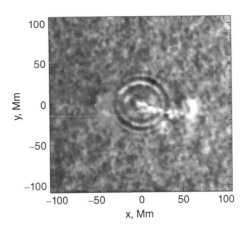

A nice example of a transitory disturbance is shown in Fig. 9.3. A *sunquake* was produced from a solar flare. The oscillations (like 'ripples') were seen to propagate over the surface, travelling about 10 Earth diameters over the course of an hour, before becoming damped and indistinguishable from the solar background. However, the waves actually travelled through the interior of the Sun and rebounded. A movie of this event can be seen at Online Resource 9.7.

Aside from sound, other types of waves can also occur within stars. A perturbation of *mass* is also possible. Convection zones, for example, are regions of instability such that if a pocket of gas is perturbed upwards, it will continue to rise due to buoyancy (Sect. 4.3.1). However, if a pocket of gas is perturbed radially in a region that is *stable* against convection, it can oscillate up and down about its equilibrium position like a simple harmonic oscillator. The natural frequency of the oscillation is called the *buoyancy frequency* (also called the *Brunt-Väisälä frequency*), and the result is a propagation of *gravity waves* in a non-radial direction. The effect is similar to perturbing the surface of water and watching the waves propagate sideways. Thus, gravity waves only exist in stable (radiative) regions – not in convective regions – and produce *non-radial* ('sideways') oscillations. Other types of waves can also exist, such as those that are induced by rotation or tidal gravitational effects from companion stars.[5] However, the important and most readily measured waves in stars are sound waves and gravity waves. The former represents an oscillation of pressure, and the latter represents an oscillation of mass.

When waves propagate through an infinite uniform medium, they maintain their original amplitude until they damp out. However, when they propagate through a finite medium, they can hit a boundary and reflect back. The 'boundary' could be a location at which no oscillation is possible, such as the center of a star, or a region in which the density changes significantly from the original medium, such as the surface of a star. This is analogous to an organ pipe that is closed at one end (a *node*) and open at the other (an *antinode*). Those waves that add coherently before and

---

[5] Other examples are the Alfvén waves introduced in Sect. 1.2.1, and fast and slow magnetohydrodynamic (MHD) waves.

after reflection (constructive interference) amplify the signal and result in *standing waves* or *resonances*. These occur at specific frequencies, given the properties of the medium and the size of the star or size of the *resonant cavity* within which the reflections are occurring. It is helpful, then, to distinguish between *waves* and *modes*. A wave propagates in both space and time, whereas a mode is a standing wave that does not propagate in space but does oscillate in time. Any wave can be obtained by a sum of modes, and any mode can be obtained by a sum of waves; however, when boundary conditions are introduced, it makes more sense to speak of modes rather than waves. Pressure waves (sound waves) produce resonances called *p-modes*, and gravity waves produce resonances called *g-modes*.

In stars, the situation can be complicated because unlike the organ pipe, the properties of a star, such as its pressure and density, change with radius. Therefore, waves that are not travelling exactly in the radial direction bend (*refract*) as they travel through the star. The frequencies of the modes that emerge at the surface also depend on the stellar properties through which the waves have travelled. But this is just the information that we want to know. *Asteroseismology* is the study of stellar oscillations in order to infer the structure and properties of the stellar interior. The process is analogous to what geophysicists do when they study the propagation of seismic waves through the Earth. Earthquakes, although destructive, are informative about the interior structure of our planet. Asteroseismology, when applied to the Sun, is called *helioseismology*.

Figure 9.4 shows regions in which sound waves and gravity waves can propagate in the Sun. The sizes of the regions may differ for different stars, but in general, p-modes have shorter periods (higher frequencies) and probe the outer regions of stars, whereas g-modes have longer periods (lower frequencies) and probe the inner regions closer to the core. We have argued elsewhere (Sect. 2.1.8 and Sect. 4.3.1) that both sound waves and mass perturbations are adiabatic. In other ways, though, these two types of oscillation are quite different and provide information on different regions in the star.

Gravity waves must propagate at or below the buoyancy frequency, which is the natural frequency of oscillation for local conditions and is designated as an angular frequency, N, in the figure. Returning to our concept of the simple harmonic oscillator, the frequency of an oscillating pocket of gas cannot be higher than what local conditions allow naturally,[6] so all g-modes fall below this curve. Throughout most of the solar interior, $N \approx 3 \times 10^{-3}$ rad s^{-1} [251], corresponding to a period of $P = 2\pi/N \sim 35$ min. So the periods of g-modes will be greater than 35 minutes. G-modes are difficult to measure on the Sun because they cannot propagate through the convective layer, so they are measured via their effect on the observed p-modes [104].

-----

[6] It can be shown that $N^2 = g\left(\frac{1}{\gamma_a}\frac{1}{P}\frac{dP}{dr} - \frac{1}{\rho}\frac{d\rho}{dr}\right)$, where g is the local acceleration of gravity, ρ is the density, P is the pressure, γ_a is the adiabatiac index and r is the radial distance from the star's center [60].

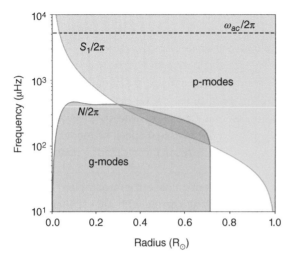

Figure 9.4 Diagram showing the regions in which g-modes (in blue) and p-modes (in orange) dominate in the Sun, from Model S described in [59]. The x-axis gives the radius in solar units, and the y-axis indicates the frequency of the modes. Note that the y-axis scale designates a frequency $\nu = \omega/2\pi$, where ω is an angular frequency. N is the buoyancy frequency, ω_{ac} is the acoustic cutoff frequency and S_1 is the Lamb frequency for this model in which the angular degree $l = 1$. Credit: Courtesy of Jørgen Christensen-Dalsgaard and Warrick Ball.

By contrast, the minimum frequency for sound waves is given by the *characteristic acoustic frequency*, or *Lamb frequency*, which is shown as S_1 (angular frequency) in the figure. Essentially, for a mode to exist, its wavelength λ must fit within the resonant cavity. For a pipe closed at one end and opened at another, for example, this can be approximated by the requirement that $\lambda/4 < r$, where r is the distance from the center. Since $c_s = \lambda \nu$, the frequency of a resonant sound wave should be $\nu > c_s/(4\,r)$. The angular frequency of the wave ($\omega = 2\pi\nu$) must then be $\omega > \pi\,c_s/(2\,r)$ for a mode to exist. The quantity on the right-hand side of this inequality represents the characteristic acoustic frequency, S_1, to within factors of a few. This example is clearly a gross simplification to reality, but it does illustrate why there is a *lower limit* to the allowed frequencies of p-modes.[7] Typical pressure mode frequencies in the Sun are of order 5 minutes. P-modes are predominantly radial oscillations [3].

At even higher frequencies, the value ω_{ac} is the *acoustic cutoff frequency*. At this point, the propagation of a p-mode is limited by the pressure scale height, h_P (Box 4.1 on page 95). If the wavelength is very small compared to h_p, then waves are no longer trapped and therefore cannot resonate. The corresponding upper angular frequency limit is $\omega_{ac} \approx c_s/2\,h_P$. At frequencies higher than ω_{ac}, acoustic disturbances propagate as traveling waves through the chromosphere and into the base of the corona [103].

---

[7] It can be shown that $S_l^2 = \frac{l(l+1)\,c_s^2}{r^2}$, where l is the angular degree, c_s is the sound speed and r is the radius [60].

Because stars are spherical and modes require resonances that only occur at specific multiples of the wavelength, asteroseismology is well-suited to the use of mathematical functions called *spherical harmonics*. References [3], [109], [60] and [181] provide details. A mode is fully characterized by three integers (*eigenvalues*), l, m and n, which describe the type of stellar oscillation that is occurring, similar to quantum numbers in an atom.

The radial order, n, is the number of *nodes* (points at which there is no oscillation) along a radius. If $n = 0$, the vibration is radial with no interior nodal point other than at the center. If $n = 1$, there is one node along a radius. A node at some radial point within a star is equivalent to a shell that is at rest, whereas gas above and below it is in motion. An illustration of this type of pulsation will be given in Sect. 9.3.

The angular degree, l, is the number of nodal lines at the surface of the sphere. If $l = 0$, there are no nodes at the surface, so the situation reduces to a purely radial pulsation. As soon as $l \geq 1$, the star becomes a *non-radial* pulsator. Examples are shown in Fig. 9.5. The simplest example is when $l = 1$, $m = 0$ ($\{l, |m|\} = \{1, 0\}$) at *Top Left*. Here, there is a single node at the surface. In the top hemisphere, the surface is advancing outwards (blue); and in the bottom hemisphere, the surface is receding inwards (red). Later in the cycle, this will reverse, and the top hemisphere will recede while the bottom hemisphere advances. The *Top Row* shows models in which there is no rotation, so 'top' and 'bottom' of a star don't really have any meaning. We might have rotated the images and specified 'right' and 'left'.

The final number, m, is the *azimuthal order*: the number of surface nodes that pass through the poles. This number only has meaning if the star is rotating (essentially all real stars), in which case the star has two poles and an equator (*Second and Third Rows* of Fig. 9.5). The nodes that pass through the poles advance around the surface with time, so this number can be positive or negative, corresponding to retrograde and prograde, respectively, with respect to the sense of rotation [109]. It is possible for all surface nodes to pass through the pole, in which case $|m| = l$ (*Second Row* in the figure), or just some of them (*Third Row*) ($|m| < l$), so m can take integer values from $-l$ to $+l$.

The final image at *Bottom Right* of Fig. 9.5 shows many modes. Generally, large numbers of modes can only be observed on the Sun because it is spatially resolved. In fact, several thousand modes of oscillation have been detected in the Sun [252]! Other stars are unresolved, though, so only a low number of modes can typically be detected.

Figure 9.5 also hints at how stellar oscillations can be detected. As the surface (or a sector of a surface) advances and recedes with respect to the equilibrium surface level, there are corresponding deviations in surface temperature and therefore luminosity. The motions can be detected via Doppler velocity measurements (Sect. 2.1.9) or measurements of the changes in brightness. Such data are very sensitive, especially for the Sun. For example, at a frequency of $\nu = 3 \times 10^{-3}$ Hz, the radial velocity of the oscillation is 17 cm s^{-1} [105, 150]. A turtle could walk as fast! Fluctuating brightness, though, is observationally most common. Since stars are unresolved, all such changes

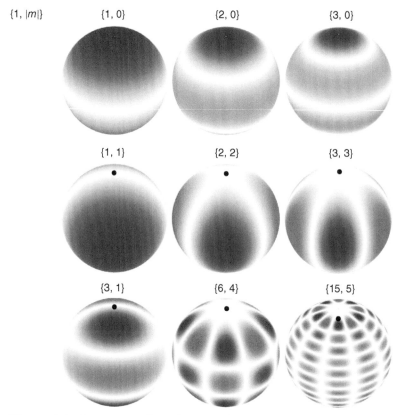

Figure 9.5 Representation of non-radial modes at one point during oscillation. The stars are tilted by 60° for clarity. White represents nodes (no motion), blue represents regions where the surface is moving outwards, and red shows where the surface is moving inwards. Modes are indicated by {l, $|m|$}: {total number of nodal lines on the surface, number of nodes that pass through a pole}. The first row shows axisymmetric models (no rotation), so there is no pole, and $m = 0$. The second and third rows have rotation, so a pole (black dot) is indicated. Various combinations of l and $|m|$ are shown. Credit: Conny Aerts; [3] / With permission of American Physical Society.

are observed together. The net result is periodic fluctuations in brightness of the star. An example is shown in Fig. 9.6, for which the displayed modes are p-modes corresponding to high-radial-overtone values (high values of n). The signal is remarkably distinct.

The information hiding in the spectrum is rich and provides a remarkable inside view of the star. Each pulsation mode tells us about the region in which the corresponding waves propagate. By starting with a physical model, the modes can be calculated and matched to the observations and then adjusted until the best fit is obtained. This *inversion*, where one begins with a physical model and predicts the

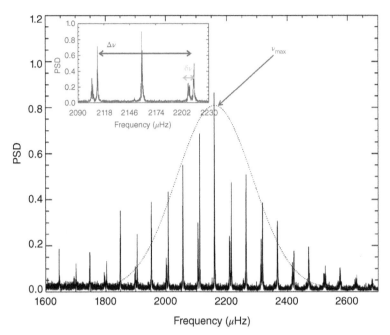

Figure 9.6 The p-modes of 16 Cyg A. *Inset:* A blow-up near the peak. The y-axis, which shows the power spectrum density (PSD), has been arbitrarily scaled. Credit: Courtesy of Rafael A. Garcia and Jérôme Ballot; [109] / Springer Nature / CC BY-4.0.

observations, is common in astrophysics. Information on the interior sound speed and interior rotation can be obtained. The structure, pressure, density, temperature, luminosity and/or combinations thereof can also be found [2]. The star's density profile depends on its composition. As a star ages, it turns more hydrogen into helium, increasing the mean molecular weight. Thus, even the star's age can be inferred. Asteroseismology is a fruitful, burgeoning subfield of astronomy and promises to reveal even more about stars in the future [109]. Examples of different kinds of pulsators are given in the footnotes to Table 9.1.

Box 9.1

Starring … Polaris (the North Star)

Distance: $d = 137$ pc; Magnitude: V $= 1.86 - 2.13$;
Variable Type: Classical Cepheid; Spectral type: F7Ib;
Effective Temperature: $T_{eff} = 6015$ K; Mass: M $= 5.4$ M$_\odot$
The point in the sky directly above the Earth's North Pole is called the North Celestial Pole (NCP), marked with a yellow plus sign in the figure. Polaris is

(continued)

(*continued*)

very close to the NCP; hence its common moniker, the North Star. As the Earth rotates daily, all stars circle about its north/south axis, making circular streaks in this time-lapse seven-hour photograph. Polaris makes only a small arc of radius 44 arcmin during this time (see arrow), so it appears almost fixed in the sky. Since ancient times, sailors and explorers have used this star to help with navigation. What isn't widely known, however, is that Polaris is a *classical Cepheid variable* and is the closest Cepheid to Earth.

Its pulsation period is 3.97 days and is increasing with time. Its amplitude also shows variations, possibly decreasing with time [8]. Polaris is losing mass at a high rate of $\dot{M} \sim 10^{-6}$ $M_{\odot}$ yr^{-1}. It also has a close companion with an orbital period of 29.6 yr [95]. Even some of the most basic parameters of this star are not accurately known, however. In short, on close inspection, this most fixed of all stars is far from fixed.

9.3 RADIAL PULSATION

The simplest oscillation mode is when the star is undergoing *radial pulsation*. As indicated in Sect. 9.2, this corresponds to $l = 0$ and $m = 0$, so only n can have a value. If $n = 0$, then the star pulsates in its *fundamental* mode, moving out and in, out and in again, etc. In the fundamental mode, the entire star expands and contracts together. This is similar to sound waves in an organ pipe, with the pipe closed at one end (node, center of the star) and open at the other (antinode, surface of the star). Low-order modes in a pipe are shown in Fig. 9.7. The *Leftmost* pictures show the fundamental mode, which occurs when the pipe has a length of one-quarter of a wavelength. It is helpful to recall that sound is a *longitudinal* wave (introduced in Sect. 1.2.1). Therefore, the out/in motion is to the right/left in the figure (see Online Resource 9.6 for a helpful animation). The *Bottom Left* picture shows an instant of time in which motions are 'out'. At the closed end, particles are unable to move, the displacement is zero and the pressure is maximum. At the open end, the displacement is maximum and the pressure is minimum, reducing to atmospheric pressure in the case of a pipe, or

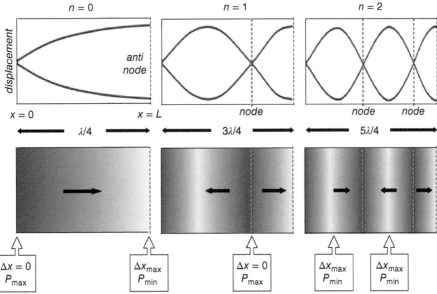

Figure 9.7 Standing waves in a pipe of length L that is open at one end (dashed vertical line on the right of each picture) and closed at the other (left of each picture). The radial order, n, is indicated at the top, representing the fundamental mode ($n = 0$) and the first and second overtones ($n = 1$, $n = 2$, respectively). The corresponding pipe length, represented by the wavelength of the standing wave is given between the upper and lower panels. **Top Row**: A plot of *longitudinal* (**right/left**) displacement (Δx). **Bottom Row**: Direction of gas motions (arrows) at one instant of time. Nodes (aside from the node at the closed left end) are shown as vertical red dashed lines. The behaviours of the gas displacement and pressure are indicated for several locations at the bottom. See Online Resource 9.6 for an animation.

reducing to zero at the surface of a star. The cases in which $n = 1$ and $n = 2$ in the figure show the first overtone and second overtone, respectively.

This simple illustration is a helpful visualization as to what happens in stars, but the fact that stars are spherical and that the density, temperature and pressure vary with radius are all key differences. The frequency ratios between the fundamental and overtones, for example, are different from the simple organ pipe, and those ratios are helpful diagnostics of interior conditions. The nodes, shown as vertical red dashed lines in the figure, are concentric shells in stars. On a nodal shell, particles have no collective velocities in the radial direction. That is, gas particles are at rest, aside from those random velocities expected from a Maxwell-Boltzmann velocity distribution in an ideal gas (Sect. 2.1.3). Stars can also exhibit higher-order overtones compared to what is shown in the figure. For example, the 3 mHz solar oscillation described in Sect. 9.2 corresponds to $n = 23$.

The most well-known and, arguably, the most important radial pulsators are the classical Cepheid variables, RR Lyrae stars and LPVs [247]. The first two types, together with the δ Scuti stars, lie within the instability strip that has been approximated by diagonal lines in Fig. 9.2. Radial pulsations occur in all of these stars, although the δ Scutis also show some non-radial modes. It is possible for the fundamental mode as well as overtones to occur simultaneously, but certain modes tend to have higher amplitudes and dominate others. In these three variable types, the fundamental mode is representative of the class. There are a variety of subclasses, though, in which $n > 0$ dominates, and it isn't always clear, from a theoretical perspective, what 'sets' the dominant mode in a star.

Classical Cepheid stars, RR Lyrae stars and δ Scuti stars are named after their prototypes: that is, stars in which this behaviour was originally identified. The corresponding constellations are Cepheus, Lyra and Scutum, respectively. If 'Cepheids' is used without a qualifier it is assumed to be a classical Cepheid, a Type I Cepheid or, equivalently, a δ Cephei type. Several other types of Cepheids are also known (e.g. β Cepheids, Type II Cepheids; see Table 9.1) but they are different from the classical Cepheids and will not be discussed here. The importance of these stars is that they obey *period-luminosity relations*, which will be discussed in Sect. 9.5.

Figure 9.8 displays what happens to a Cepheid variable when it is pulsating in the fundamental mode. This star is the prototype δ Cephei, which has a mass $M = 4.5\ M_\odot$, radius $R = 44.5\ R_\odot$ and pulsation period of 5.37 days. Consider the origin of each plot, at which time the star shows maximum compression (*Top Right*). The compression is enormous compared to the size of the Sun (orange circle). However, δ Cephei is a

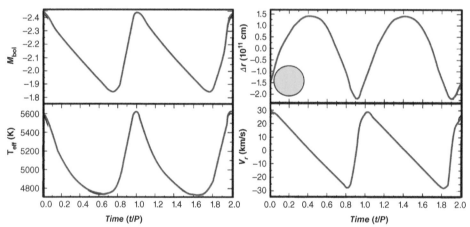

Figure 9.8 Surface properties of the Cepheid prototype δ Cephei as a function of time over two periods, P. **Top Left**: Bolometric absolute magnitude. **Bottom Left**: Effective temperature. **Top Right**: Displacement of the radius from its equilibrium value (positive is expanded, negative is compressed). **Bottom Right**: Radial velocity (positive is expansion, negative is compression). The orange circle shows the size of the Sun for comparison. The curves are approximate. Credit: Adapted from [44].

yellow supergiant, and this compression represents only ~ 5% of its equilibrium size. Near this time, the star's outwards velocity is largest (*Bottom Right*), its bolometric magnitude (*Top Left*) is most negative (the star is at its highest luminosity), and its temperature is at a maximum (*Bottom Left*). Compression has caused the gas to heat (Sect. 2.1.8), and this has a greater effect on its luminosity than its smaller radius (recall $L \propto R^2 T^4$, Eq. 2.56). Inspection of the plots at later times shows that they can also be understood in a similar way, although there are some phase lags between the various quantities. Notice the high maximum speed: 30 km s^{-1} is over 100,000 km hr^{-1}!

9.4 THE DRIVERS – THE KAPPA MECHANISM

Why don't perturbations damp out? Put another way, why is the instability stable (Sect. 9.1)? Most of the time, perturbations should indeed damp out, as suggested in Sect. 7.3.1, but there are conditions in which a persistent *driving mechanism* is present. For the small oscillations on the Sun and other solar-type stars, turbulent convective envelopes are the seeds that stochastically drive the observed p-modes [3, 60, 109]. Another process, called the ϵ mechanism, works in deep regions in which nuclear burning is taking place. The ϵ mechanism has been shown to excite instabilities in some stars [24]. However, this mechanism is insufficient to drive the pulsations seen in the instability strip that contains the Cepheids, RR Lyraes, δ Scutis and LPVs. In these regions, the driver of pulsations is the kappa (κ) mechanism, so-named to indicate that the *opacity* (cf. Sect. 4.2.1) is the crucial parameter.

Let us consider the interior of a star. Here, the opacity is largely governed by a Kramers law, for which $\kappa \propto \rho/T^{3.5}$ (Eqs. 4.11, 4.12). If a star is compressed, both ρ and T $\uparrow$ (increase). Normally, T dominates because of its higher power, so $\kappa \downarrow$ under stable conditions. With lower opacity, heat can easily escape from lower regions, there is no driver to cause expansion, and the star settles back to an equilibrium structure. By contrast, in the instability strip when a star is compressed, $\kappa \uparrow$ in certain regions rather than $\downarrow$.

It was Arthur Eddington (Fig. I.1) [86] who first suggested that a pulsating star could be thought of as a kind of heat engine (see Box 9.2 on page 226). When the star is compressed (A to B), at first the compression is adiabatic and T $\uparrow$. However, if a significant part of the compression occurs in a *partial ionization zone*, the energy from the compression goes into ionizing the particles, rather than increasing the temperature. This is equivalent to having more degrees of freedom f, and a switch from adiabatic to isothermal (T= *const*) conditions (Sect. 2.1.7, point B). Then $\kappa \uparrow$, closing the 'opacity valve', and the interior heat cannot escape. Instead, it is absorbed into the layer, and the compressed layer is forced out again (B to C). The heat reservoir, Q_h, is the interior heat produced from nuclear reactions. The trapping of interior heat from a closed opacity valve is equivalent to opening a valve to let heat in, in the engine figure. By C, the density is low enough that $\kappa \downarrow$ again and the outwards push stops. However, the star continues to expand past the equilibrium radius and continues outwards adiabatically, so T $\downarrow$. Once the ions recombine with free electrons, they supply excess heat, Q_c (D), which can easily escape from the cool expanded envelope. With heat leaving, the star contracts again to point A, and the pulsation cycle repeats. The controlling valve is the opacity, κ.

The instability strip is determined by the depth at which ionization zones are present with sufficient mass. Ionization of hydrogen (HI → HII) occurs at T = 1 → 1.5×10^4 K, and the second ionization of helium (HeII → HeIII) occurs at T ≈ 4×10^4 K, so the partial ionization zone of helium is deeper in a star than the partial ionization zone of hydrogen. While both zones can play important roles, the location of the He ionization zone is responsible for the instability strip delineated in Fig. 9.2. If the star is too hot, the He ionization zone is at a higher radius where the density is too low for the kappa mechanism to be effective. This sets the blue edge of the instability strip. If the star is too cold, the He ionization zone is at a lower radius in the star where the density is too high to be effectively driven. This sets the red edge of the instability strip.[8] Apparently the instability strip is a 'Goldilocks' region in which the temperature, and therefore location, of the ionization zone is just right for the kappa mechanism to be effective. Hydrogen ionization zones play an important role for the LPVs, shown as red dots in Fig. 9.2.

Box 9.2

The Carnot Heat Engine

The job of a heat engine is to turn heat into work. This requires that a temperature differential be maintained on some system, here taken to be an ideal gas in a piston, as was shown in Fig. 2.4.

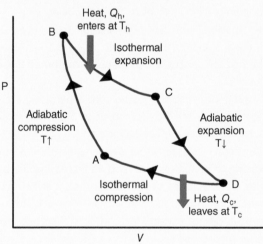

A common representation of a heat engine is to plot the pressure, P, of the gas against the volume, V, as shown in this picture for the *Carnot engine*. This is the 'ideal' engine, in the sense that for a given difference in temperature, it has the highest heat → work efficiency. Consider the gas at a cool temperature T_c at point A. It is then

---

[8] Convection also plays a very important role in determining the red edge.

compressed adiabatically (all valves closed) until its temperature rises (T ↑) to T_h at point B. Here, its pressure is maximum and volume is minimum. At B, a valve is opened that connects to a heat reservoir, and an amount of heat Q_h is added, causing the gas to expand. With expansion, V ↑ and P ↓, but the temperature remains constant because of the added heat. At C, the gas continues to expand, but the valve closes; so no more heat is added, the expansion is adiabatic, and T ↓. By D, the gas has reached its largest volume, lowest pressure and lowest temperature of T_c. At D, a different valve is then opened to a cold sink, allowing heat Q_c to leave the system. Heat continues to leave the system during compression from D to A, so T_c remains constant until A is reached again. The cycle then repeats.

9.5 PERIOD-LUMINOSITY RELATIONS

The most important practical consequence of variability is that many pulsators obey a *period-luminosity (PL) relation*. That is, the period of pulsation is directly related to the luminosity of the star. Once the relation has been calibrated for some filter band, the period, which is easily measured, can tell us the star's absolute magnitude in that band. The absolute magnitude (Eq. I.12) together with the observed apparent magnitude yield the distance (Eq. I.11). Thus, such variable stars are *standard candles* (Box 3.1 on page 60) and important as *distance indicators*.

Cepheids are Population I (Sect. 7.5) evolved stars in the mass range ∼ 5 to 10 $M_\odot$ with spectral types ∼ F6 to K2. In their post-main-sequence evolution, they are traversing the blue loop (Sect. 8.2.2.3, Fig. 8.6) where helium core burning is occurring [42]. Therefore, they can pass through the instability strip twice in the loop — coming and going. Arguably the most famous star that is a Cepheid is Polaris — the North Star (Box 9.1 on Page 221). Cepheids are shown as green dots in Fig. 9.2. These stars are very bright (M_V ∼ −1 → −6) and so are readily seen over great distances, making them exceptionally good distance indicators, even out to tens of Mpc. Figure 9.9 shows an example in the relatively nearby galaxy M 31 (the *Andromeda Galaxy*).

Historically, it was in the *Small Magellanic Cloud* (SMC), a companion galaxy to the Milky Way, that Henrietta S. Leavitt first discovered a number of Cepheids and found that they varied with periods of order tens of days. The variation in the *apparent magnitude* was found to be related to the period [189], and since these variables were all at approximately the same distance, the implication was that the variation in the *absolute magnitude* (and hence the luminosity) was also related to the period. The *PL* relation was thus inferred.

Figure 9.10 shows a modern version of this relation for the V (visual) and I (infrared) filter bands. The central wavelengths for these bands were given in

Figure 9.9 Image of the galaxy M 31 (the *Andromeda Galaxy*), which is at a distance of 0.78 Mpc. *Inset Pictures*: A Cepheid variable changing brightness with time. The optical maximum-to-minimum variation is 1.2 magnitudes. Credit: NASA/ESA and the Hubble Heritage Team (STSci/AURA).

Eqs. 1.8 and 1.10, respectively. This figure shows the intrinsic Cepheid relations from [203],

$$M_V = -2.670 \left(\log P - 1.0 \right) - 3.944 \quad (\sigma = 0.21) \tag{9.1}$$

$$M_I = -2.983 \left(\log P - 1.0 \right) - 4.706 \quad (\sigma = 0.14) \tag{9.2}$$

where P is the period in days and σ represents the standard deviation of the fit. Put simply, brighter Cepheids pulsate more slowly. These relations have undergone much scrutiny (and adjustment) over the years because of their importance to the extragalactic distance scale.

Because Cepheids are high-mass stars, they have short lifetimes and so are found near star-forming regions. This means they are confined to the thin disk of a galaxy where star formation is occurring. In the Milky Way, any line of sight along the disk suffers from location-dependent *extinction* and *reddening*[9] from interstellar dust. Corrections need to be applied to each magnitude measurement to account for this before making use of Eq. 9.1 or 9.2. The effects of extinction and reddening are far less in the infrared; hence the attraction of observing in the I-band or at other infrared wavelengths.

---

[9] *Extinction* is the attenuation of light as it passes through the dusty interstellar medium. Reddening occurs because longer-wavelength red light is less attenuated than shorter-wavelength blue light. Hence stars appear redder than they are intrinsically.

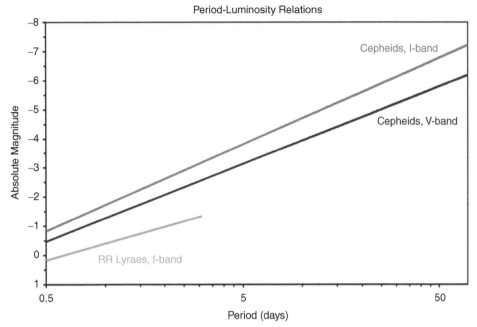

Figure 9.10 Modern *PL* relations for classical Cepheid variables in the V-band (blue, Eq. 9.1) and I-band (red, Eq. 9.2), as well as RR Lyrae variables in the I-band (turquoise, Eq. 9.4) The y-axis is the absolute magnitude in the respective filter bands. Note that RR Lyrae variables have periods that are usually less than about a day (Table 9.1).

There are other reasons for observing in the infrared. Spectra are smoother because spectral lines due to metals mainly occur in the ultraviolet, blue and visual parts of the spectrum (cf. F- to K-type stars in Fig. 3.1). Possible metallicity dependences are then also minimized. The width of the instability strip, equivalent to ΔT_{eff} between the blue and red edges, also has a smaller effect on the *PL* relation. Consequently, the net scatter in infrared *PL* relations is small compared to the optical, as is evident from the smaller value of σ in Eq. 9.2. Similarly, the brightness amplitude of the pulsations, ΔM_I, is also smaller than in optical bands. At first glance, this could seem to be a disadvantage, because a magnitude normally needs to be determined at some fiducial point (e.g. the mean of the light curve, cf. Fig. 9.8) in order to determine the *PL* relation. Thus one would have to plot out the light curve for all stars in the sample, and find the mean for each. However, in the infrared, small ΔM_I implies that at any given time, although some Cepheids will be brighter and some dimmer than their respective means, in bulk the brightnesses will average around the mean value, thus minimizing the observing effort required [202].

Improvements to the observed V-band relation can be made if a third parameter – namely, colour, as a measure of effective temperature (Sect. 3.1) – is explicitly taken into account. This third parameter is needed because of the finite width of the instability strip. The period-luminosity-colour (*PLC*) relation for the V-band is

$$M_V = -3.62 \left(\log P - 1.0 \right) + 3.03 \left(V - I \right)_0 - 6.25 \qquad (\sigma = 0.06) \qquad (9.3)$$

By including a colour term, the uncertainty, σ, decreases. The subscript 0 means the values have been dereddened.

A *PL* relation exists for many other types of variable stars. In the instability strip being considered, the other stars that pulsate in the fundamental mode are the δ Scutis and RR Lyraes. The δ Scuti stars have their own *PL* relation [331] but experience other complicating overtones. RR Lyraes are a well-known group that have been studied for many years. These are Population II (halo) stars that are often found in globular clusters (Sect. 7.5). Being Population II objects, there are no high-mass stars in the sample, so this group is of lower luminosity than Cepheids and cannot be seen over as great a distance. In evolution, like Cepheids, they are also in the helium-burning evolutionary phase. RR Lyraes are shown as closed yellow and open turquoise dots in Fig. 9.2. A *PL* relation for the fundamental mode of RR Lyraes is [225]

$$M_I = 0.17 - 1.92 \left(\log P_f + 0.30 \right) \qquad (9.4)$$

where P_f indicates the fundamental period in days. The Type II Cepheids (see Table 9.1), which are also lower-mass Population II stars, have their own *PL* relation (e.g. [211]). These various types, once clearly identified, can all be important standard candles.

Why is there a *PL* relation? Detailed arguments cannot be presented here, but a very rough approach is possible.

We have seen previously that pressure waves propagate on a dynamical timescale (e.g. Prob. 7.1). Consequently, periods of pulsating stars should also be of order τ_{dyn}. Using Eq. 7.21 and writing the period P for τ_{dyn},

$$P \sim \frac{1}{\sqrt{G \bar{\rho}}} \propto \left(\frac{R^3}{M} \right)^{1/2} \qquad (9.5)$$

where R is the radius and M is the stellar mass. From the Stefan-Boltzmann relation (Eq. 2.56), $L \propto R^2 T_{eff}^4$, so rearranging for radius and continuing with proportionalities,

$$R \propto \frac{L^{1/2}}{T_{eff}^2} \qquad (9.6)$$

Substituting Eq. 9.6 into Eq. 9.5 yields

$$P \propto \left(\frac{L^{3/4}}{M^{1/2} T_{eff}^3} \right) \qquad (9.7)$$

We now have a relationship between the period and luminosity, but the mass and temperature are still present, so we need to find relationships between the remaining parameters. Firstly, regarding the mass, we cannot use the main sequence mass-luminosity relation that we developed in Sect. 6.3 because Cepheids are evolved stars that are off the main sequence. Even off the main sequence, however, the luminosity is still a steep function of mass, similar to the steep dependences found for the main sequence (cf. Fig. 6.2). Inverting, it means the mass is only *weakly* dependent on luminosity. Consequently, for our rough arguments, the dependence on M is minor. Secondly, even though Cepheids cover a wide luminosity range from 100 to 10,000 $L_\odot$ [282], the range of temperature across the instability strip is comparatively small. Typically, $T_{eff} \approx (6 \pm 1) \times 10^3$ K, so the dependence on T_{eff} is also minor.

In summary, there are only weak dependences of P on T_{eff} and M in Eq. 9.7. If we take them into account, they modify the (3/4) power on the luminosity to about 0.87. Then

$$P \propto L^{0.87} \Rightarrow log(P) \propto 0.87\, log(L) \tag{9.8}$$

Using Eq. I.12 to convert from luminosity to bolometric magnitude, M_{bol},

$$log(P) \propto 0.87 \left(\frac{M_{bol}}{-2.5}\right) \Rightarrow M_{bol} = -2.9\, log(P) + const \tag{9.9}$$

A comparison of Eq. 9.9 with Eqs. 9.1 and 9.2 shows similar powers on the period. Although we have not made any conversions from bolometric magnitude to magnitude in a given filter, and have also presented a rather crude argument, it is still clear that the *PL* relation makes sense from a theoretical perspective. The arguments can be simplified as follows. Larger stars have lower densities and so have longer periods (Eq. 9.5). Larger radii lead to larger luminosities as long as T_{eff} does not change markedly (Eq. 9.6). Consequently, higher luminosities are associated with longer periods.

Online Resources

9.1 *Conny Aerts YouTube lecture, 'Space Asteroseismology: The Renaissance of Stellar Interiors'*:
https://www.youtube.com/watch?v=dveqFMamwyQ

9.2 *General Catalogue of Variable Stars (GCVS)*:
http://www.sai.msu.su/gcvs/gcvs

9.3 *All Sky Automated Survey for SuperNovae (ASAS-SN) Variable Stars Database*:
https://asas-sn.osu.edu/variables

9.4 *OGLE (Optical Gravitational Lensing Experiment) Atlas of Variable Star Light Curves*: https://ogle.astrouw.edu.pl/atlas/index.html

9.5 *Eclipsing Binaries Database – Timing Database at Krakow (TIDAK)*:
https://www.as.up.krakow.pl/ephem

9.6 *Animation of standing sound waves, by Daniel Russell*:
https://www.acs.psu.edu/drussell/demos/standingwaves/standingwaves.html

9.7 *Movie of a sunquake*:
http://soi.stanford.edu/press/agu05-98

9.8 *American Association of Variable Star Observers*: https://www.aavso.org/

PROBLEMS

9.1. (a) From the Stefan-Boltzmann law, show that

$$\frac{dL}{L} = 2\frac{dR}{R} + 4\frac{dT}{T} \qquad (9.10)$$

(b) From the adiabatic relation between T and V for an ideal gas (Sect. 2.1.8), find dT/T in terms of dV/V and the adiabatic constant, γ_a.

(c) Find dT/T in terms of dR/R and the adiabatic constant, γ_a.

(d) Show that

$$\frac{dL}{L} = (14 - 12\gamma_a)\frac{dR}{R} \qquad (9.11)$$

9.2. (a) Suppose that a Cepheid increases its radius by a maximum of 5%. Use Eq. 9.11 to find the percentage change in luminosity. Does the luminosity increase or decrease? (Assume that $\gamma_a \approx 5/3$.)

(b) Starting with Eq. I.12, find an equation that relates dM_{bol} to dL/L.

(c) From the *Top Left* plot in Fig. 9.8 for the star δ Cephei, measure the change in bolometric magnitude from the midpoint to the maximum/minimum and, using your result from part *(b)*, convert this to dL/L.

(d) Compare your result from part *(a)* with your result from part *(c)* and comment.

9.3. From the information given in the text regarding the advancing ripples of Fig. 9.3, estimate the sound speed in km s^{-1}. Calculations from the standard solar model (SSM) indicate that between $r = 99.5\,R_\odot$ and $r = R_\odot$, the sound speed declines from $c_s = 20.6$ km s^{-1} to $c_s = 7.98$ km s^{-1}. Is your simple estimate within this range?

9.4. Specify n, l and $|m|$ for the following cases for a rotating star. Specify whether the pulsation is radial or non-radial.

(a) There are no nodes within the star and three nodes on the surface, none of which pass through the pole.

(b) There are four nodes within the star and no nodes on the surface.

(c) There is one node within the star and five nodes on the surface, all of which pass through the poles.

(d) There are 2 nodes within the star and 10 nodes on the surface, of which 8 pass through the poles.

9.5. *(a)* A Cepheid in the galaxy NGC 1365 has a period of 10.0 days and apparent magnitudes (corrected for extinction and reddening) of $V = 27.344$ and $I = 26.606$ in the V-band and I-band, respectively.

(a) Find the distance d (Mpc) to NGC 1365.

(b) Suppose that the V-band and I-band magnitudes were not corrected properly for the effects of dust. Would the calculated distances be higher or lower? Would the V-band or I-band be affected more?

9.6. An RR Lyrae star in the globular cluster M4 is measured to have an apparent magnitude of $I = 11.462$ and a period of 0.5 days. What is the distance modulus of M4? What is the distance (kpc)?

Chapter 10

The Dying Star and Its Remnant

How often have I said to you that when you have eliminated the impossible, whatever remains, *however improbable*, must be the truth?

Sherlock Holmes, in *'The Sign of the Four'* [64]

One can hardly imagine a more dramatic transformation in the natural universe than the one that occurs when a star dies. The story of a dying star can be one of gentle mass loss and core exposure, or of a violent explosion that forms exotic remnants from the core and creates new elements. In either case, the end looks nothing like the beginning, and the dramatic change involves the complete cessation of nuclear reactions. Some of this story has already been told in Sect. 8.2. But there is much more to the saga because the remaining remnants and surrounding nebulae can, themselves, be beacons and probes that can lead us to a deeper understanding of the nature of our universe.

10.1 PLANETARY NEBULAE

Planetary nebulae (PNe) are arguably the most poorly named of all astronomical objects because they have nothing to do with planets. They were dubbed 'planetary nebulae' by Sir William Herschel in the 18th century because those that he observed were small and roundish with a slight greenish tint, similar to the planet Uranus. We now know that PNe are the ejecta of the envelopes of stars that are in the process of evolving from the asymptotic giant branch to the white dwarf stage of stellar evolution (Fig. 8.5, [185]).

Just when the stellar envelope becomes completely detached is uncertain. The transition to the left across the H-R diagram (to higher temperatures) happens quickly after a number of mass-loss episodes have already occurred at the asymptotic giant branch (AGB) phase (e.g. from instabilities such as thermal pulses, Sect. 8.2.1). Outflowing material will also sweep up any pre-existing ISM matter as part of the nebula. The beginning of a PN is more specific. It occurs when the central star (really just the contracting core of the original star) increases in temperature to $\gtrsim$20,000 K. At this temperature, the surrounding hydrogen in the nebula becomes ionized, and by a central star temperature of ~30,000 K, surrounding oxygen is double-ionized. This means the nebula 'turns on' in the sense that it can shine as an emission nebula, showing Balmer lines (Box 1.3 on page 14) and a characteristic OIII line at 5007 Å that supplies its greenish colour [183].

Stars whose initial masses are too small to produce supernovae but high enough to pass through the asymptotic giant branch (AGB) phase ($0.6 \lesssim M/M_\odot \lesssim 8$, Sect. 8.2.2) produce planetary nebulae. They constitute a significant fraction of all stars in the Galaxy. Consequently, the return of metals into the interstellar medium via PNe provides an important source of chemical enrichment in the Galaxy [185]. By the AGB phase, a variety of elements are present. Carbon is a direct result of the triple-α process, but a variety of other nuclear burning products are also present (see several reaction examples in Eqs. 5.30 to 5.32). Additional heavier metals are produced via *neutron capture*. This is when a free neutron is captured by a nucleus and then *β-decays*, when the new neutron in the nucleus converts into a proton and ejects a negative β particle (an electron). During the AGB, where surface temperatures are cooler, abundant molecular species as well as dust are also formed. These constituents are also observed in PNe. It may be that most of the carbon necessary for life as we know it is produced in this fashion [209].

Planetary nebulae include some of the most intricate and exquisite nebulosities in existence. Figure 10.1 shows some examples. It is clear that these objects are far from spherical! The resulting shapes can result from multiple factors. Apparently, at the AGB stage, some minor asymmetries are present deep in the star, even if the star itself is spherical at the surface [241]. When one includes magnetic fields, rotation and a fast wind, these asymmetries can be magnified [183]. For example, a toroidal magnetic field can collimate a jet along the symmetry axis, so jets can exist within the outflowing wind and have been observed in a number of PNe [18]. The multiple

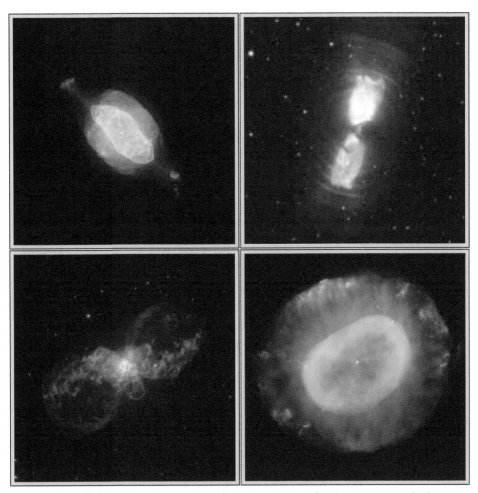

Figure 10.1 A selection of planetary nebulae. **Top Left**: The Saturn Nebula (NGC 7009) in Aquarius. This Hubble Space Telescope (HST) image shows a pair of knots along its major axis in a jet-like structure. **Top Right**: The Cotton Candy Nebula (IRAS17150-3224), discovered by Sun Kwok and Bruce Hrivnak using the HST. This object is evolving from the red giant stage to become a planetary nebula. A series of concentric arcs can be seen on both lobes, resulting from previous multiple mass-loss episodes. **Bottom Left**: Hubble 5 in Sagittarius, taken with the HST. Winds produce a bipolar nebula because of a small constricting disk. **Bottom Right**: NGC 7662 has a ring-shaped shell and an outer 'crown' showing many microstructures. Credit: [183] Kwok (2021), Cambridge University Press.

mass-loss events that occur prior to the complete ejection of the envelope can result in complex PNe structures if subsequent ejecta interact with each other. Companion stars as well as an uneven surrounding ISM can also play a part. PNe are thus rich test beds for studying the final stages in the lives of low and intermediate mass stars.

10.2 WHITE DWARFS – STELLAR CINDERS

White dwarfs (WDs) are the remnants of stars whose initial masses are M $\lesssim$ 8 M$_\odot$. Since over 99% of all stars fall into this category, WDs are the most common stellar remnant. They represent the exposed dense degenerate cores of highly evolved stars once the outer layers have been expelled (Sect. 8.2.2). Internal energy is transported by electron conduction, and electron degeneracy supplies the necessary pressure to hold the remnant up (Sect. 4.4). Such objects are polytropes (Sect. 6.2.2.2) whose equation of state links the pressure to the density but is independent of temperature. While the temperature does not factor into the structure of the star, it has important observational consequences because WDs cool over time and therefore follow an evolutionary path on an H-R diagram (see Sect. 10.2.2).

It is difficult to imagine what a white dwarf is actually like. A typical white dwarf mass is M$_{WD}$ ~ 0.6 M$_\odot$ [304], and as we saw in Box 8.3 on page 192, a remnant of this mass would be left behind from a star whose initial mass is ~1 M$_\odot$. A typical luminosity is only $L \sim 10^{-3}$ $L_\odot$ (see Fig. 8.10), and a typical density is $\rho \sim 10^6$ g cm^{-3}. The densities alone are far outside of human experience, since the densest material on Earth is the manufactured element hassium (atomic number 108) at a mere $\rho = 40.7$ g cm^{-3}. With these typical values of mass and density, the WD radius is only $R_{WD} = 6{,}580$ km, which is comparable to the Earth! Figure 10.2 shows an artist's concept of a white dwarf with the Earth behind it, for size comparison. Gaseous hydrogen is at the surface, followed by helium and then carbon and oxygen

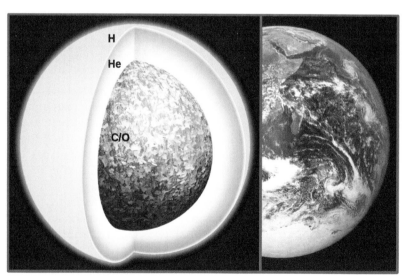

Figure 10.2 Artist's conception of a typical white dwarf with the Earth behind it for size comparison. Credit for the white dwarf drawing: Mark Garlick; University of Warwick; European Research Council. Credit for the Earth picture: NASA.

(cf. Sect. 8.2.1), which, in this example, has crystallized into a solid state. The mass of the hydrogen and helium layers is $\sim 10^{-4}$ and $\sim 10^{-2}$ of the mass of the WD, respectively [270]. An artist's conception is all that is available because WDs are so small that none have yet been spatially resolved.

There are various types of white dwarfs and also a variety of subtypes. Observationally, most WDs are 'DA' white dwarfs that show hydrogen absorption lines in their spectra, but about 20% are 'DB' types that show no (or very little) hydrogen [102]. DBs are thought to be 'failed' DAs in the sense that they have lost their outer envelope of hydrogen, revealing a predominantly helium atmosphere. The hydrogen may have been lost at the post-asymptotic giant phase (Fig. 8.5) with a very late helium shell flash or from a final AGB thermal pulse [320].

WDs can show variability from g-mode pulsations (Sect. 9.2). Pulsating DA WDs are called ZZ Ceti stars (black dots in Fig. 9.2), and the pulsations are driven by the kappa mechanism, which involves the ionization/recombination zone of hydrogen in the atmosphere (Sect. 9.4). The narrow ZZ Ceti instability strip is centered around an effective temperature of $\sim 11,800$ K, which is about the temperature that H recombines in the gaseous envelope. Periods are of order hundreds of seconds, and light curve variations are small, measured in mmag (milli-magnitudes). There is a corresponding g-mode pulsation for DB WDs that involves helium ionization/recombination, and the corresponding instability strip is at a higher temperature (around 25,000 K) because helium has a higher ionization potential than hydrogen (Fig. 2.2) [102]. As a WD cools, it can pass through these WD instability strips just as we saw for ageing stars that pass through their own instability strips to the right of the main sequence (Sect. 9.1).

A typical WD interior temperature is T $\sim 10^{7}$ K, which describes the random motions of the ions, since electrons are decoupled from their parent atoms in degenerate matter. The high-speed free electrons conduct energy quite efficiently (the matter has a high *thermal conductivity*), and there is no source of internal nuclear energy that could introduce a temperature gradient, so the interior of a white dwarf is essentially isothermal. On the other hand, the effective temperature at the surface is high (roughly 5000 K to 30,000 K, Fig. 8.10), but it is still much lower than the interior. Therefore, in a narrow atmospheric layer, the temperature must decline steeply. This favours convection, as we have seen before (Sect. 4.3), and this convection plays a role in the pulsation described in the previous paragraph [e.g. 71]. Thus, WDs have narrow convection zones that are just a few km deep and contain a very small fraction of the total stellar mass. For example, for DA WDs with $log(g) = 8$ (g is the acceleration of gravity) and $T_{eff} = 11,500$ K, the convection zone contains only 10^{-14} of the mass of the WD [180].

10.2.1 The Mass-Radius Relation for White Dwarfs

In Eqs. 6.23 and 6.24, we provided polytropic relations for white dwarfs when the electrons are non-relativistic, in which case the interior pressure is related to the average density by $P \propto \rho^{5/3}$ (polytropic index of $n = 3/2$), and relativistic, in which case $P \propto \rho^{4/3}$ (polytropic index of $n = 3$). Let us now fill in the missing constant in the proportionality. The electron degeneracy pressure, P_{ed} (dyn cm^{-2}), is

Non-relativistic:

$$P_{ed} = \left(\frac{3}{\pi}\right)^{2/3} \frac{h^2}{20\,m_e} \left(\frac{Z\,\rho}{A\,m_H}\right)^{5/3}$$

$$= 3.1 \times 10^{12}\,\rho^{5/3} \tag{10.1}$$

Relativistic:

$$P_{ed} = \left(\frac{3}{8\,\pi}\right)^{1/3} \frac{h\,c}{4} \left(\frac{Z\,\rho}{A\,m_H}\right)^{4/3}$$

$$= 4.9 \times 10^{14}\,\rho^{4/3} \tag{10.2}$$

where Z is the atomic number (number of protons), A is the atomic weight (number of protons plus neutrons), m_H is the mass of hydrogen, m_e is the mass of the electron, h is Planck's constant, and c is the speed of light (all cgs units). (For a derivation, see [44].) The quantities Z and A refer to any nuclei in the WD, and since there are twice as many nucleons as protons, $Z/A = 1/2$ for the evaluation of the constant. Notice that the temperature does not enter into these equations, as expected for a polytrope. Notice also that Planck's constant, h, is present, indicating that quantum mechanical considerations have entered into the derivation. The transition from non-relativistic to relativistic velocities is not abrupt, of course. Generally, non-relativistic velocities occur when $\rho << 10^6$ g cm^{-3} and relativistic velocities occur when $\rho >> 10^6$ g cm^{-3}. The two equations give equivalent results when $\rho = 3.9 \times 10^6$ g cm^{-3}. The corresponding transitional mass is ~ 0.6 M$_\odot$.

Although the temperature is $\sim$ constant in a WD interior, the density and pressure are not. Figure 10.3 shows a plot of the variation in density as a function of interior mass, for WDs of different masses. The canonical WD mass of 0.6 M$_\odot$ (red arrow) would have a *central* density between 10^6 and 10^7 g cm^{-3}. The corresponding pressures can be calculated using either Eq. 10.1 or 10.2. For example, a WD with M = 1.46 M$_\odot$ has a central density of $\rho = 10^{11}$ g cm^{-3}, which, from Eq. 10.2, corresponds to a central pressure of $P_{ed} = 2.3 \times 10^{29}$ dyn cm^{-2}. This is 12 orders of magnitude greater than the central pressure of the Sun (2.3×10^{17} dyn cm^{-2})!

We can now explore some interesting relations for WDs, starting with the equation of hydrostatic equilibrium (Eq. 6.4). Recall that we did not specify any particular type of pressure in order to use this equation – only that internal pressure must exist to counteract gravity for the star to be stable. As we did in Sect. 6.3, we use proportionalities, star-wide gradients and relations between the average density, total mass, and total radius and find

$$P_{ed} \propto \frac{M^2}{R^4} \tag{10.3}$$

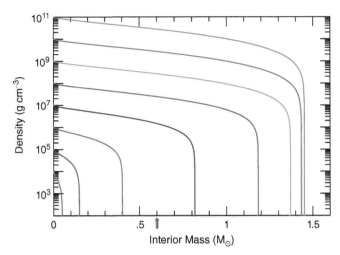

Figure 10.3 Interior density profiles of white dwarfs as a function of interior mass, M_r, for WDs of different masses, represented by different colours. For example, a WD of mass, 0.82 $M_\odot$ (purple curve) has a central density of 10^7 g cm^{-3}. The red arrow marks the location of a 0.6 $M_\odot$ WD. Notice the upper mass cutoff on the far right at the Chandrasekhar mass of 1.46 $M_\odot$. Credit: Frank Timmes.

where R and M are the total radius and total mass of the WD, respectively, and P is the central pressure. This is the same proportionality that we found in Eq. 6.27. Let us look at the non-relativistic and relativistic regimes separately.

Non-relativistic: From Eqs. 10.1 and 10.3,

$$P_{ed} \propto \rho^{5/3} \propto \frac{M^2}{R^4} \tag{10.4}$$

$$\Rightarrow \left(\frac{M}{R^3}\right)^{5/3} \propto \frac{M^2}{R^4} \tag{10.5}$$

$$\Rightarrow R \propto M^{-1/3} \tag{10.6}$$

Therefore, *more massive white dwarfs are smaller*! This rather counter-intuitive result has important consequences. For example, suppose a WD were accreting mass from a nearby companion with no significant mass-loss mechanisms at work. Then it would become smaller, its internal density and pressure would increase, and non-relativistic electrons could become relativistic, depending on the initial state of the WD and amount of accretion. The reality is more complex, though, because the accretion triggers surface nuclear reactions on the WD that cause energetic outbursts and mass *loss*. Such an event is observationally identified as a *nova*, abbreviated from *stella nova*, meaning 'new star' [83] (see Box 10.1 on page 241).

Relativistic: From Eqs. 10.2 and 10.3,

$$P_{ed} \propto \rho^{4/3} \propto \frac{M^2}{R^4} \tag{10.7}$$

$$\Rightarrow \left(\frac{M}{R^3}\right)^{4/3} \propto \frac{M^2}{R^4} \tag{10.8}$$

$$\Rightarrow \left(M^{4/3}\right) \propto M^2 \tag{10.9}$$

$$\Rightarrow M \approx constant \tag{10.10}$$

We now see that the mass is independent of radius. What does this mean? Once the mass has reached a critical upper limit (the constant), electron degeneracy pressure can no longer support the star, and hydrostatic equilibrium breaks down. In practical terms, imagine again that a WD in the relativistic regime is in a binary system, and mass is accreted from a companion without appreciable mass-loss. The consequences are dramatic, resulting in a *supernova*. This type of supernova will be discussed in Sect. 10.3, but at this point, it suffices to note that there is an upper limit to the mass of any WD. This limit is called the *Chandrasekhar mass* after Subrahmanyan Chandrasekhar [48]. In modern form, it can be expressed as [186]

$$M_{ch} = 1.46 \left(\frac{2}{\mu_e}\right)^2 M_\odot \tag{10.11}$$

where μ_e is the mean molecular weight of electrons, given by Eq. 2.65. For the most common C/O WDs (Sects. 8.2.2.2, 8.2.2.3), one can approximate $X = Y = 0$ and $Z = 1$, so $\mu_e = 2$. For the less common He WDs (Sect. 8.2.2.1), $X = Z = 0$ and $Y = 1$, so again, $\mu_e = 2$. This places the upper limit on WD masses at $M_{ch} = 1.46\ M_\odot$.

Figure 10.4 shows the mass-radius relation for WDs. It is clear that the observations (light grey points) roughly follow the $R \propto M^{-1/3}$ relation. The dashed curve, R_{ch}, corresponds to an 'ideal white dwarf'. Real WDs follow a relation that includes

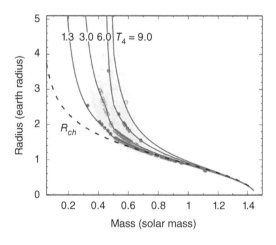

Figure 10.4 The mass-radius relation for type DA white dwarfs, from [285]. The reddish circles trace different effective temperature values, T_4 (labelled in units of 10^4 K at top), along with their fits as blue curves. The light grey circles with error bars represent data. The dashed line corresponds to the Chandrasekhar model for ideal WDs. Credit: Courtesy of Elvis do A. Soares.

corrections to the ideal electron degenerate equation of state as well as finite temperature effects. Higher T_{eff} as well as increased hydrogen layer thickness produce models that have systematically larger radii than the ideal WD curve [161]. Curves for different effective temperatures (labelled) are shown in blue. It is encouraging that the observations match the theoretically expected relation as well as they do. At the high-mass end, we simply do not observe WDs that exceed M_{ch}.

Box 10.1

Novae – The 'Guests'

Observationally, a nova is the sudden brightening of a star, a type of *cataclysmic variable (CV)* shown in magenta in Fig. 9.1. A nova results from a binary system in which one star is a WD and its companion is a non-degenerate star. Material flows from the star through an *accretion disk* onto the WD, with mass-transfer rates of $\dot{M} \sim 10^{-10}$ to 10^{-8} $M_{\odot}$ yr^{-1}. When sufficient material has built up, nuclear reactions start on the *surface* of the WD. This nuclear burning is unstable, causing a *thermonuclear runaway*. The result is a powerful ejection of mass, typically $M_{ej} \lesssim 10^{-4}$ $M_{\odot}$ with outflow speeds from 200 to 7,000 km s^{-1}. The event is seen as a brightening over many wavelengths. Optical magnitudes can increase by 8 to 15 magnitudes within just a few days, maximum luminosities are typically a few $\times 10^{38}$ erg s^{-1} [276], and radio, X-ray and even γ-ray emission may be observed. Light curves can be complex, but eventually the light fades over timescales of days to months. If matter continues to accrete, the outburst can occur again, resulting in a *recurrent nova* [57].

(continued)

(*continued*)

This artist's concept is of nova Z Camelopardalis, which is a recurrent nova and one in which the ejected matter has been directly observed as an expanding shell. It is possible that this nova was seen in outburst as early as 77 BCE by Chinese imperial astrologers [277]. These ancient astrologers referred to novae (as well as supernovae) as 'guest stars' [83].

10.2.2 The Cooling Curve – A Cosmic Clock

The cooling sequence for WDs is shown and labelled in Fig. 8.10. It stretches along a region that is to the lower left of (fainter and hotter than) the main sequence. A WD that lands on this curve at upper left will eventually slide down the curve to the lower right as it cools. The cooling curve has subtle substructure, mainly a bifurcation that splits the curve into two parts with the upper branch corresponding to hydrogen-rich atmospheres and the lower branch representing helium-rich atmospheres [304], although they are somewhat blended in the displayed figure. Compared to the main sequence, WDs are rather simple objects because the interior is not continually adjusting to the consumption of nuclear fuel. The Stefan-Boltzmann law (Eq. 2.56), for example, can be expressed as $L = const\ T_{eff}^4$ because the radius is essentially a constant. Thus, in an appropriate H-R diagram that plots $log\ L - log\ T_{eff}$, the slope is 4.

As time passes, individual WDs cool, meaning the thermal energy of the *ions* diminishes. The radiation comes from stored heat that is not replenished. A simple luminosity-age analytic relation of $L \propto t^{-7/5}$ was developed early on by treating the ions as an ideal gas [171], which would be appropriate prior to the stage at which there are phase changes. However, modern approaches are more sophisticated and must take into account a number of additional contributions to cooling. Examples are the release of latent heat during a phase transition from a fluid to a crystalline state in the interior, and the separation of elements in which heavier elements (e.g. oxygen or, for massive WDs, neon) shift closer to the center than carbon, releasing gravitational energy [110, 155]. Such additional sources of energy have the effect of slowing the cooling rate. In addition, about 20% of WDs are known to be magnetic, with fields on the surface ranging from 10^4 to 10^8 G and possible higher values inside. Strong magnetic fields could also modify the cooling of the star [81], although the exact effects are not yet known.

Figure 10.5 shows a modern relation from numerical calculations for massive WDs which have with hydrogen-rich atmospheres (red curves) and corresponding hydrogen-deficient atmospheres (dashed curves). The hydrogen-deficient curves show more rapid cooling because helium is more transparent to escaping photons than

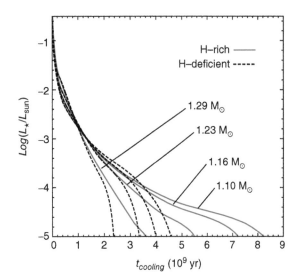

Figure 10.5 White dwarf cooling curves for massive WDs with hydrogen-rich envelopes. Credit: Maria E. Camisassa, [40]/With permission of EDP Sciences.

hydrogen. In all cases, lower-mass WDs take longer to cool, and this trend continues to masses that are lower than shown in the figure. Most WDs are of lower mass, ~0.6 $M_\odot$, so characteristic cooling times are very long, of order $t_{cool} \sim 10^{10}$ years.

Since WDs cool over a timescale that is comparable to the age of the universe (13.7 Gyr), they have been used as *cosmic clocks* to date individual stellar populations, to date the Milky Way, and even to reconstruct the star formation history of the Milky Way. In the most straightforward application of this concept, the oldest WD in any population cannot be older than the population itself. In the Milky Way, the halo (Population II stars, Sect. 7.5) consists of the oldest population of stars in our galaxy. Figure 10.6 plots the luminosity function (Sect. 8.1.5) for this population from a variety of data sets. Data from the Gaia satellite (Box 3.1 on page 60) are shown in black with relatively small error bars. This data set makes it clear that the number of white dwarfs drops off dramatically at low luminosity (to the right in the figure). The bolometric magnitude of the drop-off can be converted to a luminosity (Eq. I.12) and then converted to an age using a relation like the one shown in Fig. 10.5. The resulting estimated age of the Milky Way is 12 ± 0.5 Gyr, and fits to the data suggest that there was a burst of star formation between 10 and 12 Gyr ago that ceased about 8 Gyr ago [300].

Although we do not know all possible effects that could occur, it is intriguing to extrapolate our current knowledge of WDs into the far future. They should continue to cool, fade and crystallize. Under high pressures, the most stable form of crystallized carbon is diamond. Eventually, the temperature will decline to absolute zero, unless some other source of heating is discovered, and no light will be emitted at all. We call these futuristic objects *black dwarfs*. Imagine a dark galaxy filled with stellar remnants of black diamond!

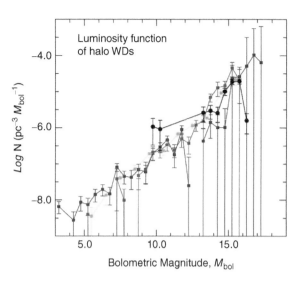

Figure 10.6 Luminosity function of white dwarfs in the halo of the Milky Way plotted against bolometric magnitude. The different colours represent different data sets, and vertical lines represent error bars. Black dots and lines specify WDs observed by the Gaia satellite, from [300]. Notice the drop-off seen in the Gaia data at the right. Credit: Courtesy of Santiago Torres.

10.3 SUPERNOVAE

A supernova (SN) is, essentially, an exploding star – but there is more than one way to blow up a star, as we will soon see. Figure 10.7 shows some examples. On the *Left* are pre-SN images of two galaxies, and on the *Right* are post-SN images of the same galaxies with supernovae (SNe) marked. Observationally, a supernova is seen as a rapid (days long [101, 123]) brightening followed by a slower (months to years) decline, similar to novae (Box 10.1 on page 241), but supernovae are intrinsically more luminous, typically 10^{43} erg s^{-1}, some five orders of magnitude brighter than novae. For a short time, such a luminosity rivals that of an entire galaxy! Like novae, early Chinese astrologers referred to supernovae as 'guest stars'. Unlike novae, there is no such thing as a 'recurring supernova' since the original star is destroyed in the process.

Observationally, supernovae are divided into two broad classes based on what lines are detected in their spectra: those that show hydrogen lines and those that don't (reminiscent of the classification of WDs, Sect. 10.2). Supernovae that *do* are called *Type II Supernovae*, and supernovae that *do not* are called *Type I Supernovae*. Additional spectral lines assist in the classification, but the presence or absence of hydrogen is a distinguishing feature. There are also a number of subtypes. For example, a typical SN Ia has no hydrogen or helium lines but does show a Si absorption feature. SN Ib and SN Ic also show no hydrogen but differ in the presence or absence, respectively, of helium lines [284]. Although the before-and-after images of Fig. 10.7 look similar, their spectra reveal that SN 1987A is of Type II and SN2011fe is of Type Ia. The underlying origins of these two types are quite different.

Type II supernovae are *core collapse* supernovae. These result from high-mass stars: that is, those with initial masses greater than approximately 8 M$_\odot$ that have

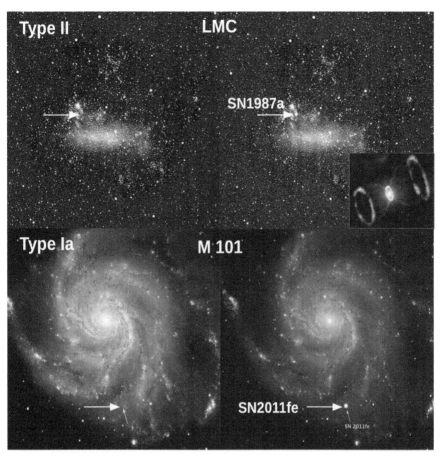

Figure 10.7 **Top Row**: Before (**Left**) and after (**Right**) picture of SN 1987A, a Type II supernova that occurred in 1987 in the Large Magellanic Cloud (LMC). Credit for top row: European Southern Observatory/CC BY 4.0. *Inset*: Artist's drawing of the observed inner and outer rings of the SN ejecta. Credit: European Southern Observatory/Wikimedia Commons/CC BY 3.0. **Bottom Row**: Before (**Left**) and after (**Right**) picture of SN 2011fe, a Type Ia supernova that occurred in 2011 in the galaxy M 101. Credit for bottom row: Courtesy B.J. Fulton and PTF.

evolved naturally to the point at which the core consists of iron, as shown in Fig. 8.8.[1] Any nuclear fusion of iron is endothermic (Sect. 5.3), so as the core becomes hotter and denser, iron burning cannot provide pressure to support the star. The consequences are rapid and catastrophic. Since abundant hydrogen was in the envelope of

---

[1] Stars in the lower-mass end (possibly $8 \lesssim M/M_\odot \lesssim 10$, though the range is uncertain) may have cores of oxygen, neon and magnesium instead of iron but are still expected to produce SNe [208].

the pre-supernova object, called the *progenitor*, it is present in the spectrum of the supernova as well. In addition, there are some cases in which the hydrogen-rich outer envelope of the progenitor may have been stripped prior to the explosion. The stripping could have occurred from massive stellar winds or mass loss onto a companion star. If that has occurred, then a core-collapse SN would likely be classified as Type I. Types SN Ib and SN Ic are thought to fall into this category. Because these SNe are associated with high-mass stars that have short lifetimes, they are only found in regions or galaxies in which active star formation is occurring. All told, core-collapse supernovae are the most numerous type of SNe. In a volume-limited sample, an estimate of the fraction of SNe that are of core-collapse type (i.e. Type SN II, Type SNIb, plus Type SNIc) is 76% [194, 279]. We will look at core-collapse supernovae in more detail in Sect. 10.3.1.

Type Ia supernovae have an entirely different origin involving a binary (or possibly multiple) star system, of which at least one component is a degenerate WD. Mass transfer onto the WD ultimately causes its destruction. In the past, it was thought that the trigger was from driving the mass of the WD over the Chandrasekhar limit (M_{ch}, Sect. 10.2.1). However, modern observations suggest that the WD could be close to the Chandrasekhar mass or could be lower than M_{ch}, i.e. *sub-Chandrasekhar*. These are called *thermonuclear supernovae*. The resulting supernova shows no hydrogen in the spectrum because the WD precursor also had no hydrogen, other than in a very thin atmospheric layer. The regions within which SNe Ia are found must be old enough to harbour WDs, so SNe Ia are observed, on average, in older stellar populations than SNe II. Of all supernovae, 24% are estimated to fall into the Type Ia category. Thermonuclear supernovae are important standard candles for distance determination. More information will be presented in Sect. 10.1.

The most obvious outcome from the supernova is the presence of rapidly expanding ejecta from the explosion, called a *supernova remnant* (SNR). Here, the use of the word 'remnant' should not be confused with the meaning in the title of this chapter. A SNR refers to the expanding nebulosity centered on the supernova, whereas the central remnant from the star will be either a neutron star (Sect. 10.4) or a black hole (Sect. 10.5). Figure 10.8 shows SNRs whose precursors were a SN II (RCW 103) and a SN Ia (SN 0509-67.5). These SNRs look circular in projection because of their spherical shapes. Many SNRs, however, are not so symmetrical, often because of their interaction with the surrounding interstellar medium. A catalogue of SNRs in the Milky Way can be found in Online Resource 10.1.

It is possible to estimate the SN rate in our Milky Way from the initial mass function (Box 8.2 on page 183) and our knowledge of massive star evolution (Sect. 8.2.2.4). Additionally, one can measure radioactive decay products from elements that are produced via *explosive nucleosynthesis* at the time of the SN. Elements of intermediate mass, Fe-group and heavier elements are readily produced in this way [e.g. 294] and returned to the ISM in the ejecta. Those products that decay on time periods comparable to the lifetimes of massive stars are especially useful for fine-tuning the SN

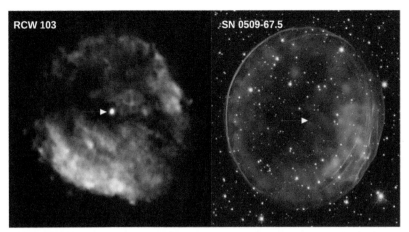

Figure 10.8 Two SNRs shown in X-rays. The colours red, green and blue show increasing X-ray energies, respectively. **Left**: RCW 103 results from a Type II SN that occurred in our Milky Way ~2,000 years ago. It is at a distance of 3.3 kpc and has a diameter of 8.6 pc. The arrow points to the neutron star (possibly a binary) called 1E 161348-5055 [72]. Credit: NASA/CXC/Penn State/G.Garmire et al. **Right**: SNR 0509-67.5, located in the Large Magellanic Cloud (distance ~50 kpc), resulted from a Type Ia supernova that went off ~400 years ago [15]. The SNR has a diameter of 7.2 pc [174]. The arrow points to the location of the explosion. No remnant has been found. Credit: J. Hughes et al/NASA.

rate. An example is ^{26}Al, which emits a γ-ray with a half-life[2] of 7.2×10^5 yrs. This high-energy radiation can penetrate over large distances in the Milky Way because it is not strongly affected by ISM dust absorption. Measuring the Galaxy-wide flux of the ^{26}Al γ-ray leads to a Galaxy-wide mass-estimate of ^{26}Al. Together with a knowledge of the IMF and the expected yield of ^{26}Al per massive star, the SN rate in the Milky Way can be calculated. An estimate is 1.9 ± 1.1 core-collapse SNe per century [78], or 53^{+72}_{-20} years between SNe, on average. The youngest SNR detected (so far) in our Galaxy is called G1.9+03, with an age of ≈ 100 yrs [260], which is within the estimated range.

The oldest SN that has been observed historically and linked to a currently observed SNR is one that was seen by the ancient Chinese in the year 185 AD [312]. The resulting SNR is RCW 86, and the likely progenitor was a Type 1a SN [36]. Historically, numerous SNe have been observed by eye in various cultures [e.g. 133].

---

[2] The *half-life* is the time it takes for the quantity of a sample to reduce by one-half.

10.3.1 Core-Collapse Supernovae

How does a core collapse produce a supernova explosion? There are many steps to this complicated process, and active research is on-going to understand a number of non-trivial issues. Nevertheless, a consensus as to the major ingredients has emerged [38] and will be briefly touched on here.

We begin with the progenitor, which is a high-mass (M $\gtrsim$ 8 M$_\odot$) star in the latest stage of its evolution. Differentiation of the elements with radius has occurred, like an onion skin, as suggested by the stylized representation of Fig. 8.8. The layers of the 'onion' consist of shells of various elements intermixed with shell burning and overlaid with a very extended outer envelope.

The final nuclear reaction phases proceed quickly, largely due to the increased importance of neutrinos. Energy losses by neutrinos, L_ν, dominate the photon luminosity, L_{phot}, at temperatures $\gtrsim 10^9$ K. While neutrinos are a product of some nuclear reactions, most neutrinos in SN progenitors are produced from other types of reactions such as electron-positron pair annihilation, the interaction of photons with electrons or positrons, and plasmon decay (Sect. 8.2.1). This extra loss of energy adds an extra term to the denominator of Eq. 7.18 ($\dot{x} = L_{phot} + L_\nu$), shortening the timescale and so accelerating the pace of the evolution. For example, in a 25 M$_\odot$ star, the steps involving silicon burning, which lead to iron, take only a single day [90]! In a typical high-mass star, the end result in the progenitor is an inert iron core (Sect. 5.5.2).

Figure 10.9 shows a more realistic view of the distribution of elements in a SN progenitor. The *Left* image shows a model for the blue supergiant progenitor star of

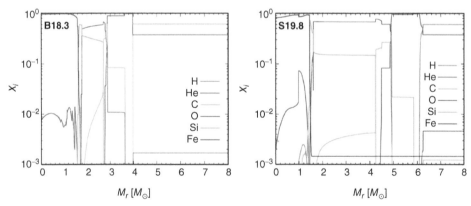

Figure 10.9 Mass fraction, X_i, of various elements, i, as a function of stellar mass for the interior of a SN II progenitor star. The elements are indicated in the legend on the right of each plot. **Left**: Model B18.3 shows a mass distribution of one model for the progenitor of SN 1987A, which was a blue supergiant of mass 18.3 M$_\odot$. **Right**: Model S19.8 shows the distribution for a red supergiant progenitor of mass 19.8 M$_\odot$. Note that only the interior region out to 8 M$_\odot$ has been plotted. Credit: Salvatore Orlando. [235]/With permission of EDP Sciences.

SN 1987A, which was pictured at the top of Fig. 10.7.[3] The *Right* image shows the interior of a red supergiant for comparison. Only the inner 8 $M_\odot$ has been plotted. The remaining 12 $M_\odot$ contains mainly hydrogen and helium in an extended envelope. Notice the iron core.

As pointed out in Sects. 5.3, 8.2.2.4 and 10.3, iron cannot fuse into heavier elements with the release of energy. Once an iron core has been produced, increased densities and temperatures do not start reactions that can produce sufficient pressure to stop core contraction. When the core reaches temperatures of $T > 10^{10}$ K, photons are energetic enough to break up iron nuclei via *photodisintegration* [186], i.e.

$$^{56}\text{Fe} + \gamma \Rightarrow 13\ ^4\text{He} + 4\text{n} \qquad (10.12)$$

This helium contribution can be seen in the cores of the models shown in Fig. 10.9. However, even this photodisintegration reaction is endothermic (Sect. 5.3), so iron 'consumes' energy rather than produces it.

Progenitors immediately prior to the supernova explosion have central densities that range from a few $\times 10^9$ g cm^{-3} for a 27 $M_\odot$ star to a few $\times 10^{10}$ g cm^{-3} for a 9 $M_\odot$ star [38]. These densities place the iron core firmly in the regime of electron degeneracy pressure, so the pre-SN iron core pressure is supplied by electron degeneracy. As Fig. 10.9 shows, the core mass is close to the Chandrasekhar limit (Eq. 10.11), and such a condition is gravitationally unstable. Once the mass exceeds the effective Chandrasekhar limit for the interior conditions, the core *implodes*. The imploding core releases an enormous amount of gravitational potential energy (Prob. 10.7). In less than a second, the core reaches neutron star densities at which neutron degeneracy pressure (see Sect. 10.4) suddenly stops the collapse. Copious numbers of electron neutrinos, ν_e (Box 5.2 on page 121), result from electron capture onto protons in the iron nuclei,

$$\text{p}^+ + \text{e}^- \rightarrow \text{n} + \nu_e \qquad (10.13)$$

This condensed object is a *proto-neutron star*.

When neutron degeneracy pressure suddenly stops the core collapse, the core rebounds like a piston and sends a shock wave outwards [38]. (For a helpful visualization, see Online Resource 10.2.) This outwards-moving *bounce shock* is powerful, but all modern models suggest that it is insufficient to drive the explosion. The bounce shock starts in the inner core in a region that is so dense ($\rho \sim 10^{11}$ g cm^{-3}) that it is opaque even to neutrinos, but it emerges within milliseconds to larger radii where the neutrinos can escape. Because of this energy loss, as well as the fact that the shock is advancing into material that is still infalling, the shock now *stalls* at a radius of around 100 to 200 km. Figure 10.10 *Left* shows how the shock levels off around this radius for a short time.

---

[3] This particular progenitor model results from the merger of two massive stars. Other models also exist.

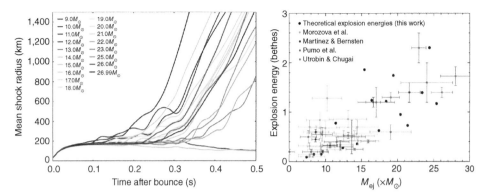

Figure 10.10   Results from two-dimensional models. **Left**: Plot of the location of the bounce shock in a supernova as a function of time for different progenitor masses shown in the legend. **Right**: Plot of the explosion energy, in 'bethes' (1 bethe = 10^{51} erg), as a function of the mass of the ejecta. The progenitor star mass is equal to the ejecta mass plus an assumed ~1.6 $M_\odot$ mass that belongs to the neutron star remnant. Black dots give the theoretical expectations, and coloured dots show observations from different data sets. Mass loss from the envelope has not been taken into account. Credit: Adam Burrows [38]/With permission of Springer Nature.

The region just inside the stalled shock is called the *gain region*. A proto-neutron star emits ~3×10^{53} ergs in neutrinos, whereas a typical supernova's kinetic energy is ~10^{51} ergs. Neutrino energies surpass the energy required to explode the star, so only a fraction of the neutrinos need to interact with matter in the gain region to supply the necessary impulse required to drive the explosion. Neutrino-generated turbulence in the gain region supplies additional pressure. In short, energy that originates from gravitational collapse leads to a strong neutrino flux, which, aided by turbulence, is sufficient to cause the SN explosion.

Figure 10.10 *Right* plots the explosion energies for a variety of progenitor masses. The x-axis plots the ejecta mass, which is the progenitor mass minus an assumed ~1.6 $M_\odot$ mass that belongs to the neutron star remnant. There is good agreement between theoretical models (black dots) and data (coloured dots). Three-dimensional modelling is necessary to include certain asymmetries, though. For example, effects such as convection or instabilities can explain asymmetries such as those observed in the ejecta of SN 1987A (Fig. 10.7 *Inset*) [235]. Online resource 10.7 also provides a creative visualization of asymmetric explosions.

Which supernovae progenitors produce neutron star remnants, and which produce black holes? The boundary between these two outcomes is not exactly known, but a modern consensus suggests that high-mass stars with initial masses up to ~20 $M_\odot$ end up forming neutron stars, and stars $\gtrsim$30 $M_\odot$ result in black holes (BHs). However, even if a BH is the ultimate outcome, a supernova must pass through the proto-neutron star phase, in which case the neutrino flux still occurs and a supernova is still expected [38].

10.3.2 Thermonuclear Supernovae

Thermonuclear supernovae are explosions of WD stars in interacting binary (or multiple) systems. The general picture is similar to the nova shown in Box 10.1 on page 241, except that the accretion drives a detonation that ultimately involves the entire WD. For example, unstable mass transfer from a companion can produce a detonation in the helium envelope of the WD, which then can lead to the detonation of the C/O core. This is called *double detonation* and is a way of explaining the explosion of sub-Chandrasekhar-mass WDs. There are various other models, though, and little consensus so far as to the nature of the system. For example, there could be two WDs, called a *doubly degenerate* (DD) case, or just one, called *singly degenerate* (SD). It isn't clear whether the companion survives the explosion, how the mass transfer occurs, how close the WD mass is to M_{ch} or whether some progenitors could actually be triple systems rather than binaries [268, 286]. The proliferation of models is, in part, a result of more and better observations that have uncovered fainter Type Ia supernovae as well as a variety of subtypes. The consequence, nevertheless, is a supernova and the complete destruction of the WD (e.g. Fig. 10.8, *Right*).

Here, we will focus on 'normal' SNe Ia, which are taken to be the explosion of a C/O WD in a binary system [292] and whose light curves look similar to each other. This similarity is what makes normal SNe Ia so useful. The intrinsic luminosities at the peaks of the light curves are similar, and so are their shapes, suggesting that similar physics are at work. To find the intrinsic luminosity, of course, requires calibration. That is, the distance to a set of SNe Ia first needs to be known in order to establish this relation. Those distances are usually determined from the period-luminosity relation of Cepheid variables (Sect. 9.5) that are in the same galaxy. Once the intrinsic luminosity is known for the standards, the relation can be extrapolated to other SNe Ia to find their (otherwise-unknown) distances. Thus, Type Ia supernovae are important standard candles. Moreover, SNe Ia are $\sim 10^6$ times brighter than Cepheids and therefore can be seen $\sim 10^3$ times farther away, allowing higher 'rungs' to be put on the distance ladder (Box 3.1 on page 60). SNe Ia are therefore crucial to the study of the large-scale structure and behaviour of our universe (see Box 10.2 on page 251).

Box 10.2

The Accelerating Universe

The large-scale structure of our universe can be probed with the *Hubble-Lemaître (H-L)* diagram (see picture). This is a plot of a measure of distance of a galaxy against its *redshift*. The redshift is $z \equiv (\lambda - \lambda_0)/\lambda_0$, where λ_0 is the rest wavelength of a spectral line from the object and λ is the measured longer wavelength of the line. The value $c\,z$, where c is the speed of light, has units

(continued)

(continued)

of velocity, but it is not a Doppler shift (cf. Eq. 2.45) because it measures the expansion of space itself rather than the motion of galaxies through space. That the universe is expanding has been known since the 1920s, but modern H-L diagrams show something unexpected, originally discovered from SNe Ia distances. At early times, the universe was expanding at a rate that is lower than it is today. In other words, the universe is now *accelerating*. Just what is causing this acceleration is not known, but it is referred to as *dark energy*, and it takes up ~70% of the total energy in our universe. For more information, see Sect. 9.1.2 of [154].

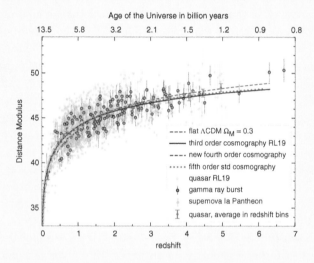

This H-L diagram plots the distance modulus, DM (Eq. I.11), against z. The origin (0,0) is 'here and now', and the cyan dots represent SNe Ia. If the universe were expanding at a constant rate (distance $\propto c z$), then $DM = 5\ log(z) + const.$ However, the displayed curve actually departs from this pure logarithmic shape in a sense that reveals the acceleration.

Credit: Elisabeta Lusso, from [198].

A thermonuclear explosion of a C/O WD whose mass is at the Chandrasekhar limit can produce ~0.6 $M_\odot$ of ^{56}Ni plus about the same amount of mass of other elements, including those of intermediate masses, elements around the iron peak (Fig. 5.2) and residual unburnt carbon and oxygen. The energy released from the production of ^{56}Ni is ~10^{51} ergs, and a comparable amount of energy is released from the other reactions, so this is sufficient to unbind the WD and give the ejecta the ~10^4 km s^{-1} outflow speeds that are observed [207].

^{56}Ni is radioactive and follows a decay chain of $^{56}Ni \rightarrow {}^{56}Co \rightarrow {}^{56}Fe$. The first step has a half-life of 6.6 days, the second has a half-life of 77.1 days, and in both steps, a γ-ray is released. This decay chain is the principal driver of the early light curve. The γ-rays quickly diffuse and degrade into the optical light that is observed. The rise to maximum is due to the increasing size of the rapidly expanding optically thick material. At maximum light, the release of energy is what is expected from the decay of ^{56}Ni and its decay product, ^{56}Co, combined [191]. Declining abundances from these decays and diminishing optical depths then drive the light curve decline. Additional factors can affect the light curve shape, such as circumstellar material, the propagation of a shock, the recombination of ionized material, dust formation and others. Nevertheless, both the peak luminosity as well as the light curve decline are connected via ^{56}Ni, and this makes SNe Ia a roughly unified group.

In reality, however, the luminosities and decay curves of SNe Ia are not exactly identical. Apparently the quantity of synthesized radioactive nickel can vary between SNe by a factor of ~ 8, for reasons that are not yet clear [267]. In short, brighter supernovae have broader light curves and bluer colours, and dimmer supernovae have narrower light curves and redder colours. This is known as the *Phillips relation*. The peak luminosity depends on the mass of synthesized nickel. But if there is more ^{56}Ni, then there will also be more of its product, ^{56}Fe, and iron is the major source of opacity in SNe. The higher opacity slows the evolution of the SN, thereby stretching out the curve.

The *Top* plot of Fig. 10.11 illustrates the Phillips relation. It is possible to correct for these effects, though, with proper calibration. For example, one can fit standard

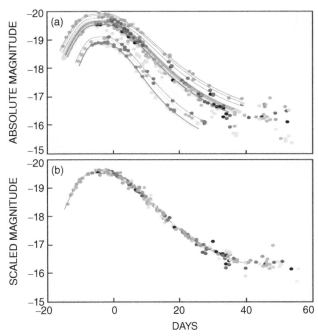

Figure 10.11 Sampling of SN Ia light curves before (**Top**) and after (**Bottom**) correcting for luminosity and duration. Credit: [249]/AIP Publishing.

templates of SNe Ia that have firm values for distance and luminosity (e.g. [141, 265]). An example of the resulting corrected light curves is shown in the *Bottom* plot of Fig. 10.11. While the uncorrected maxima vary by $\approx$1 magnitude, the corrected peaks vary by a small fraction of a magnitude.

10.4 THE DENSEST REMNANTS – NEUTRON STARS AND PULSARS

A neutron star (NS) is the remnant of a high-mass star that has undergone a supernova explosion. It is held up by *neutron degeneracy pressure* and is the densest material known. As suggested by Eq. 10.13, the remnant represents a 'sea' of neutrons, like a giant neutron nucleus, with a typical mass between 1 and 2 $M_\odot$ and a typical radius of $\sim$ 10 km [291]. The average density of a 1.4 $M_\odot$ NS is then $\bar{\rho}_{ns} \sim 7 \times 10^{14}$ g cm^{-3}. This can be compared to 'nuclear densities' (essentially, the density of a neutron) of $\sim$2 $\times 10^{14}$ g cm^{-3}. Consequently, the densest regions of NSs surpass nuclear densities and may probe sub-nuclear physics, delving into the exotic realm of *quarks, hyperons, hadrons* and *mesons*.[4]

Figure 10.12 provides a schematic of the modelled interior structure of a NS. Densities are expressed in units of a number density, n_0, called the *saturation density*, where the energy per nucleon is a minimum and the interior density of heavy nuclei would be a constant, i.e. nucleons are 'packed together'. The corresponding mass density is $\rho_0 = n_0 \, m_n = 2.67 \times 10^{14}$ g cm^{-3} (m_n is the mass of the neutron),[5] so 'saturation density' provides a more specific definition for 'nuclear density'. A NS's density changes by many orders of magnitude between the surface crust and the center. Near the surface, stable nuclei are possible, mostly ^{56}Fe, along with free electrons. The outer crust is likely to be crystalline [147]. By $\sim$10^{-8} n_0 in the outer crust, the electron capture reaction of Eq. 10.13 becomes energetically favoured, and neutrons are abundant. When the inner crust is reached ($\sim$10^{-3} n_0), *neutron drip* occurs, where there are so many neutrons that they begin to 'drip' out of nuclei. Just below the inner crust is a region called *nuclear pasta* because structures in this region resemble various forms of pasta – long, thin features called *spaghetti*, flattened ones called *lasagne* and other shapes. In this region, where $\rho \sim 10^{14}$ g cm^{-3}, nuclear attractive forces and

---

[4] Quarks, of which there are six flavours – 'up', 'down', 'charm', 'strange', 'top' and 'bottom' – are the constituents of protons and neutrons. The neutron is composed of two down quarks and one up quark, whereas the proton has two up quarks and one down quark. Protons and neutrons are examples of *baryons*, which have an odd number of quarks. A hadron is either a baryon (odd number of quarks) or a meson (even number of quarks). A hyperon also has three quarks, but none of them are charm or bottom, and one or more of them are strange.

[5] According to [148], $n_0 = 0.150 \pm 0.010$ fm^{-3}, or $\rho_0 = (2.5 \pm 0.2) \times 10^{14}$ g cm^{-3}. This agrees with the value of n_0 used in the figure.

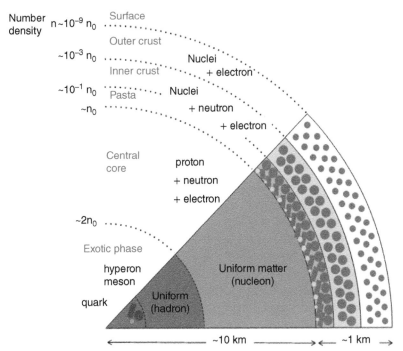

Figure 10.12 Schematic diagram of the interior of a neutron star. The quantity n_0 is a number density with a value $n_0 = 0.16 \, \text{fm}^{-3} = 1.6 \times 10^{38} \, \text{cm}^{-3}$, so the corresponding mass density is $\rho_0 = n_0 \, m_n = 2.67 \times 10^{14} \, \text{g cm}^{-3}$, where m_n is the mass of a neutron. Credit: [291]/Kohsuke Sumiyoshi.

Coulomb repulsive forces have about the same magnitude and so are in competition in a kind of 'tug-of-war', resulting in pasta and other complex morphologies [147]. Then comes the central core at the saturation density in which nuclei have dissolved into neutrons and some protons and electrons, and the matter is of uniform density. The inner core is in the 'exotic phase' in which sub-nuclear particles should exist [291].

At such high densities, Newtonian physics is no longer adequate and the equation of state of a NS must be found in the framework of general relativity. Electromagnetic interactions in addition to those of the strong force (responsible for binding nuclei together, Sect. 5.1) and weak force (responsible for radioactive decay) must also be taken into account. Thus, general relativity as well as subatomic physics are required to understand a neutron 'fluid', and it is not surprising that the equation of state for NSs is not well known.

However, as we have done before (e.g. Sect. 10.2.1), we can find some simple relations by taking a more straightforward approach. For example, a neutron degenerate gas is a polytrope, and that will lead us to some important conclusions about NSs.

10.4.1 The Mass-Radius Relation for Neutron Stars

The polytropic equation of state of a NS in which the neutrons are non-relativistic follows $P \propto \rho^{5/3}$ (Eq. 6.23), and for the relativistic case, it is $P \propto \rho^{4/3}$ (Eq. 6.24). These relations are the same as for WDs (Eqs. 10.1 and 10.2, respectively) except for the value of the constant. Consequently, we can follow the same development as in Sect. 10.2.1 and find the same kind of mass-radius relation for NSs as we did for WDs (cf. Eqs. 10.6 and 10.10), i.e.

$$R \propto M^{-1/3} \quad (Non\text{-}relativistic) \tag{10.14}$$

$$M \approx constant \quad (Relativistic) \tag{10.15}$$

We see that more massive NSs will be smaller, and there is an *upper limit to the mass* beyond which neutron degeneracy pressure can no longer support the remnant. This upper limit is called the *Tolman-Oppenheimer-Volkoff (TOV)* mass. It is not known as well as the Chandrasekhar mass, but an estimate is $M_{TOV} \sim 2.16\ M_\odot$ [261].

Any uncertainties in the equation of state will translate into uncertainties in the mass-radius relation. This is revealed in Fig. 10.13, in which a large variety of equations of states have been used to determine the mass-radius relation. An increasing radius with decreasing mass is seen, similar to Eq. 10.14, along with an upper mass cutoff that varies depending on the model.

The value of the upper mass limit can be approached from an observational point of view as well. Figure 10.14 shows a list of NSs in binary systems for which the masses have been measured (see Sect. 3.4). Here, we see some high-mass NSs that would suggest a higher upper-mass limit than suggested earlier. The vertical line represents the Chandrasekhar mass, and the presence of NSs to the left of this line shows that there is some overlap between the masses of WDs and NSs. All told, the mass range of NSs is very small. Only a few solar masses separate NSs from WDs at the low end and black holes at the high end.

The radii of NSs also exist over a narrow range (Fig. 10.13). A 10 km radius NS, for example, is about the size of *Phobos*, the small moon of Mars, or any number of large cities on Earth. But if radii are so small, how bright are NSs? Following a supernova explosion, the NS cools from a combination of neutrino emission from the interior and photon emission from the surface. The thermal emission from the surface behaves like a black body, and an effective temperature of $T_{eff} \sim 10^6$ K is maintained for $t \sim 10^5$ years [237]. Using the Stefan-Boltzmann law (Eq. 2.56) and a radius of 10 km, the luminosity will be low, $L \sim 0.2\ L_\odot$. Moreover, by Wien's displacement law (Eq. 2.50), the peak of the Planck curve occurs in the X-ray regime, and X-rays must be observed from space-based telescopes. These limitations have made it difficult, but not impossible, to obtain direct measurements of thermal emission from the surfaces of NSs [e.g. 237]. However, there is another way of seeing NSs, and it relates to the fact that these remnants have strong magnetic fields – they are seen as *pulsars*.

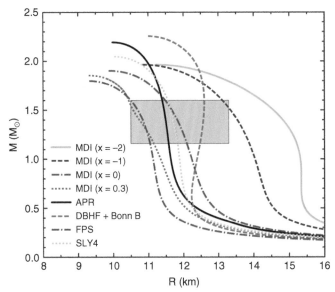

Figure 10.13 Mass-radius relation for neutron stars using different equations of state that are given by the acronyms on the left. The shaded region, enclosing masses from 1.16 and 1.6 $M_\odot$ and radii between 10.5 and 13.3 km, specifies a region for which a gravitational wave constraint has been included. Credit: Plamen G. Krastev, from [176].

10.4.2 Stellar Beacons – Pulsars

'Little green men'. That is what the discoverers of pulsars jokingly called the source of the new radio pulses that they detected in the year 1967. These repeating signals were so surprisingly regular that they seemed to be coming from an intelligent source. They were first detected by Jocelyn Bell at the Mullard Radio Astronomy Observatory in England; she had discovered the pulsar PSR B1919+21 using radio observations at a frequency of 81.5 MHz. It wasn't long before more such signals were detected at a variety of positions across the sky, laying to rest the 'little green men hypothesis'. The fact that the signal was emitted over a wide range of frequencies was also typical of natural signals.

From the first measurements, it was realized how precisely the signal period could be measured. With periods on the order of seconds or fractions of a second, repetitions of a great many pulses led to exceptionally high-precision measurements. The first paper in 1968, for example, quoted $P = 1.3372795 \pm 0.0000020$ s for PSR B1919+21 [145], an uncertainty of only 0.00015%!

What source in nature could be producing a signal that regularly ticked like a clock? The repetitive nature, short regular periods, high inferred energies of the

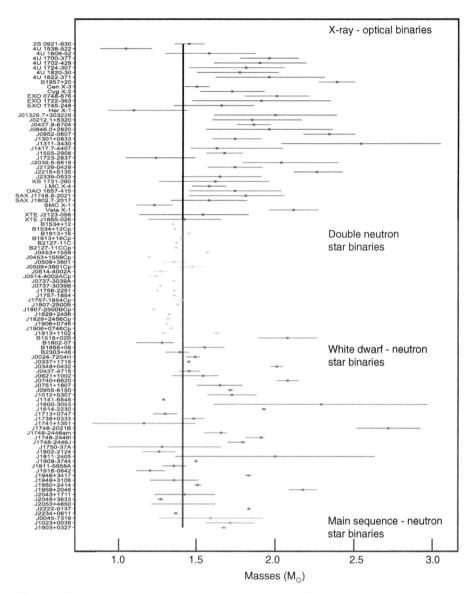

Figure 10.14 Neutron star masses as determined from binary systems. The name of the neutron star is listed at left, and the mass is at the bottom. The type of binary is labelled in the different colour zones. The vertical line shows the Chandrasekhar mass. Credit: Adapted from Horvath and Valentim 2017.

pulses- and strong magnetic fields- were all clues. These clues pointed to rotation of something small and dense with strong gravity [121], leading to the picture of a magnetized rotating NS whose magnetic axis is inclined to the rotation axis, as shown in Fig. 10.15. This is called the *oblique rotator model*, also called the *lighthouse model*,

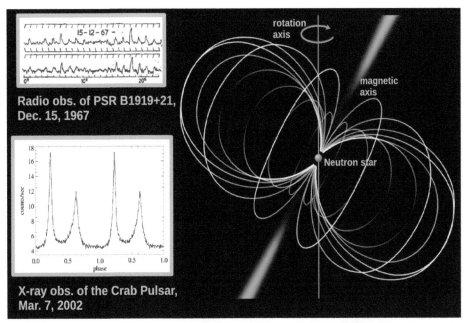

Radio obs. of PSR B1919+21,
Dec. 15, 1967

X-ray obs. of the Crab Pulsar,
Mar. 7, 2002

Figure 10.15 Sketch of the pulsar 'lighthouse' model showing how the magnetic axis is tilted from the rotation axis, as an explanation for the radio emission. As the pulsar rotates, the jet (blue) produces pulses when it sweeps by the line of sight. Curves represent magnetic field lines. Credit: Roy Smits, GNU Free Documentation License. **Top Inset**: Recorded radio pulses from the first pulsar detected. Credit: [145]/Springer Nature. **Bottom Inset**: Recorded X-ray pulses from the Crab Pulsar (period of 33.5 ms). Credit: [168]/With permission of SPIE.

because the observed radio pulses come from emission along the magnetic axis as it regularly sweeps through the observer's line of sight, like a lighthouse.

Today, well over 3,000 pulsars are known (Online Resource 10.5). A histogram of pulsar periods (*Top*) and pulsar magnetic fields (*Bottom*) is shown in Fig. 10.16. What could be called 'normal pulsars' have periods between 0.3 and 3 s, but there is also a distinct group of *millisecond pulsars (MSPs)* with periods between about 1 and 10 ms. Most MSPs are known to be in binary systems (light grey) and are thought to have been spun up by accretion[6] from a companion star [205]. There are also pulsars, called *magnetars*, with extremely high magnetic fields of order $\sim 10^{15}$ Gauss; they are shown in red in Fig. 10.16. These high-magnetic-field pulsars have long periods and have likely been slowed down by magnetic braking (see Box 8.5 on page 202).

Radio emission from pulsars is *non-thermal*, which means the luminosity is independent of temperature. It has been interpreted as *curvature radiation*, which results

---

[6] Recall that if mass M ↑, the radius R ↓ (Eq. 10.14). The rotation rate must then increase as R ↓ from conservation of angular momentum.

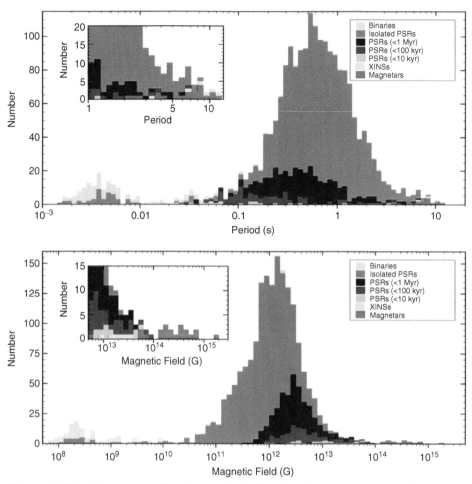

Figure 10.16 Histograms of pulsar periods (**Top**) and pulsar magnetic fields (**Bottom**). Credit: [233]/IOP Publishing. Colour scheme: light grey – binary pulsars, dark grey – isolated pulsars, shades of blue – young isolated binaries of different ages (see legend), yellow – X-ray emitting isolated neutron stars and red – magnetars. The *Insets* show blow-ups of the magnetar region. Credit: S.A. Olausen and V.M. Kaspi.

from relativistic electrons streaming along curved magnetic field lines [317], but other mechanisms have been proposed [165]. Pulsars also emit in other wavebands, but the lighthouse model of Fig. 10.15, which works so well at radio wavelengths, may not work so well as the waveband changes. This is because the rotating NS can produce emission from different locations in the pulsar *magnetosphere* (the region around the NS that has a significant magnetic field) rather than just along the magnetic poles [127]. The locations can be inside or outside of the *light cylinder*, which marks the boundary beyond which the field lines cannot co-rotate with the star; otherwise they would be moving faster than the speed of light. The light cylinder radius is

$$R_{LC} = \frac{c}{\omega} \tag{10.16}$$

where c is the speed of light and ω is the angular rotation frequency. Outside of the light cylinder, the field lines spiral outwards.

These exotic objects show even more extreme and astonishing properties. For example, the surface gravity, g (Eq. 6.5), of a NS of radius 10 km and solar mass is $g_{NS} = 1.3 \times 10^{14}$ cm s^{-2}, some five *billion* times what it is on the Sun. It follows that forces at the surface of a NS are that much stronger than those on the surface of the Sun. The escape velocity (Eq. 1.27) at the surface of such a NS is $v_{esc} = 1.6 \times 10^{10}$ cm s^{-1}, or 0.54 c! Observational consequences of general relativity are manifest, including the *gravitational redshift*. This is when the wavelength of a photon that leaves the surface is stretched out because of curved space-time around the NS (see Sect. 9.1.3 of [154]). In fact, pulsars *gravitationally lens* their own radiation! Photons that exit at an angle on the rear side of the pulsar can curve around into the line of sight [e.g. 227]. For more on gravitational lenses, see Sect. 9.2.1 of [154].

10.4.2.1 Scaling Down to the Pulsar

There is nothing simple about pulsars. However, some straightforward scaling relations and physical arguments can help us arrive at the picture shown in Fig. 10.15. While not rigorous, they do illustrate *feasibility*, or provide *order-of-magnitude estimates*, in order to rule out alternative explanations.

Let us start by asking what kind of mechanism could produce such strong, fast pulses. A pulsating WD, for example, produces varying light, but the magnitudes of the light curve variations are tiny (Sect. 10.2), and there is no match between the WD and pulsar light curves in other ways, such as the waveband. The most important difficulty, however, is the timescale. Pulsation should occur on a dynamical timescale ($\tau_{dyn} \sim 1/\sqrt{G\,\rho}$, Eq. 7.21), which, for a WD, is a few seconds – longer than observed for many pulsars. There are higher-frequency overtones, but they should be weaker than the fundamental, and one would have to explain why the fundamental is missing. Moreover, pulsar periods are observed over a wide range of values, from milliseconds to a few seconds (Fig. 10.16). If the pulses resulted from intrinsic variability, this would suggest that there is a wide range of densities in the objects. For example, the range of WD densities ($\rho \sim 10^{6-7}$ g cm^{-3}) results in only a factor of three variation in periods. The range of NS densities is even smaller. Consequently, we can rule out pulsation of WDs and NSs as the pulsar mechanism.

This leads us to *rotation*. Some basic relations for rotation and acceleration are

$$\nu = \frac{\omega}{2\,\pi} = \frac{1}{P} \tag{10.17}$$

$$a_c = \frac{v^2}{R} = \omega^2\,R \tag{10.18}$$

where ν (s^{-1}) is the rotational frequency, P (s) is the period, ω (rad s^{-1}) is the angular frequency, R (cm) is the radius, υ (cm s^{-1}) is the equatorial rotation speed and a_c (cm s^{-2}) is the centripetal (or its reaction, 'centrifugal') acceleration.

If the rotation rate is too high, there is insufficient gravity at the surface to retain material, and the star will start to tear itself apart. For stability, then, we require that the accelerations

$$\omega^2 R < \frac{GM}{R^2} \tag{10.19}$$

$$\left(\frac{2\pi}{P}\right)^2 R < \left(\frac{4\pi}{3}\right) G R \left(\frac{M}{\frac{4\pi}{3}R^3}\right) \tag{10.20}$$

$$P > \sqrt{\frac{3\pi}{G\overline{\rho}}} \tag{10.21}$$

where we have used Eqs. 10.17 and 10.18 and rewritten the expression in terms of the average density, $\overline{\rho}$, in the last step.[7] It is not surprising that the resulting period is a dynamical timescale, as we have seen before (cf. Eq. 7.21). Rearranging Eq. 10.21, we find

$$\overline{\rho} > \frac{3\pi}{G P^2} \tag{10.22}$$

Equation 10.22 tells us the minimum density needed for a rotating object to be stable. For the pulsar PSR B1919+21 (Fig. 10.15 *Top Inset*), with a period of 1.34 s, the average density must be $> 8 \times 10^7$ g cm^{-3}. The Crab Pulsar, with a 33.5 ms period (Fig. 10.15 *Bottom Inset*), requires an average density $>1.3 \times 10^{11}$ g cm^{-3}. These values are greater than average WD densities (Sect. 10.2.1). The numbers point to rotating NSs.

If NSs are the origin of the pulses, then most of them must have originated from the cores of massive stars that went through a supernova explosion. Certain NS properties might then be related to those of the pre-collapse, pre-supernova core of a red supergiant star. We will just call this the *core*. For example, in the absence of other forces or mass loss, the *angular momentum*, J (g cm^2 s^{-1}), should be conserved between the core and the NS. Also, the *specific angular momentum*, j (cm^2 s^{-1}), which is the angular momentum per unit mass, should also be conserved. The expressions for these quantities are

$$J = I \omega \tag{10.23}$$

$$j = \frac{(I \omega)}{M} \tag{10.24}$$

where I (g cm^2) is the *moment of inertia*. Equating the core and NS angular momenta,

$$J_{core} = J_{NS} \tag{10.25}$$

$$= I_{NS} \, \omega_{NS} \tag{10.26}$$

---

[7] This is the same argument as used in Eq. 8.20, in which we set $\omega^2 R = a_c = \frac{\upsilon^2}{R}$.

For the moment of inertia, I, we use

$$I = [0.8]\left(\frac{2}{5} M R_{NS}^2\right) = 0.32 M R_{NS}^2 \tag{10.27}$$

where the quantity in parentheses is the moment of inertia for a uniform solid sphere and the quantity in brackets is a modification for the fact that the sphere is not uniform [142].[8] Then, using Eq. 10.17,

$$J_{core} = 0.32 M R_{NS}^2 \frac{2\pi}{P} \tag{10.28}$$

$$j_{core} = \left(0.32 R_{NS}^2\right)\frac{2\pi}{P} \tag{10.29}$$

The properties of the core are not well known, but post-main-sequence stellar evolution models can provide some information (e.g. see the Online Resources of Chapter 8), and in some cases, the core region can be probed using asteroseismology (Sect. 9.2). A sample model is provided by [169], in which a star of initial mass $M_{ZAMS} = 13\ M_\odot$ evolves a core mass of $M_{core} \sim 1.7\ M_\odot$ immediately prior to the SN explosion. The specific angular momentum of this core is $j_{core} = 5 \times 10^{13}$ cm^2 s^{-1}, so by Eq. 10.29, the period of the resulting NS of radius $R_{NS} = 10$ km is 0.04 s, or 40 ms. This simple conservation of angular momentum argument, which does not include any other effects, suggests that fast rotation rates are feasible. Sophisticated numerical models that include convection and magnetic stresses that can couple the core to the outer layers of the star give values closer to ~1 s, which is more typical of normal pulsars [169].

What about the magnetic fields? The magnetic flux, Φ_B, is a measure of the density of field lines passing through a surface. If the flux is *frozen in* – that is, the magnetic field and plasma move together – then the magnetic field strength, B, of a collapsing object should increase according to $B \propto \frac{1}{R^2}$, by geometry. Then the magnetic field of the pulsar, B_{NS}, should be

$$B_{NS} = \left(\frac{R_{core}}{R_{NS}}\right)^2 B_{core} \tag{10.30}$$

As with rotation, the magnetic field strengths of cores are also not well known, but asteroseismic signatures suggest some values to be of order $B_{core} \sim 10^6$ G in a core of size $R_{core} \sim 0.5\ R_\odot$ [41]. With these values, the magnetic field strength of a NS could be as high as $B_{NS} \sim 10^{15}$ Gauss, typical of magnetars. There are alternatives to generating B through flux freezing, though. For example, convective regions are expected in proto-neutron stars shortly after bounce (Sect. 10.3.1). This can amplify magnetic fields via dynamo action (cf. Sect. 1.4) and may produce both the field strengths seen in pulsars as well as magnetars [321].

To arrive at the picture of the lighthouse (Fig. 10.15) requires the presence of radiation that is beamed along the magnetic poles. Misalignment between the rotational

---

[8] The formula for I given by [326] gives a similar result. Reference [169] uses $I = 0.35\ M R^2$.

and magnetic poles is then required from geometrical considerations, in order for the jets to be detected. The mechanism for producing this beaming is not completely understood, but magnetic fields are essential and known to be both present and strong.

We have pieced together some simple arguments that point to pulsars as rotating magnetic NSs with jets. The arguments involving angular momentum and magnetic fields are, in fact, more likely to be reversed. That is, the observed properties of pulsars can be brought to bear on the internal structure of the pre-supernova cores of red giants in order to hone existing stellar evolution codes. As is typically the case in astronomy, observations and theory work hand in hand to advance our understanding, and pulsars are a good example of that process.

10.4.3 The $P\dot{P}$ Relation and Characteristic Age

A pulsar is a rotating magnetic dipole (Fig. 10.15) and therefore emits electromagnetic radiation whose frequency is the frequency of the rotation (Eq. 10.17). This radiation is too low to be directly observed (e.g. $\nu = 1/P \sim 1/(1\ \mathrm{s}) \sim 1$ Hz), but the consequence of the energy loss is the slowing, or *spin-down*, of the pulsar. Spin-down rates are very small, of order tenths of microseconds per year, yet they are easily measurable [165], and the study of this spin-down leads to important information about the nature of pulsars.

The radiative energy loss rate from spin-down, L (erg s^{-1}) [e.g. 236], is

$$L = -\frac{2\mu_\perp^2 \omega^4}{3c^3} = -\frac{32\ \pi^4\ \mu_\perp^2}{3\ c^3\ P^4} \tag{10.31}$$

where $\mu_\perp$ (erg G^{-1}) is the component of the magnetic dipole moment that is perpendicular to the axis of rotation, c is the speed of light, and the right-hand expression makes use of Eq. 10.17. The negative sign means radiation is leaving the spinning body. We see that $L \propto 1/P^4$, revealing that short-period, fast-rotating pulsars lose significantly more energy per unit time than long-period pulsars. The magnetic dipole moment of a uniformly magnetized sphere with average surface magnetic field B and radius R can be taken as [65][9]

$$\mu = R^3\ B \Rightarrow \mu_\perp = R^3\ B_\perp \tag{10.32}$$

where $B_\perp$ is the component of the magnetic field that is perpendicular to the axis of rotation.

The energy loss manifests as a slowing-down of the rotation rate. The rotational kinetic energy of a spinning sphere is

$$E = \frac{1}{2} I\ \omega^2 \tag{10.33}$$

---

[9] Specifically, the magnetic moment (cgs units) is $\mu_\perp = (B\ R^3/2)\ \sin \alpha$, where B is the magnetic field of the uniformly magnetized sphere and α is the angle between the axis of rotation and the magnetic axis [156, Prof. R. N. Henriksen (private communication)].

where I is the moment of inertia. The loss of rotational kinetic energy during spin-down (using 'dot notion' for the time derivative) will then be

$$\dot{E} = \frac{d}{dt}\left(\frac{1}{2}I\,\omega^2\right) = I\,\omega\dot{\omega} = -4\pi^2\,I\,\frac{\dot{P}}{P^3} \tag{10.34}$$

where we have made use of Eq. 10.17 for ω (along with its differential with time, $\dot{\omega}$). Equation 10.34 describes the energy loss rate due to spin-down, regardless of the reason for the spin-down.

Let us now assume that there are no other causes of spin-down other than magnetic dipole radiation.[10] We can therefore set L from Eq. 10.31 equal to $\dot{E}$ from Eq. 10.34. Rearranging and making use of Eq. 10.32, we can solve for the magnetic field [e.g. 165],

$$B_\perp = \left[\frac{3\,c^3\,I}{8\,\pi^2\,R^6}\,(P\dot{P})\right]^{1/2} \tag{10.35}$$

for P in seconds and $\dot{P}$ in seconds per second, i.e. unitless. The *canonical* pulsar can be considered to have $M = 1.4\,M_\odot$ and $R = 10$ km. Then an estimate of the moment of inertia is $I \sim 10^{45}$ g cm^2 (Eq. 10.27). However, this value may not vary by much. That is because $I \propto M\,R^2$ (Eq. 10.27), and if the mass-radius relation follows $R \propto M^{-1/3}$ (Eq. 10.14), then $I \propto M^{1/3}$. If M varies by a factor of ~ 2 (Fig. 10.14), then $I^{1/2}$ varies by only 12%. Adopting $R = 10$ km,

$$B_\perp = 3.2 \times 10^{19}\,(P\dot{P})^{1/2}\ \ (\mathrm{G}) \tag{10.36}$$

Notice that provided the pulsar's radius and moment of inertia do not change during spin-down, the perpendicular component of the magnetic field can be inferred based only on two very easily measured pulsar properties, namely P and $\dot{P}$. One would expect $B_\perp$ to be of the same order of magnitude as B for most pulsars in a random sample, so with observations of a statistically significant number of pulsars, measurements of P and $\dot{P}$ can provide information about the strength of pulsar magnetic fields. For any given pulsar, we have $B_\perp$, i.e. a lower limit to B. The field strengths of Fig. 10.16 were found this way.

Moreover, for any given pulsar, as long as B also does not change during spin-down,

$$P\dot{P} = C \tag{10.37}$$

where C is a constant that depends on the magnetic field strength of the pulsar. This is referred to as the *$P\dot{P}$ relation* and indicates that as the pulsar slows ($P \uparrow$), the rate of spin-down decreases ($\dot{P} \downarrow$).

We can take this development a step farther. From Eq. 10.37,

$$P\frac{dP}{dt} = C \Rightarrow \int_{P_0}^{P} P\,dP = C\int_0^t dt \Rightarrow P = (2\,C\,t)^{1/2} \tag{10.38}$$

---

[10] An additional contributor to pulsar spin-down could be *magnetic braking* when the field lines are embedded in surrounding gas, which 'brakes' the pulsar spin (Box 8.5 on page 202).

where we have started the clock when the pulsar is formed ($t = 0$, called the *birth time*), t is the current time (the age of the pulsar), and we have assumed that the initial period was much less than the current period ($P_0 \ll P$). The period derivative is

$$\dot{P} = \frac{1}{2} (2 C)^{1/2} (t)^{-1/2} \qquad (10.39)$$

We can then define a *characteristic age* of a pulsar to be

$$\tau_c \equiv \frac{P}{2\dot{P}} \qquad (10.40)$$

Notice that, again, a measurement of τ is quite straightforward, depending only on measured quantities. The true age could depart from the characteristic age if, for example, the pulsar is very young and $P_0 \ll P$ (Prob. 10.10).

There are forms of braking other than simple magnetic dipole braking in a vacuum as described earlier. For example, there could be coupling to surrounding material, producing additional torques (see [196] for other examples). If one assumes that $\dot{P} \propto P^{2-n}$, rather than $\dot{P} \propto P^{-1}$ as in Eq. 10.37, then [e.g. 164]

$$\tau = \frac{P}{(n-1)\,\dot{P}} \left[1 - \left(\frac{P_0}{P} \right)^{n-1} \right] \qquad (10.41)$$

where n is called the *braking index*. For simple magnetic dipole braking with $P_0 \ll P$, as described in this section, n = 3 and Eq. 10.41 reduces to Eq. 10.40.

The $P\,\dot{P}$ relation, observationally, is shown in Fig. 10.17. A given isolated pulsar should slide down and to the right along a diagonal line of constant B with time. Notice the high spin-down rates of the magnetars, which cluster towards the right on the diagram. Although not marked, the youngest pulsars, shown in blue, are often located in supernova remnants, consistent with their birth mechanism. The MSPs are clustered in the lower left. These are in binary systems and have been spun up by mass-accretion from a companion [e.g. 165]. Their low values of $\dot{P}$ are consistent with this model. The $P\,\dot{P}$ figure is sometimes referred to as the 'H-R diagram for pulsars'.

10.5 THE ULTIMATE STELLAR REMNANTS – BLACK HOLES

A black hole (BH) is a region of space within which the gravity is so strong that nothing can escape – not even light itself. The size of a non-rotating BH is specified solely by the Schwarzschild radius, R_s,

$$R_s = \frac{2 G M}{c^2} \qquad (10.42)$$

where G is the universal gravitational constant, M is the mass, and c is the speed of light. This is not a physical boundary but specifies the region within which all matter and light are trapped. The 'surface' at this radius is called the *event horizon*. Any object that passes through the event horizon from the outside to the inside will

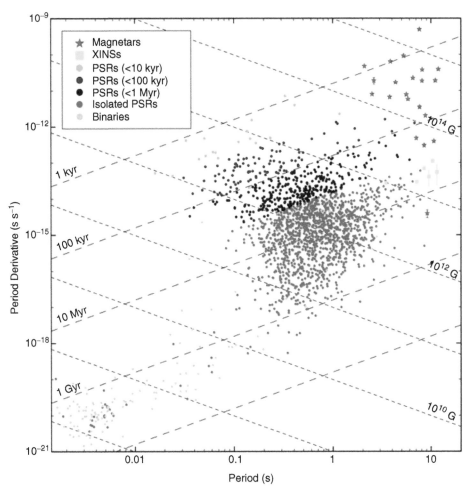

Figure 10.17 Distribution of periods, P, and spin-down rates, Ṗ for pulsars. Lines of constant B (from Eq. 10.36, labelled on the inside at right) and constant τ_c (from Eq. 10.40, labelled on the inside at left) are indicated. The colour scheme is the same as in Fig. 10.16. Credit: S.A. Olausen and V.M. Kaspi; [233]/IOP Publishing.

descend to a point-like location at the center called the *singularity*. We do not yet have physics that can adequately describe the singularity. General relativity is required to understand the physics of BHs. Such an approach is beyond the scope of this text, but several relevant points will be summarized in this section, especially from an observational point of view.

There are different classes of BHs, but the difference is only because of different mass regimes. For example, at the centers of galaxies are *supermassive black holes* (SMBHs), which likely formed at the beginning of the formation of the galaxy itself and grew thereafter. These SMBHs are in the range of M $\sim 10^6 \rightarrow 10^9$ M$_\odot$ and have now been imaged [92, 93]. The second class is *stellar mass black holes* of much lower mass, and these will be discussed shortly. A possible third mass range would be *intermediate mass black holes* (IMBHs) with M $\sim 10 \rightarrow 10^5$ M$_\odot$, which are more elusive and for which there is comparatively scant observational evidence [126].

We have seen that massive stars that go through the core-collapse supernova phase produce remnants that are either NSs or BHs. The mass boundary between these two fates is not known exactly, but a BH is the likely outcome if the initial stellar mass is $\gtrsim$30 M$_\odot$ (Sect. 10.3.1). In addition, if accretion onto a NS is sufficient to push it above the TOV upper limit (Sect. 10.4.1), then neutron degeneracy pressure is no longer sufficient to hold up the NS, and it will collapse to a BH. This suggests that the masses of stellar mass BHs should be of order a few M$_\odot$ at the low-mass end and could go up to approximately 10 M$_\odot$, depending on the value of the highest-mass stars (Sect. 7.4.2). More recently, however, signals from *gravitational wave*[11] detectors have shown signatures of the mergers of BHs in binary systems whose individual masses are of order $\sim$30 M$_\odot$ and whose final BH masses are $\sim$60 M$_\odot$ or more (see Table 2.2 from [154]). This extends the mass range of stellar-mass BHs on the high end.

Once a BHs is formed, there is no turning it into something else. It can only become a bigger BH, provided there is sufficient fuel for growth. Although massive stars are fewer in number than low-mass stars (Sect. 8.2), they live short lives, so many generations of stars can contribute to the population of BHs. An estimate for our Milky Way alone is 10^8 to 10^9 BHs [259]. Effort has also gone into determining the *black hole mass function* (total mass contained in BHs as a function of BH mass) over cosmic time. An estimate of the mass contained in stellar-mass BHs within a spherical volume of $\sim$1 Mpc^{-3} in the local universe is $\sim$5 $\times 10^7$ M$_\odot$ [281].

How can we 'see' a BH when the interior is inaccessible to us and the event horizon is not an emitting surface like other stellar remnants? The only way is to understand how the BH affects other objects or signals. For example, light from background objects bends around a BH. This produces either observable distortions or observable fluctuations in the background light. This process is called *gravitational lensing*, and the BH is the lens (see Sect. 9.2 of [154]). The key to 'observing a BH' more directly, though, is *accretion*. A BH is 'lit up' via material from a companion that falls down towards the event horizon, forming an accretion disk as material funnels inwards. It is emission from the region of the accretion disk outside of the event horizon that is observed. Figure 10.18 illustrates this process. Notice that a jet is formed as a natural outcome from accretion, as we have seen before (Sects. 8.1.3, 10.1, 10.4.2). Aside from the presence of a companion star, the image could just as easily apply to *active galactic nuclei* (AGN), which are produced from accretion onto a SMBH at the centers

---

[11] A gravitational wave is a distortion of space due to the acceleration of a mass in a system that is not symmetrical, such as a binary. See Chapter 2 of [154] for more information.

Figure 10.18 Artist's impression of the black hole Cygnus X-1, which is in a binary system where material is flowing from the companion (large blue star) through an accretion disk (orange/yellow) and into the black hole at the center. Jets are seen up and down, centered on the BH. Credit: NASA/CXC/M.Weiss.

of galaxies. AGNs that show jets are called *quasars*, so stellar-massed objects that resemble them are sometimes called *microquasars*.

The accretion disk is a high-energy object. Consider, for example, a 2 $M_\odot$ BH with a radius of 6 km (Eq. 10.42). The gain in potential energy of infalling material per unit mass is $\sim G\,M/R_s \sim 4.5 \times 10^{20}$ erg g^{-1}. If the mass of infalling material from a companion star amounts to ~ 0.1 $M_\odot$, then the amount of energy available at R_s is $\sim 10^{53}$ ergs. This is 100 times the typical kinetic energy released by a core-collapse supernova (Sect. 10.3.1). Only a fraction of this energy will be converted into observable radiation, but this rough estimate shows the high-energy nature of the region around a BH – not because the BH is massive, but because it is small.

10.5.1 Observational Evidence

There was a period of time when BHs were simply theoretical objects, though predicted from Einstein's general theory of relativity. Karl Schwarzschild's solution to Einstein's equation in 1916 [274] for a 'mass point' became known as the *Schwarzschild metric* and is applicable to a non-rotating BH. For the next 60 years, though, the existence of BHs would have to await observational verification. This was finally provided in 1975 by Tom Bolton at the University of Toronto when he observed the source, Cygnus X-1 [33], illustrated in Fig. 10.18. Let us review the arguments as presented in this paper.

Cygnus X-1 is in a known binary system, and its name reveals the X-ray-emitting nature of the source. The X-rays showed variability on timescales of $\Delta t \sim 1$ ms. By causality arguments,[12] the source size would have to be *less than* $c\Delta t \sim 300$ km. This suggests a compact source but does not rule out a NS. The high-energy X-rays suggested mass accretion onto the compact source via an accretion disk, but there were other creative explanations for this emission.

At the position of the X-ray object, only the companion star, HDE 226868, was observed optically, and it turned out to be a normal B0 supergiant at a distance of ~ 3 kpc. This star had a mass of ~ 30 $M_{\odot}$ and was too big for any accretion onto it to be the source of the X-rays. Spectroscopy was the key step. The system turned out to be a single-line spectroscopic binary (Sect. 3.4.2) with an orbital period of 5.6 days. For such a system, only the mass function (Eq. 3.17) can be found, which is a function of both masses, m_1, m_2, and the inclination of the orbit, i. Although the inclination was not known, a limit could be put on this value because of the observation that there were no X-ray eclipses, as well as some other geometrical arguments related to the ellipsoidal shape of HDE 226868. With knowledge of the observed star's mass plus its uncertainties, as well as limits on the inclination, the mass of the unseen object could be pinned down to a *minimum* of $M = 5.5$ $M_{\odot}$. As this value was significantly higher than the NS upper mass limit, the conclusion was inescapable that the unseen object must be a BH.

Today, the mass of the Cygnus X-1 BH is known to be between 13 and 23 $M_{\odot}$ [332]. The early work on this system, though, reveals the kind of sleuthing required to piece together a compelling story, from ruling out what doesn't work to gathering evidence of what does.

Box 10.3

Starring … SS 433 (V1343 Aquilae)

Distance: $d = 5.5$ kpc
Primary: Black hole; Secondary: Giant of spectral Type A7Ib
Mass$_{BH}$: >7 $M_{\odot}$; Mass$_{secondary}$: > 12 $M_{\odot}$

SS 433 ('SS' stands for 'Stephenson-Sanduleak') is an eclipsing X-ray binary system in which a black hole and an A-type giant star are orbiting their center

---

[12] The light from an unresolved source can only be seen to vary if all parts of the source vary together. Otherwise the variations from different parts would cancel and the source would shine steadily. But to vary together requires that the source be small enough that a light signal travelling at c could propagate from one side to the other in the time of interest, Δt.

of mass with a period of 13.1 days. The system is located at the center of the supernova remnant W50. The star loses mass to the BH, forming an accretion disk around the BH as well as two jets, similar to the diagram of Fig. 10.18. What is exceptional about SS 433, though, is that the jets are *precessing*, like the wobble of a child's top. Over a precession period of 162.3 days [53], the outflow from the jet forms a helical spray, like water from a rotating garden sprinkler. The jet speed is up to 80,000 km s^{-1}, or over 25% of the speed of light. The picture shows radio emission at 4.85 GHz and reveals the helical structure of the jet.

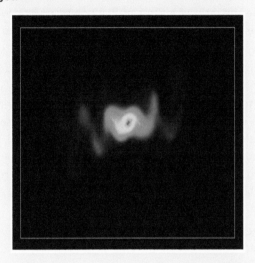

The box is 10 arcsec accross, from [32]. Credit: Blundell and Bowler, NRAO/AUI/NSF.

Online Resources

10.1 *A Catalogue of Galactic Supernova Remnants, by D.A. Green*: https://www.mrao .cam.ac.uk/surveys/snrs

10.2 *Supernova core-bounce demonstration*: https://www.youtube.com/watch? v=HQ8yvGN2VcY

10.3 *A montage of planetary nebula pictures*: https://skyandtelescope.org/astronomy-news/how-planetary-nebulae-get-their-shapes

10.4 *A montage of supernova remnants, Figure 1 from Lopez and Fesen, 2018 [197]*: https://arxiv.org/pdf/1804.00024.pdf

10.5 *The Australia Telescope National Facility (ATNF) pulsar catalogue [206]*: https:// www.atnf.csiro.au/research/pulsar/psrcat

10.6 *The Sounds of Pulsars*: https://www.jb.man.ac.uk/research/pulsar/Education/ Sounds

10.7   *Thierry Foglizzo, 'Aspherical Explosions Illustrated by Dynamics of a Water Fountain'*: https://www.youtube.com/watch?v=cB2GhQT5fls (see timestamp 40:35 to 51:55)

PROBLEMS

10.1.   (a)   Consider a non-relativistic white dwarf and, similar to the proportionalities found in Sect. 10.2.1, find proportionalities between the average density, ρ, and mass, M, as well as between ρ and the radius, R.

(b)   Suppose the ratio of masses of two non-relativistic WDs is $M_2/M_1 = 2$. Find the ratio of their average densities and the ratio of their radii.

10.2.   In Section 10.2.2, we suggested that the ions in a WD could be approximated as an ideal gas until there are phase changes. In Eq. 2.5, we also provided a guideline as to when to consider whether a gas is ideal. For a WD of mass of density $\rho = 10^6$ g cm^{-3} and T $= 10^7$ K, determine whether the ion gas is ideal. Assume that the size of carbon and oxygen can be represented by the de Broglie wavelength using the appropriate momentum (Eq. 2.6).

10.3.   Consider a completely degenerate carbon/oxygen (C/O) WD, and compute the following quantities for both the non-relativistic and relativistic cases, where possible. Justify your choices of input parameters, and state any assumptions.

(a)   The electron degeneracy pressure, P_{ed}.

(b)   The thermal pressure of the ions, $P_{i,th}$, assuming an ideal gas.

(c)   The radiation pressure, P_{rad}.

(d)   What pressure dominates? What pressure would you use in the equation of hydrostatic equilibrium?

10.4.   (a)   Calculate the ratio of the escape velocity (Eq. 1.27) of a 1 $M_\odot$ WD compared to that of the Sun. [HINT: Figure 10.4 will be helpful.]

(b)   Suppose some matter started from rest at infinity and fell to the surface of these two objects. What would be the ratio of the kinetic energies of the matter between the two masses by the time they reached the surface?

(c)   *Briefly* comment on the energetics of any activity on the WD as a result of accretion, compared to a normal star.

10.5.   The *isothermal compressibility* is a measure of the compressibility of an object and is defined as

$$\kappa_T \equiv -\frac{1}{v}\frac{\partial v}{\partial P}\bigg|_T \tag{10.43}$$

where v is the *specific volume* (volume per unit mass, cm^3 g^{-1}), P is the pressure, and the units of κ_T are cm^2 dyn^{-1} (Do not confuse this with the opacity.). When an object is compressed, ∂v is negative, so κ_T is a positive number, and a harder object will have a smaller value of κ_T.

(a) Rewrite κ_T with a change of variable so that it is a function of $\frac{\partial \rho}{\partial P}$ and ρ rather than $\frac{\partial v}{\partial P}$ and v.

(b) Find expressions for κ_T in terms of ρ, for a non-relativistic and relativistic C/O white dwarf. [HINT: Equations 10.1 and 10.2 will be of assistance.]

(c) Evaluate κ_T for a typical non-relativistic and relativistic C/O WD. Look up κ_T for the hardest substance on Earth, and compare.

10.6. In 2020, a headline announced, 'Compressibility measurements reach white dwarf pressures'. The reference was to a laboratory measurement of up to 450 Mbar [219]. Do you agree with this statement?

10.7. (a) Estimate the amount of gravitational potential energy that is released when an iron core at the Chandrasekhar mass within a supernova progenitor collapses to form a proto-neutron star of size $R = 20$ km. Assume that the initial radius is much greater than the final radius, and state any other assumptions.

(b) For the collapsing object of part (a), what is the free-fall time?

10.8. Online Resources 10.3 and 10.4 show many pictures of planetary nebulae and supernova remnants, respectively. It is not always obvious which is which, from morphology alone. If you had a wealth of observational data available to you, provide a list of ways that you could distinguish between the two.

10.9. Do a dimensional analysis on Eq. 10.31, and verify that the units match. Repeat for Eq. 10.32, and verify that the units are [erg/G].

10.10. (a) Show that if $P_0 \ll P$ for a pulsar, then

$$\tau_c \equiv \frac{P}{2\dot{P}} = t + \frac{P_0^2}{2C} \tag{10.44}$$

where t is the true age of the pulsar and the other quantities are defined in Sect. 10.4.3.

(b) If you calculate the age of a young pulsar using τ_c (Eq. 10.40), will the pulsar be younger or older than your value?

10.11. (a) Plot (with axes in logarithmic form) the $P\dot{P}$ diagram using Online Resource 10.5. [HINT: P0 is the period, and P1 is the period derivative.]

(b) From this resource, look up P and $\dot{P}$ for the Crab Pulsar (M = 1.4 $M_\odot$), also known as PSR B0531+21. Compute its characteristic age, τ_c.

(c) The supernova that gave birth to the Crab pulsar was observed by Chinese astronomers in the year 1054 AD. Compare τ_c to the true age. If there is a difference, offer an explanation.

(d) What is the magnetic field of the Crab Pulsar?

10.12. For a black hole mass range from 10 to 80 $M_\odot$, what is the range of radii? Compare this to the estimated radius range of neutron stars.

Chapter 11
A Stellar Invitational

I imagine that each star represents a single thought.
Star Trek Voyager, Season 7, episode 15, 'The Void'

Several stars have approached me and indicated their disappointment that they have not been featured more prominently in this text. As a result, I have invited them to this meeting so that they can introduce themselves and indicate how they are special. We all invite you to do an in-depth study of any one of these remarkable stars, or any others that might not have made it to this meeting.

– My name is ζ **Ophiuchi**, and I am a **runaway star**. I was kicked out of a binary system when my companion went supernova and suddenly lost a great deal of mass. Some of my cousins have also been expelled at high speed from close dynamical interactions with other stars in stellar clusters. My fast motion makes a shock wave as I travel through the interstellar medium.

– Hi, I'm a **Population III star**, and I have essentially no metals – just hydrogen and helium and trace amounts of lithium, berillium and boron that were formed in the Big Bang. No one has convincingly observed me, but I must surely exist in the early universe. That's because even Population II stars, which have very few metals (Sect. 7.5), do have *some metals*, and those metals must have been formed in earlier generations of stars. I guess that makes me the oldest star in this meeting.

– I'm called **S4716**, and I am orbiting the center of the Galaxy. The center is called Sgr A*, where there is a supermassive black hole. I am so close that I orbit the Galactic center in only four years and reach a speed of 8,000 km s^{-1}. At closest approach, I am only 98 AU from the black hole. I'm a bit embarrassed by how ellipsoidal I appear at this meeting, but on the other hand, it is an important distinguishing feature.

– Hi everyone. Happy to be here. I am a **polluted white dwarf**. Please don't look at me so negatively just because I am polluted. In fact, it is because of this pollution that all kinds of interesting information can be found about my system. You see, I normally would have a hydrogen or helium envelope, but if other chemicals are found in my atmosphere, they could be signatures of infalling planets. Heavier elements should sink very quickly (astronomically speaking), so the infall must have occurred recently.

– I am a **quark star**, and I have to admit, I'm a bit of an imposter because I am only hypothetical. I consist only of quarks under extreme pressure, possibly like the centers of neutron stars (Fig. 10.12). People aren't too sure if I'm stable.

– Well I'm definitely not hypothetical. I am **CW Leonis**, and I am a **carbon star**. On the H-R diagram, I sit on the asymptotic giant branch, and I have so much carbon in my atmosphere because of dredge-ups from the interior. I'm losing a lot of mass each year, so I have a thick envelope of carbon-rich material around me. I'm looking forward to the future when I'll shed my envelope and become a white dwarf.

– My name is somewhat mundane, **WOCS11005**. I am a **blue lurker**, and I hide out in globular clusters. I'm a main sequence star, but I don't end up in the right place on the colour-magnitude diagram because I have a white dwarf companion that skews the colours. I hope you don't mind that I've brought along my little white dwarf companion to this meeting. Perhaps he can give CW Leonis some tips on what it's like.

– Hello everybody! My name is θ^1 **Ori C**. I'm a hot O7 main sequence star and the primary source of ionization of the Orion Nebula. I'm interesting because I have a **magnetic field**, which is unusual for hot stars because we lack a surface convection region that can generate fields via dynamo action. So the magnetic fields are probably *fossil fields* that are remnants of an earlier time in my formation. My magnetic field is going to be important as I evolve because it will affect mass loss, rotation and eventually the characteristics of my future supernova explosion [315].

– Move over. I am a γ-**ray burster (GRB)**. I am the most powerful of all supernova explosions, emitting high-energy γ-rays with energies up to 10^{54} ergs. I come in two flavours: 'long' duration events (> 2 s) and 'short' duration events (< 2 s). My long version is thought to be from the core collapse of massive stars, and my short version is from mergers of compact binaries. I'm so luminous that I can be seen over great distances and can extend the Hubble diagram to higher redshifts. Check out my contribution to the Hubble-Lemaître plot in Box 10.2 on page 251.

– Well, you're not the only burster. My name is **FRB 20121102A**, and I am a **fast radio burst (FRB)**. Not only am I an FRB, but I am an FRB *repeater*. I exhibit sub-second radio pulses from cosmological distances. In other words, I live in another galaxy, so I could be a useful cosmological probe, too. I might be related to magnetars (Sect. 10.4.2), but since I was only discovered as recently as 2007, there's a lot that people don't know about me.

– Hi, I'm **TRAPPIST 1**. I'm a cool red dwarf star in the constellation of Aquarius, but what's really cool about me is that I'm surrounded by a planetary system. In fact, seven planets orbit me, a few of which could even be habitable.

– I am a **pair-instability supernova**. I'm also a hypothetical object – so far, anyway. But I'm certainly a theoretical possibility for stars that are very massive (well over 100 $M_\odot$) and with low metallicity (kind of like our Population III friend at the other side of the room). When that is the case, pair-production occurs, which is the creation of electrons and positrons when γ-rays and atomic nuclei interact. A chain of events leads to a supernova explosion that completely annihilates the star, leaving behind *no* remnant. Let's wrap up this meeting, please, because I won't be here for long.

– My name is **HV 2112**, and I'm a candidate for a **Thorne-Żytkow object (TZO)**. Okay, yes, TZOs are also hypothetical, but they could occur if a neutron star collides with a red giant. So I'm a red giant that has essentially 'eaten' the NS. The NS is still inside me, but eventually we should evolve together into either a single NS or BH. Sorry for taking up so much room here.

– I am **the Sun**. I know that a chapter was devoted to me in this text, but really, my importance cannot be overstated. I rule over a planetary system that contains a remarkable water- and land-covered planet that is filled with life. The bipeds on this planet have wondered about me since they opened their child-like eyes at the dawn of their existence. Today, they can view my active surface in all its glory, and they can also glimpse, by physical inference or by observation, into my interior. They have some understanding as to where I came from and what my future holds. And they understand the connection between me and my myriad cousins that sprinkle the night sky. But there is much more work to do before their active curiosity can be satisfied. If I may re-quote Eddington from the beginning of the Introduction, let us hope that 'in a not too distant future we shall be competent to understand so simple a thing as a star'.

Appendix A
Physical and Astronomical Data

Table A.1 Fundamental physical data

Symbol	Meaning	Value
Physical constants		
c	Speed of light in vacuum	$2.99792458 \times 10^{10}$ cm s^{-1}
G	Universal gravitational constant	$6.6742(10) \times 10^{-8}$ cm^3 g^{-1} s^{-2}
g	Standard gravitational acceleration (Earth)	9.80665×10^{2} cm s^{-2}
k	Boltzmann constant	$1.3806505(24) \times 10^{-16}$ erg K^{-1}
		$8.617343(15) \times 10^{-5}$ eV K^{-1}
h	Planck constant	$6.6260693(11) \times 10^{-27}$ erg s
$N_A{}^a$	Avogadro's constant	$6.0221415(10) \times 10^{23}$ mol^{-1}
$\mathcal{R}$	Molar gas constant	$8.314472(15) \times 10^{7}$ erg mol^{-1} K^{-1}
R_∞	Rydberg constant	$109737.31568525(73)$ cm^{-1}
Atomic and nuclear data		
e^b	Atomic unit of charge	$4.803250(21) \times 10^{-10}$ esu
eV	Electron volt	$1.602176634 \times 10^{-12}$ erg
m_e	Electron mass	$9.1093837015(28) \times 10^{-28}$ g
m_n	Neutron mass	$1.67492749804(95) \times 10^{-24}$ g
m_p	Proton mass	$1.67262192369(51) \times 10^{-24}$ g
u	Unified atomic mass unit	$1.66053906660(50) \times 10^{-24}$ g
$r_e{}^c$	Classical electron radius	$2.817940325(28) \times 10^{-13}$ cm
λ_C	Compton wavelength	$2.426310238(16) \times 10^{-10}$ cm
λ_{C_n}	Neutron Compton wavelength	$1.3195909067(88) \times 10^{-13}$ cm

(Continued)

Astrophysics: Decoding the Stars, First Edition. Judith Irwin.
© 2023 John Wiley & Sons Ltd. Published 2023 by John Wiley & Sons Ltd.

Table A.1 *(Continued)*

Symbol	Meaning	Value
λ_{C_p}	Proton Compton wavelength	$1.321\,409\,8555(88) \times 10^{-13}$ cm
a_0	Bohr radius	$0.529\,177\,2108(18) \times 10^{-8}$ cm
σ_T	Thomson cross section	$0.665\,245\,873(13) \times 10^{-24}$ cm^2
Radiation constants		
σ	Stefan-Boltzmann constant	$5.670\,400(40) \times 10^{-5}$ erg s^{-1} cm^{-2} K^{-4}
b	Wien's displacement law constant	$2.897\,7685(51) \times 10^{-1}$ cm K
a^d	Radiation constant	$7.565\,91(25) \times 10^{-15}$ erg cm^{-3} K^{-4}

[a] This is the number of particles in one *mole* (mol) of material.
[b] From [187]. An esu is an 'electrostatic unit'. The corresponding SI unit of charge is 1.602×10^{-19} coulomb. Note that the force equation in cgs units, $F = (q_1 q_2)/r^2$, does not have a constant of proportionality for F in dynes, q_1, q_2 in esu and r in cm.
[c] Here, $r_e = e^2/(m_e c^2)$.
[d] [66].
Credit: [293], converted to cgs units, unless otherwise indicated. The values in parentheses, where provided, represent the standard error of the last digits that are not in parentheses.

Table A.2 Astronomical data

Symbol	Meaning	Value
Distance		
AU	Astronomical unit	$1.495\,978\,706(2) \times 10^{13}$ cm
pc	Parsec	$3.085\,677\,582 \times 10^{18}$ cm
ly	Light year	$9.460\,730\,47 \times 10^{17}$ cm
Time		
y	Tropical year[a]	$3.155\,692\,58 \times 10^{7}$ s
y	Sidereal year[b]	$3.155\,814\,98 \times 10^{7}$ s
d	Mean sidereal day[b]	$23^h\,56^m\,04.090\,524^s$ s
Earth		
$M_\oplus$	Earth mass	$5.973\,70(76) \times 10^{27}$ g
$R_\oplus$	Mean Earth radius[c]	$6.371\,000 \times 10^{8}$ cm
R_{equ}	Earth equatorial radius[c]	$6.378\,136 \times 10^{8}$ cm
R_{pol}	Earth polar radius[c]	$6.356\,753 \times 10^{8}$ cm
$\rho_\oplus$	Mean Earth density[c]	5.515 g cm^{-3}
Moon		
M_m	Moon mass[d]	7.353×10^{25} g
R_m	Mean Moon radius[c]	$1.738\,2 \times 10^{8}$ cm
ρ_m	Mean Moon density[c]	3.341 g cm^{-3}
r_m	Mean Earth–Moon distance[c]	$3.844\,01(1) \times 10^{10}$ cm

Table A.2 *(Continued)*

Symbol	Meaning	Value
Sun		
$M_\odot$	Solar mass[d]	$1.98892(13) \times 10^{33}$ g
$R_\odot$	Solar radius[d]	$6.9599(1) \times 10^{10}$ cm
$\rho_\odot$	Mean solar density	1.41 g cm^{-3}
$L_\odot$	Solar luminosity[d]	$3.8275(14) \times 10^{33}$ erg s^{-1}
$T_{eff\odot}$	Solar effective temperature[d]	5772.0(8) K
$S_\odot$	Solar irradiance[e]	1.361×10^6 erg s^{-1} cm^{-2}
$m_{V\,\odot}$	Solar apparent visual magnitude[e]	-26.832
$M_{V\,\odot}$	Solar absolute visual magnitude[f]	4.81
$M_{bol\,\odot}$	Solar absolute bolometric magnitude[e]	4.74
$\theta_\odot$	Angular diameter of Sun from Earth	32.0'

[a] Equinox to equinox.
[b] Fixed star to fixed star.
[c] [66].
[d] [76].
[e] [204].
[f] [324].
Credit: [187] unless otherwise indicated. Values in parentheses indicate the uncertainties in the last digits.

Appendix B
The Solar Atmosphere

The next four plots show the probability density functions of the (base-10 logarithm of the) temperature, T, mass density, ρ, electron density, n_e, and pressure, P, respectively, in the solar atmosphere, as a function of height. The zero point on the x-axis is set at an optical depth of $\tau_{\lambda 500nm} = 1$. Figures B.1, B.2, B.3, and B.4 were kindly provided by Matts Carlsson using the three-dimensional 'Bifrost' code (see [43, 130]). These plots were introduced in Sect. 1.2.1.

Note the SI units of these plots. For an example conversion to cgs units, consider the density plot of Fig. B.2 at an x-axis position of 10 Mm (10,000 km). Then $log(\rho) \sim -12.5$, $\rho \sim 3.16 \times 10^{-13}$ kg m^{-3} = 3.16×10^{-16} g cm^{-3}.

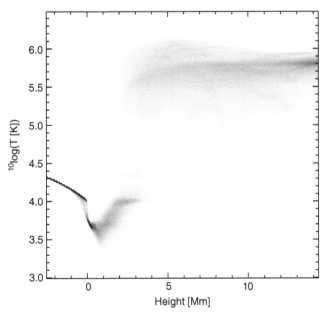

Figure B.1 Log of the temperature as a function of height.

Astrophysics: Decoding the Stars, First Edition. Judith Irwin.
© 2023 John Wiley & Sons Ltd. Published 2023 by John Wiley & Sons Ltd.

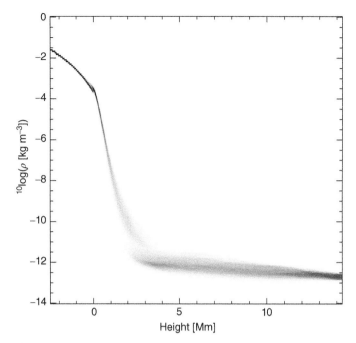

Figure B.2 Log of the density as a function of height.

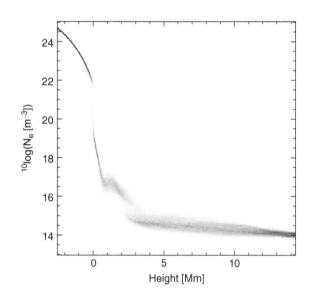

Figure B.3 Log of the electron density as a function of height.

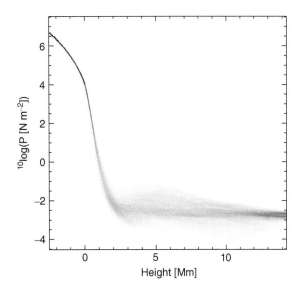

Figure B.4 Log of the pressure as a function of height.

Appendix C
The Standard Solar Model

Table C.1 presents a subset of data read from the standard solar model (SSM) data provided at https://www.ice.csic.es/personal/aldos/Solar_Data.html (GS98, Online Resource 1.12). The scientific description can be found in [313]. The GS98 model assumes a metal-to-hydrogen ratio at the solar surface of $(Z/X)_\odot = 0.0229$. The surface helium abundance is $Y_S = 0.2426$, and the surface metallicity is $Z_S = 0.0170$. The reference radius and luminosity of the Sun are $R_\odot = 6.9598 \times 10^{10}$ cm and $L_\odot = 3.8418 \times 10^{33}$ erg s^{-1}, respectively.[1] The mixing length parameter in this model is $\alpha_{ML} = 2.18$. The plots in this appendix were made from the same data. The SSM was introduced in Sect. 1.3.1.

Table C.1 Standard solar model data

Mass (M/M$\odot$)	Radius ($R/R_\odot$)	Tempera- ture (K)	Density (g/cm^3)	Pressure (dyn/cm^2)	Luminosity ($L/L_\odot$)	X (HI)^a	Y (^{4}He)^b
0.0000001	0.0005	15600000	150.8000000	2.3140E+17	0.00000	0.34715	0.63281
0.0027783	0.0300	15310000	141.2000000	2.1830E+17	0.02284	0.37276	0.60720
0.0186333	0.0585	14580000	120.0000000	1.8760E+17	0.13897	0.43401	0.54602
0.0533556	0.0870	13580000	96.6400000	1.5070E+17	0.34170	0.50856	0.47157
0.1066305	0.1155	12470000	76.0200000	1.1540E+17	0.56365	0.57679	0.40344
0.1750299	0.1440	11350000	59.1400000	8.5250E+16	0.74603	0.62890	0.35140
0.2538202	0.1725	10300000	45.5400000	6.1150E+16	0.86832	0.66405	0.31658
0.3378189	0.2010	9335000	34.6200000	4.2790E+16	0.93787	0.68528	0.29532
0.4220877	0.2295	8479000	25.9500000	2.9360E+16	0.97306	0.69720	0.28291
0.5025619	0.2580	7724000	19.1900000	1.9850E+16	0.98966	0.70355	0.27539
0.5764478	0.2865	7061000	14.0400000	1.3300E+16	0.99635	0.70710	0.27122

(continued)

[1] This value, whose error bar is $\pm0.015 \times 10^{33}$ erg s^{-1} 17, agrees with the value given in Eq. 1.1.

Table C.1 *(continued)*

Mass (M/M$_\odot$)	Radius (R/R$_\odot$)	Tempera- ture (K)	Density (g/cm^3)	Pressure (dyn/cm^2)	Luminosity (L/L$_\odot$)	X (HI)[a]	Y (^{4}He)[b]
0.7483754	0.3720	5516000	5.3670000	3.9820E+15	0.99888	0.71299	0.26811
0.7898626	0.4005	5113000	3.9080000	2.6890E+15	0.99895	0.71400	0.26734
0.8247964	0.4290	4751000	2.8600000	1.8300E+15	0.99897	0.71486	0.26662
0.8540857	0.4575	4424000	2.1070000	1.2560E+15	0.99897	0.71564	0.26591
0.8786204	0.4860	4127000	1.5630000	8.6970E+14	0.99896	0.71637	0.26524
0.8991576	0.5145	3855000	1.1680000	6.0730E+14	0.99896	0.71704	0.26460
0.9163628	0.5430	3602000	0.8791000	4.2750E+14	0.99895	0.71771	0.26395
0.9308017	0.5715	3364000	0.6669000	3.0310E+14	0.99894	0.71836	0.26333
0.9429529	0.6000	3137000	0.5097000	2.1610E+14	0.99894	0.71888	0.26281
0.9532039	0.6285	2915000	0.3926000	1.5470E+14	0.99893	0.71915	0.26250
0.9618818	0.6570	2691000	0.3048000	1.1100E+14	0.99893	0.72037	0.26116
0.9692548	0.6855	2451000	0.2383000	7.9450E+13	0.99892	0.72746	0.25447
0.9755531	0.7140	2177000	0.1890000	5.6490E+13	0.99892	0.74039	0.24260
0.9810244	0.7425	1887000	0.1522000	3.9400E+13	0.99891	0.74039	0.24260
0.9857436	0.7710	1617000	0.1205000	2.6690E+13	0.99891	0.74039	0.24260
0.9897228	0.7995	1365000	0.0931700	1.7400E+13	0.99891	0.74039	0.24260
0.9929813	0.8280	1130000	0.0698800	1.0780E+13	0.99891	0.74039	0.24260
0.9955504	0.8565	908700	0.0502200	6.2160E+12	0.99891	0.74039	0.24260
0.9974693	0.8850	701000	0.0338700	3.2240E+12	0.99891	0.74039	0.24260
0.9987908	0.9135	505700	0.0206100	1.4090E+12	0.99891	0.74039	0.24260
0.9995834	0.9420	321800	0.0103200	4.4480E+11	0.99891	0.74039	0.24260
0.9999385	0.9705	150700	0.0030770	6.0250E+10	0.99891	0.74039	0.24260
1.0000000	0.9990	12830	0.0000011	1.0330E+06	0.99891	0.74039	0.24260

[a] Mass fraction of neutral hydrogen (HI).
[b] Mass fraction of ^{4}He.
Credit: https://www.ice.csic.es/personal/aldos/Solar_Data.html

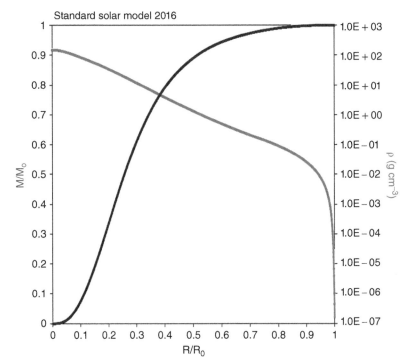

Figure C.1 The x-axis gives the solar radius as a fraction of the total solar radius. Blue curve/**Left** scale: mass as a fraction of the total solar mass. Red curve/**Right** scale: density (g cm^{-3}).

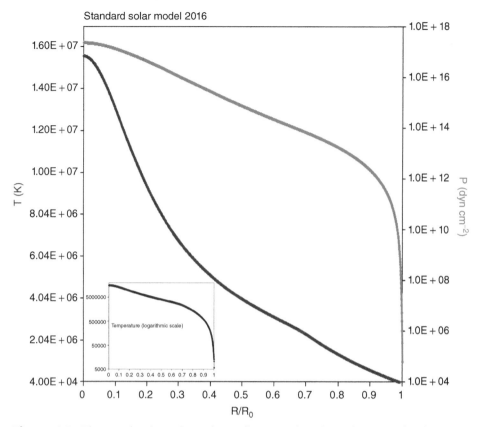

Figure C.2 The x-axis gives the solar radius as a fraction of the total solar radius. Blue curve/**Left** scale: temperature (K). Red curve/**Right** scale: pressure (dyn cm^{-2}). *Inset*: Temperature curve shown in logarithmic scale.

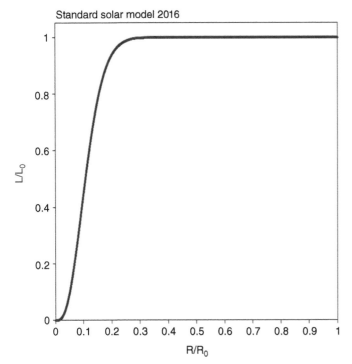

Figure C.3 The x-axis gives the solar radius as a fraction of the total solar radius. Blue curve/**Left** scale: luminosity as a fraction of the total solar luminosity.

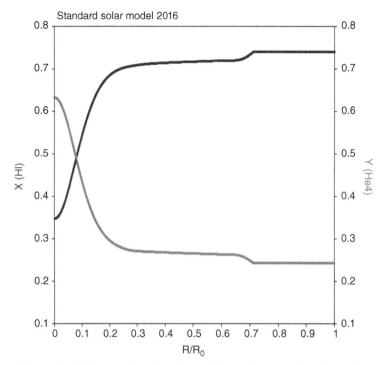

Figure C.4 The x-axis gives the solar radius as a fraction of the total solar radius. Blue curve/**Left** scale: mass fraction of neutral hydrogen (HI). Red curve/**Right** scale: mass fraction of ^{4}He.

Appendix D
Taylor Expansions for the Center of a Star

The Taylor expansion about radius $r = 0$ (i.e. a *Maclaurin expansion*) for any function $f(r)$ is

$$f = (f)_c + \left(\frac{df}{dr}\right)_c r + \frac{1}{2}\left(\frac{d^2f}{dr^2}\right)_c r^2 + \frac{1}{6}\left(\frac{d^3f}{dr^3}\right)_c r^3 + ... \qquad (D.1)$$

where c designates 'center'. This expression will give us values *near* the center of a star when the expansion is centered at $r = 0$.

Mass, m(r):

From the conservation of mass equation (Eq. 6.1) we know how mass varies with radius: $dm/dr = 4\pi r^2 \rho$. Then each term in parentheses in Eq. D.1 becomes

$$(m)_c = 0 \qquad (D.2)$$

$$\left(\frac{dm}{dr}\right)_c = (4\pi r^2 \rho)_c = 0 \qquad (D.3)$$

$$\left(\frac{d^2m}{dr^2}\right)_c = 4\pi \left(2r\rho + r^2 \frac{d\rho}{dr}\right)_c = 0 \qquad (D.4)$$

$$\left(\frac{d^3m}{dr^3}\right)_c = 4\pi \left(2\rho + 4r\frac{d\rho}{dr} + r^2 \frac{d^2\rho}{dr^2}\right)_c = 8\pi \rho_c \qquad (D.5)$$

The first three equations are zero because $r = 0$ at the center and so is the mass. The fourth equation is not zero because $\rho \neq 0$ at the center. The final expression for the mass *near* the center within radius r (inserting Eqs. D.2 through D.5 into Eq. D.1) is then

$$m(r) = \frac{4}{3}\pi \rho_c r^3 \qquad (D.6)$$

Astrophysics: Decoding the Stars, First Edition. Judith Irwin.
© 2023 John Wiley & Sons Ltd. Published 2023 by John Wiley & Sons Ltd.

This result is simply the amount of mass in a sphere of uniform density, ρ_c. 'Near the center', then, means 'until some radius at which the density varies so much that the approximation is no longer valid'.

Pressure, P(r):

For pressure, we need the hydrostatic equilibrium equation (Eq. 6.4: $dP/dr = -Gm\rho/r^2$), where we use small m for the variable, mass. Our Taylor terms then become

$$(P)_c = P_c \tag{D.7}$$

$$\left(\frac{dP}{dr}\right)_c = -\left(\frac{Gm\rho}{r^2}\right)_c = -\left(\frac{4\pi G\rho^2 r}{3}\right)_c = 0 \tag{D.8}$$

$$\left(\frac{d^2P}{dr^2}\right)_c = -\left(\frac{d\rho}{dr}\frac{Gm}{r^2} + \rho\frac{G}{r^2}\frac{dm}{dr} - 2\rho\frac{Gm}{r^3}\right)_c = -\frac{4\pi G\rho_c^2}{3} \tag{D.9}$$

We do not need the third term in the Taylor series because we already have an expression by the second term and we know the third term should be smaller.

In Eq. D.8, we have substituted Eq. D.6 for m. This is necessary because $dP/dr \propto m/r^2$, and both m → 0 and r → 0, so it is not immediately clear which term dominates. Once the substitution for m is made, it is clear that we are left with r in the numerator, leading to our conclusion that the first derivative in the Taylor series is zero at the center. In Eq. D.9, we again substitute for m and also substitute for dm/dr, leading to the stated result. Finally, we put our Taylor series terms into Eq. D.1 to find near the center,

$$P(r) = P_c - \frac{2}{3}\pi G\rho_c^2 r^2 \tag{D.10}$$

The pressure near the center is the pressure at the center less a term that includes the central density. Notice that this result already shows that near the center of a star, the pressure decreases outwards (cf. Appendix E).

Luminosity, L_r and Temperature, T(r):

It is possible to repeat this process for both luminosity and temperature, L_r and $T(r)$, respectively, near the center of any star. The results are

$$L_r = \frac{4}{3}\pi \rho_c \epsilon_c r^3 \tag{D.11}$$

where ϵ_c is the energy generation rate (erg $s^{-1}g^{-1}$) as used in Eq. 6.8.

$$T(r) = T_c - \frac{1}{8ac}\frac{\kappa_c \rho_c^2 \epsilon_c}{T_c^3} r^2 \tag{D.12}$$

where a is the radiation constant, c is the speed of light, and κ_c is the central opacity. A dimensional analysis will verify that the term on the right has units of K.

Appendix E
Chandrasekhar's Argument for a Declining Pressure Distribution in a Star

An early development, introduced by Chandrasekhar in 'An Introduction to the Study of Stellar Structure' (first published in 1939, [49]),[1] involves putting upper or lower limits on various physical quantities *based only on the equations of stellar structure and constitutive relations*. The goal is to infer limits to central values (or mean values) from only global observational properties. Here, we present a modified version of his 'Theorem 1', which argues for declining pressure in a star from the center outwards. We have already seen that the pressure declines near the star's center (Eq. D.10) but not yet throughout the entire star.

Let us define a function describing a 'modified pressure',

$$P_{mod} \equiv P + \frac{GM_r^2}{8\pi r^4} \tag{E.1}$$

where $P = P(r)$ is the pressure at some radius r in a star and M_r is the interior mass. A dimensional analysis shows that the second term on the right-hand side has dimensions of pressure.

What is the behaviour of the second term near the center of a star? Here $r \to 0$ and also $M_r \to 0$, so it is not immediately obvious. We need to work out the limit. However, from a Taylor expansion near the center (Appendix D), we know that the mass near the center can be expressed simply as $4/3\pi \rho_c r^3$. Thus near the center,

$$\frac{GM_r^2}{8\pi r^4} \propto \frac{(\rho_c r^3)^2}{r^4} \propto r^2 \tag{E.2}$$

---

[1] Several theorems are originally from Milne or Ritter, as noted by Chandrasekar in his book.

Astrophysics: Decoding the Stars, First Edition. Judith Irwin.
© 2023 John Wiley & Sons Ltd. Published 2023 by John Wiley & Sons Ltd.

and hence the second term in Eq. E.1 goes to zero at $r = 0$, which means $P_{mod} = P_c$ at the center.

We can combine the equations of stellar structure – conservation of mass (Eq. 6.2)[2] and hydrostatic equilibrium (Eq. 6.4) – to find

$$\frac{dP}{dr} = -\frac{GM_r}{4\pi r^4}\frac{dM_r}{dr} \implies \frac{dP}{dr} + \frac{GM_r}{4\pi r^4}\frac{dM_r}{dr} = 0 \tag{E.3}$$

We differentiate Eq. E.1 to find

$$\frac{dP_{mod}}{dr} = \frac{d}{dr}\left[P + \frac{GM_r^2}{8\pi r^4}\right] = \left[\frac{dP}{dr} + \frac{GM_r}{4\pi r^4}\frac{dM_r}{dr}\right] - \frac{GM_r^2}{2\pi r^5} = 0 - \frac{GM_r^2}{2\pi r^5} < 0 \tag{E.4}$$

where we have made use of Eq. E.3 for the terms within the brackets. Therefore, we have now shown that the gradient of the function P_{mod} is negative, i.e. P_{mod} is a decreasing function of r. This result is not restricted to a location near the center. Notice that the exact form of P or M_r need not be known or specified!

Let us now consider P_{mod} again. Since we have shown that P_{mod} declines with radius and $P_{mod} = P_c$ at the center, the following inequalities must hold for P_{mod}:

$$(\text{center}) \; P_c + 0 > P + \frac{GM_r^2}{8\pi r^4} > 0 + \frac{GM^2}{8\pi R^4} \; (\text{surface}) \tag{E.5}$$

where M is the total mass of the star, and R is its radius. The terms containing r in Eq. E.5 apply to any radius in the star, and the term on the right-hand side refers to the stellar surface where P = 0. Thus, we now know a limit for the pressure at the center of a star:

$$P_c > \frac{GM^2}{8\pi R^4} \tag{E.6}$$

We already know that this inequality is correct for a uniform density distribution, as given in Eq. 6.18, but now we see that it is true generally. Evaluating, we find

$$P_c > 4.48 \times 10^{14}\left(\frac{M}{M_\odot}\right)^2\left(\frac{R_\odot}{R}\right)^4 \text{ dynes cm}^{-2} \tag{E.7}$$

According to the standard solar model (SSM), $P_c = 2.3 \times 10^{17}$, so Chandrasekhar's inequality is quite correct, though it would have been more satisfying had the numerical constant in Eq. E.7 been a few orders of magnitude higher. Still, given how little went into this development, just the total mass, the total radius, the equation of hydrostatic equilibrium and continuity, to conclude that the central pressure must be *at least* 14 orders of magnitude greater than the vanishing surface value is impressive indeed! See Chandrasekhar's book for other examples.

---

[2] Here, we use the nomenclature dM_r/dr rather than dm/dr for consistency throughout this development.

Appendix F
Stellar Data

Tables F.1 through F.3 present data for stars of different spectral classes. The columns are:

1: Spectral type
2: B-V: colour index for the B filter (centered at λ 438 nm) and V filter (centered at λ 545 nm)
3: V-I: colour index for the V filter (centered at λ 545 nm) and I filter (centered at λ 798 nm)
4: M_V: absolute V magnitude
5: BC: bolometric correction (see Eq. 1.18)
6: T_{eff}: effective temperature (Eq. 2.55)
7: $R/R_\odot$: stellar radius in units of the solar radius
8: $M/M_\odot$: stellar mass in units of the solar mass

For definitions of magnitudes and colours, see Box I.2 on page xxiii. Data are from [124]. These values should be considered typical, but individual stars may differ from the values quoted here. Blanks or parentheses imply that the data are not well known. See Online Resource 3.10 for a somewhat modified table.

To obtain the stellar *luminosity*, first convert the absolute V magnitude to a bolometric magnitude using the bolometric correction (e.g. Eq. 1.18). For example, consider a B0 star on the main sequence (Table F.1):

$$M_{bol} = M_V + BC = -3.3 - 3.34 = -6.64 \tag{F.1}$$

We then use Eq. I.12 to convert to a luminosity, making use of Eq. 1.17,

$$L = 10^{-\frac{1}{2.5}\left(M_{bol} - M_{bol\odot}\right)} = 10^{-\frac{1}{2.5}(-6.64 - 4.74)} = 3.56 \times 10^4 \, L_\odot \tag{F.2}$$

and so throughout.

Astrophysics: Decoding the Stars, First Edition. Judith Irwin.
© 2023 John Wiley & Sons Ltd. Published 2023 by John Wiley & Sons Ltd.

Table F.1 Luminosity Class V (dwarfs, or main sequence)

Spectral type	B-V	V-R	M_V	BC	T_{eff} (K)	$R/R_\odot$	$M/M_\odot$
O3				(−4.3)	48000		(50)
O5			(−5)	(−4.3)	44000		(30)
O6	−0.32	−0.15	−4.8	−4.25	43000	(12)	(25)
O8	−0.31	−0.15	−4.1	−3.93	37000	10	(20)
B0	−0.29	−0.13	−3.3	−3.34	31000	7.2	17
B1	−0.26	−0.11	−2.9	−2.6	24100	5.3	10.7
B2	−0.24	−0.1	−2.5	−2.2	21080	4.7	8.3
B3	−0.21	−0.08	−2	−1.69	18000	3.5	6.3
B4	−0.18	−0.07	−1.5	−1.29	15870	3	5
B5	−0.16	−0.06	−1.1	−1.08	14720	2.9	4.3
B8	−0.1	−0.02	0	−0.6	11950	2.3	3
A0	0	0.02	0.7	−0.14	9572	1.8	2.34
A2	0.06	0.08	1.3	0	8985	1.75	2.21
A5	0.14	0.16	1.9	0.02	8306	1.69	2.04
A7	0.19	0.19	2.3	0.02	7935	1.68	1.93
F0	0.31	0.3	2.7	0.04	7178	1.62	1.66
F2	0.36	0.35	3	0.04	6909	1.48	1.56
F5	0.44	0.4	3.5	0.02	6528	1.4	1.41
F8	0.53	0.47	4	−0.01	6160	1.2	1.25
G0	0.59	0.5	4.4	−0.05	5943	1.12	1.16
G2	0.63	0.53	4.7	−0.08	5811	1.08	1.11
G5	0.68	0.54	5.1	−0.11	5657	0.95	1.05
G8	0.74	0.58	5.6	−0.16	5486	0.91	0.97
K0	0.82	0.64	6	−0.23	5282	0.83	0.9
K2	0.92	0.74	6.5	−0.3	5055	0.75	0.81
K3	0.96	0.81	6.8	−0.33	4973	0.73	0.79
K5	1.15	0.99	7.5	−0.43	4623	0.64	0.65
K7	1.3	1.15	8	−0.54	4380	0.54	0.54
M0	1.41	1.28	8.8	−0.72	4212	0.48	0.46
M2	1.5	1.5	9.8	−0.99	4076	0.43	0.4
M5	1.6	1.8	12	−1.52	3923	0.38	0.34

Table F.2 Luminosity Class III (giants)

Spectral type	B-V	V-R	M_V	BC	T_{eff} (K)	$R/R_\odot$
F0	0.31		1	0.04	7178	7
F2	0.36		0.9	0.04	6909	
F5	0.44		0.8	0.02	6528	
F8	0.54		0.7	−0.02	6160	8
G0	0.64		0.6	−0.09	5943	9
G2	0.76		0.5	−0.17	5811	10
G5	0.9	0.69	0.4	−0.29	5657	11
G8	0.96	0.7	0.3	−0.33	5486	12
K0	1.03	0.77	0.2	−0.37	5282	14
K2	1.18	0.84	0.1	−0.45	5055	17
K3	1.29	0.96	0.1	−0.53	4973	21
K5	1.44	1.2	0	−0.81	4623	40
K7	1.53		−0.1	−1.15	4380	60
M0	1.57	1.23	−0.2	−1.36	4212	100
M2	1.6	1.34	−0.2	−1.52	4076	130
M5	1.58	2.18	−0.2		3923	

Table F.3 Luminosity Class Ib (supergiants)

Spectral type	B-V	V-R	M_V	BC	T_{eff} (K)	$R/R_\odot$
A0	0.01	0.03	−5	−0.12	9550	70
A2	0.05	0.07	−5	−0.02	9000	
A5	0.1	0.12	−5	0.01	8500	
A7	0.13		−4.9	0.02	8300	
F0	0.16	0.21	−4.8	0.02	8030	80
F2	0.21	0.26	−4.8	0.02	7780	
F5	0.33	0.35	−4.7	0.04	7020	
F8	0.55	0.45	−4.6	−0.02	6080	
G0	0.76	0.51	−4.6	−0.18	5450	100
G2	0.87	0.58	−4.6	−0.26	5080	
G5	1	0.67	−4.5	−0.35	4850	
G8	1.13	0.69	−4.5	−0.41	4700	
K0	1.2	0.76	−4.5	−0.47	4500	110
K2	1.29	0.85	−4.5	−0.53	4400	
K3	1.38	0.94	−4.5	−0.68	4230	130
K5	1.6	1.2	−4.5	−1.52	3900	200
K7	1.62		−4.5		3870	
M0	1.65	1.23			3850	230
M2	1.65	1.34			3800	

Bibliography

1. Adams, F.C., Bodenheimer, P. and Laughlin, G. M dwarfs: planet formation and long term evolution. *Astronomische Nachrichten*, 326 (10): 913–919, December 2005. doi: 10.1002/asna.200510440.

2. Aerts, C. Asteroseismology. *Phys. Today*, 68 (5): 36, May 2015. doi: 10.1063/PT.3.2783.

3. Aerts, C. Probing the interior physics of stars through asteroseismology. *Rev. Mod. Phys.*, 93 (1): 015001, January 2021. doi: 10.1103/RevModPhys.93.015001.

4. Ahn, S.-H. Revisiting the epoch of the earliest Chinese star catalog titled 'Shi Shi Xing Jing'. *PASJ*, 72 (5): 87, October 2020. doi: 10.1093/pasj/psaa080.

5. Alighieri, D. *Dante's Inferno*. 1472.

6. Allers, R. and Minkoff, R. *The Lion King*. Movie, 1994. Produced by Buena Vista Pictures.

7. Althaus, L.G., De Gerónimo, F., Córsico, A. et al. The evolution of white dwarfs resulting from helium-enhanced, low-metallicity progenitor stars. *A&A*, 597: A67, January 2017. doi: 10.1051/0004-6361/201629909.

8. Anderson, R.I. Probing Polaris' puzzling radial velocity signals. Pulsational (in-)stability, orbital motion, and bisector variations. *A&A*, 623: A146, March 2019. doi: 10.1051/0004-6361/201834703.

9. Andrae, R., Fouesneau, M., Creevey, O. et al. Gaia data release 2. First stellar parameters from Apsis. *A&A*, 616: A8, August 2018. doi: 10.1051/0004-6361/201732516.

10. Arlt, R. and Vaquero, J.M. Historical sunspot records. *Living Rev. Sol. Phys.*, 17 (1): 1, February 2020. doi: 10.1007/s41116-020-0023-y.

11. Aschwanden, M.J. *Physics of the Solar Corona. An Introduction*. 2004.

12. Asplund, M., Grevesse, N., Sauval, A. and Scott, P. The chemical composition of the sun. *ARA&A*, 47 (1): 481–522, September 2009. doi: 10.1146/annurev.astro.46.060407.145222.

13. Augustson, K.C., Brown, B.P., Brun, A.S. et al. Convection and differential rotation in F-type stars. *ApJ*, 756 (2): 169, September 2012. doi: 10.1088/0004-637X/756/2/169.

Astrophysics: Decoding the Stars, First Edition. Judith Irwin.
© 2023 John Wiley & Sons Ltd. Published 2023 by John Wiley & Sons Ltd.

14. Ayres, T.R. Stellar and solar chromospheres and attendant phenomena. In Engvold, O., Vial, J.-C. and Skumanich, A., editors, *The Sun as a Guide to Stellar Physics*, pages 27–57, 2019. doi: 10.1016/B978-0-12-814334-6.00002-9.

15. Badenes, C., Hughes, J.P., Cassam-Chenaï, G. and Bravo, E. The persistence of memory, or how the x-ray spectrum of SNR 0509-67.5 reveals the brightness of its parent type ia supernova. *ApJ*, 680 (2): 1149–1157, June 2008. doi: 10.1086/524700.

16. Badnell, N.R., Bautista, M.A., Butler, K. et al. Updated opacities from the Opacity Project. *MNRAS*, 360 (2): 458–464, June 2005. doi: 10.1111/j.1365-2966.2005. 08991.x.

17. Bahcall, J.N., Serenelli, A. and Basu, S. 10,000 standard solar models: A Monte Carlo simulation. *ApJs*, 165 (1): 400–431, July 2006. doi: 10.1086/504043.

18. Balick, B., Frank, A. and Liu, B. Models of the mass-ejection histories of pre-planetary nebulae. IV. Magnetized winds and the origins of jets, bullets, and FLIERs. *ApJ*, 889 (1): 13, January 2020. doi: 10.3847/1538-4357/ab5651.

19. Ballai, I. and Marcu, A. The effect of anisotropy on the propagation of linear compressional waves in magnetic flux tubes: Applications to astrophysical plasmas. *A&A*, 415: 691–703, February 2004. doi: 10.1051/0004-6361:20034625.

20. Bally, J. Protostellar outflows. *ARA&A*, 54: 491–528, September 2016. doi: 10.1146/annurev-astro-081915-023341.

21. Barnes, S.A. Ages for illustrative field stars using gyrochronology: Viability, limitations, and errors. *ApJ*, 669 (2): 1167–1189, November 2007. doi: 10.1086/ 519295.

22. Bastian, N., Covey, K. and Meyer, M.R. A universal stellar initial mass function? A critical look at variations. *ARA&A*, 48: 339–389, September 2010. doi: 10.1146/annurev-astro-082708-101642.

23. Basu, S. and Antia, H.M. Helioseismology and solar abundances. *Phys. Rep.*, 457 (5–6): 217–283, March 2008. doi: 10.1016/j.physrep.2007.12.002.

24. Battich, T., Miller Bertolami, M.M., Córsico, A.H. and Althaus, L.G. Pulsational instabilities driven by the ϵ mechanism in hot pre-horizontal branch stars. I. The hot-flasher scenario. *A&A*, 614: A136, June 2018. doi: 10.1051/0004-6361/ 201731463.

25. Bergin, E. and Tafalla, M. Cold dark clouds: The initial conditions for star formation. *ARA&A*, 45 (1): 339–396, September 2007. doi: 10.1146/annurev.astro.45. 071206.100404.

26. Bergström, J., Gonzalez-Garcia, M.C., Maltoni, M. et al. Updated determination of the solar neutrino fluxes from solar neutrino data. *J. High. Energy Phys.*, 2016: 132, March 2016. doi: 10.1007/JHEP03(2016)132.

27. Bernath, P.F. Molecular astronomy of cool stars and sub-stellar objects. *Int. Rev. Phys. Chem.*, 28 (4): 681–709, January 2009. doi: 10.1080/01442350903292442.

28. Bestenlehner, J.M., Crowther, P.A., Caballero-Nieves, S.M. et al. The R136 star cluster dissected with Hubble Space Telescope/STIS - II. Physical properties of

the most massive stars in R136. *MNRAS*, 499 (2): 1918–1936, December 2020. doi: 10.1093/mnras/staa2801.

29. Bhatnagar, A. and Livingston, W. *Fundamentals of Solar Astronomy*, volume 6. World Scientific, 2005. doi: 10.1142/5171.

30. Bhattacharyya, I. A review of the neutrino emission processes in the late stages of the stellar evolutions. 2015. URL https://arxiv.org/pdf/1510.02678.pdf.

31. Bluhm, P., Jones, M.I., Vanzi, L. et al. New spectroscopic binary companions of giant stars and updated metallicity distribution for binary systems. *A&A*, 593: A133, October 2016. doi: 10.1051/0004-6361/201628459.

32. Blundell, K. and Bowler, M.G. Symmetry in the changing jets of SS 433 and its true distance from us. *ApJ*, 616 (2): L159–L162, December 2004. doi: 10.1086/426542.

33. Bolton, C.T. Orbital elements and an analysis of models for HDE 226868 = Cygnus X-1. *ApJ*, 200: 269–277, September 1975. doi: 10.1086/153785.

34. Borexino Collaboration. Experimental evidence of neutrinos produced in the CNO fusion cycle in the Sun. *Nature*, 587 (7835): 577–582, November 2020. doi: 10.1038/s41586-020-2934-0.

35. Borexino Collaboration, Agostini, M., Altenmüller, K. et al. Comprehensive measurement of pp-chain solar neutrinos. *Nature*, 562 (7728): 505–510, October 2018. doi: 10.1038/s41586-018-0624-y.

36. Broersen, S., Chiotellis, A., Vink, J. and Bamba, A. The many sides of RCW 86: a Type Ia supernova remnant evolving in its progenitor's wind bubble. *MNRAS*, 441 (4): 3040–3054, July 2014. doi: 10.1093/mnras/stu667.

37. Brun, A.S., Miesch, M. and Toomre, J. Global-scale turbulent convection and magnetic dynamo action in the solar envelope. *ApJ*, 614 (2): 1073–1098, October 2004. doi: 10.1086/423835.

38. Burrows, A. and Vartanyan, D. Core-collapse supernova explosion theory. *Nature*, 589 (7840): 29–39, January 2021. doi: 10.1038/s41586-020-03059-w.

39. Cadmus, R.R. The long-term spectroscopic and photometric behavior of the carbon star RS Cygni. *AJ*, 163 (6): 265, June 2022. doi: 10.3847/1538-3881/ac6590.

40. Camisassa, M.E., Althaus, L.G., Córsico, A.H. et al. The evolution of ultra-massive white dwarfs. *A&A*, 625: A87, May 2019. doi: 10.1051/0004-6361/201833822.

41. Cantiello, M., Fuller, J. and Bildsten, L. Asteroseismic signatures of evolving internal stellar magnetic fields. *ApJ*, 824 (1): 14, June 2016. doi: 10.3847/0004-637X/824/1/14.

42. Caputo, F., Bono, G., Fiorentino, G. et al. Pulsation and evolutionary masses of classical cepheids. I. Milky Way variables. *ApJ*, 629 (2): 1021–1033, August 2005. doi: 10.1086/431641.

43. Carlsson, M., De Pontieu, B. and Hansteen, V.H. New view of the solar chromosphere. *ARA&A*, 57: 189–226, August 2019. doi: 10.1146/annurev-astro-081817-052044.

44. Carroll, B. and Ostlie, D.A. *An Introduction to Modern Astrophysics and Cosmology*. 2006.

45. Casagrande, L. and VandenBerg, D.A. On the use of Gaia magnitudes and new tables of bolometric corrections. *MNRAS*, 479 (1): L102–L107, September 2018. doi: 10.1093/mnrasl/sly104.

46. Casagrande, L., Lin, J., Rains, A.D. et al. The GALAH survey: effective temperature calibration from the infrared flux method in the Gaia system. *MNRAS*, 507 (2): 2684–2696, October 2021. doi: 10.1093/mnras/stab2304.

47. Chabrier, G., Baraffe, I., Leconte, J. et al. The Mass-Radius Relationship from Solar-Type Stars to Terrestrial Planets: A Review. In Stempels, E., editor, *15th Cambridge Workshop on Cool Stars, Stellar Systems, and the Sun*, volume 1094 of *American Institute of Physics Conference Series*, pages 102–111, February 2009. doi: 10.1063/1.3099078.

48. Chandrasekhar, S. The maximum mass of ideal white dwarfs. *ApJ*, 74: 81, July 1931. doi: 10.1086/143324.

49. Chandrasekhar, S. *An Introduction to the Study of Stellar Structure*. 1939.

50. Charbonneau, P. Dynamo models of the solar cycle. *Living Rev. Sol. Phys.*, 17 (1): 4, June 2020. doi: 10.1007/s41116-020-00025-6.

51. Chen, Y., Girardi, L., Bressan, A. et al. Improving PARSEC models for very low mass stars. *MNRAS*, 444 (3): 2525–2543, November 2014. doi: 10.1093/mnras/stu1605.

52. Cherchneff, I. The formation of polycyclic aromatic hydrocarbons in evolved circumstellar environments. In Joblin, C. and Tielens, A.M., editors, *EAS Publications Series*, volume 46 of *EAS Publications Series*, pages 177–189, March 2011. doi: 10.1051/eas/1146019.

53. Cherepashchuk, A., Postnov, K., Molkov, S. et al. SS433: A massive X-ray binary in an advanced evolutionary stage. *New A. Rev.*, 89: 101542, September 2020. doi: 10.1016/j.newar.2020.101542.

54. Chevance, M., Krumholz, M.R., McLeod, A.F. et al. The life and times of giant molecular clouds. *arXiv e-prints*, art. arXiv:2203.09570, March 2022.

55. Cho, I.H., Cho, K.S., Bong, S.C. et al. Determination of the Alfvén speed and plasma-beta using the seismology of sunspot umbra. *ApJ*, 837 (1): L11, March 2017. doi: 10.3847/2041-8213/aa611b.

56. Choi, J., Dotter, A., Conroy, C. et al. Mesa isochrones and stellar tracks (MIST). I. Solar-scaled models. *ApJ*, 823 (2): 102, June 2016. doi: 10.3847/0004-637X/823/2/102.

57. Chomiuk, L., Metzger, B. and Shen, K.J. New insights into classical novae. *ARA&A*, 59: 391–444, September 2021. doi: 10.1146/annurev-astro-112420-114502.

58. Choudhuri, A.R. The meridional circulation of the Sun: Observations, theory and connections with the solar dynamo. *Sci. China Phys. Mech. Astron.*, 64 (3): 239601, March 2021. doi: 10.1007/s11433-020-1628-1.

59. Christensen-Dalsgaard, J., Dappen, W., Ajukov, S.V. et al. The current state of solar modeling. *Science*, 272 (5266): 1286–1292, May 1996. doi: 10.1126/science.272.5266.1286.

60. Christensen-Dalsgaard, J. Helioseismology. *Rev. Mod. Phys.*, 74 (4): 1073–1129, November 2002. doi: 10.1103/RevModPhys.74.1073.

61. Clark, D.B. The ideal gas law at the center of the sun. *J. Chem. Educ.*, 66: 826, 1989. doi: 10.1021/ed066p826.

62. Clementel, N., Madura, T.I., Kruip, C.J.H. et al. 3D radiative transfer simulations of Eta Carinae's inner colliding winds - I. Ionization structure of helium at apastron. *MNRAS*, 447 (3): 2445–2458, March 2015. doi: 10.1093/mnras/stu2614.

63. Colombo, S., Orlando, S., Peres, G. et al. New view of the corona of classical T Tauri stars: Effects of flaring activity in circumstellar disks. *A&A*, 624: A50, April 2019. doi: 10.1051/0004-6361/201834342.

64. Conan Doyle, A. *The Sign of Four*. Penguin Classics, reprinted from 1890 original, 2014. ISBN 9780141395487.

65. Condon, J. and Ransom, S.M. *Essential Radio Astronomy*. 2016.

66. Cox, A.N. *Allen's Astrophysical Quantities*. 2000.

67. Cranmer, S. and Winebarger, A.R. The properties of the solar corona and its connection to the solar wind. *ARA&A*, 57: 157–187, August 2019. doi: 10.1146/annurev-astro-091918-104416.

68. Crowther, P.A., Caballero-Nieves, S.M., Bostroem, K.A. et al. The R136 star cluster dissected with Hubble Space Telescope/STIS. I. Far-ultraviolet spectroscopic census and the origin of He II λ1640 in young star clusters. *MNRAS*, 458 (1): 624–659, May 2016. doi: 10.1093/mnras/stw273.

69. Cummings, J., Kalirai, J., Tremblay, P. and Ramírez-Ruiz, E. White dwarfs in star clusters: the initial-final mass relation for stars from 0.85 to 8 $M_\odot$. In *American Astronomical Society Meeting Abstracts #231*, volume 231 of *American Astronomical Society Meeting Abstracts*, page 233.01, January 2018.

70. Curtis, Jason Lee, Agüeros, M.A., Matt, S.P. et al. When do stalled stars resume spinning down? Advancing gyrochronology with Ruprecht 147. *ApJ*, 904 (2): 140, December 2020. doi: 10.3847/1538-4357/abbf58.

71. De Gerónimo, F.C., Althaus, L.G., Córsico, A.H. et al. Asteroseismology of ZZ Ceti stars with fully evolutionary white dwarf models. I. The impact of the uncertainties from prior evolution on the period spectrum. *A&A*, 599: A21, March 2017. doi: 10.1051/0004-6361/201629806.

72. De Luca, A., Caraveo, P.A., Mereghetti, S. et al. A long-period, violently variable x-ray source in a young supernova remnant. *Science*, 313 (5788): 814–817, August 2006. doi: 10.1126/science.1129185.

73. de Mink, S.E., Langer, N., Izzard, R.G. et al. The rotation rates of massive stars: the role of binary interaction through tides, mass transfer, and mergers. *ApJ*, 764 (2): 166, February 2013. doi: 10.1088/0004-637X/764/2/166.

74. De Rosa, M. and Toomre, J. Evolution of solar supergranulation. *ApJ*, 616 (2): 1242–1260, December 2004. doi: 10.1086/424920.

75. DeForest, C.E., Howard, T. and McComas, D.J. Inbound waves in the solar corona: a direct indicator of Alfvén surface location. *ApJ*, 787 (2): 124, June 2014. doi: 10.1088/0004-637X/787/2/124.

76. Degl'Innocenti, S. Stellar evolution and the standard solar model. In *European Physical Journal Web of Conferences*, volume 227 of *European Physical Journal Web of Conferences*, page 01004, June 2020. doi: 10.1051/epjconf/202022701004.

77. Di Francesco, J., Myers, P.C., Wilner, D.J. et al. Infall, outflow, rotation, and turbulent motions of dense gas within NGC 1333 IRAS 4. *ApJ*, 562 (2): 770–789, December 2001. doi: 10.1086/323854.

78. Diehl, R., Halloin, H., Kretschmer, K. et al. Radioactive ^{26}Al from massive stars in the galaxy. *Nature*, 439 (7072): 45–47, January 2006. doi: 10.1038/nature04364.

79. Dieterich, S.B., Henry, T.J., Jao, W. et al. The solar neighborhood. XXXII. The hydrogen burning limit. *AJ*, 147 (5): 94, May 2014. doi: 10.1088/0004-6256/147/5/94.

80. Dotter, A. MESA isochrones and stellar tracks (MIST) 0: Methods for the construction of stellar isochrones. *ApJs*, 222 (1): 8, January 2016. doi: 10.3847/0067-0049/222/1/8.

81. Drewes, M., McDonald, J., Sablon, L. and Vitagliano, E. Neutrino emissivities as a probe of the internal magnetic fields of white dwarfs. *ApJ*, 934 (2): 99, August 2022. doi: 10.3847/1538-4357/ac7874.

82. Duchêne, G. and Kraus, A. Stellar multiplicity. *ARA&A*, 51 (1): 269–310, August 2013. doi: 10.1146/annurev-astro-081710-102602.

83. Duerbeck, H.W. Novae: an historical perspective. In Bode, Michael F. and Evans, Aneurin, editors, *Classical Novae*, pages 1–13. Cambridge University Press, 2nd edition, 2008. An optional note.

84. Dunham, M.M., Stutz, A.M., Allen, L.E. et al. The evolution of protostars: insights from ten years of infrared surveys with Spitzer and Herschel. In Beuther, H., Klessen, R.S., Dullemond, C.P., and Henning, T. editors, *Protostars and Planets VI*, page 195, January 2014. doi: 10.2458/azu_uapress_9780816531240-ch009.

85. Duvert, G. VizieR online data catalog: JMDC: JMMC measured stellar diameters catalogue (Duvert, 2016). Art. II/345, November 2016.

86. Eddington, A.S. *The Internal Constitution of the Stars* (Reissued in the Cambridge Science Classics Series from the original 1926 edition). Cambridge University Press, 1988. ISBN 0521337089.

87. Edgar, J.S., editor. *Observer's Handbook 2021. The Royal Astronomical Society of Canada*, 2021.

88. Eker, Z., Soydugan, F., Soydugan, E. et al. Main-sequence effective temperatures from a revised mass-luminosity relation based on accurate properties. *AJ*, 149 (4): 131, April 2015. doi: 10.1088/0004-6256/149/4/131.

89. Eker, Z., Bakış, V., Bilir, S. et al. Interrelated main-sequence mass-luminosity, mass-radius, and mass-effective temperature relations. *MNRAS*, 479 (4): 5491–5511, October 2018. doi: 10.1093/mnras/sty1834.

90. Ekström, S. Massive star modelling and nucleosynthesis. *FSPAS*, 8: 53, April 2021. doi: 10.3389/fspas.2021.617765.

91. Elliott, A., Richardson, N.D., Pablo, H. et al. 5 yr of BRITE-Constellation photometry of the luminous blue variable P Cygni: properties of the stochastic low-frequency variability. *MNRAS*, 509 (3): 4246–4255, January 2022. doi: 10.1093/mnras/stab3112.

92. Event Horizon Telescope Collaboration, Akiyama, K., Alberdi, A. et al. First M87 event horizon telescope results. VI. The shadow and mass of the central black hole. *ApJ*, 875 (1): L6, April 2019. doi: 10.3847/2041-8213/ab1141.

93. Event Horizon Telescope Collaboration, Akiyama, K. Alberdi, A. et al. First Sagittarius A* event horizon telescope results. I. The shadow of the supermassive black hole in the center of the Milky Way. *ApJ*, 930 (2): L12, May 2022. doi: 10.3847/2041-8213/ac6674.

94. Eyer, L. and Mowlavi, N. Variable stars across the observational HR diagram. In *Journal of Physics Conference Series*, volume 118 of *Journal of Physics Conference Series*, page 012010, October 2008. doi: 10.1088/1742-6596/118/1/012010.

95. Fadeyev, Y.A. Evolutionary status of Polaris. *MNRAS*, 449 (1): 1011–1017, May 2015. doi: 10.1093/mnras/stv412.

96. Falceta-Gonçalves, D., Kowal, G., Falgarone, E. and Chian, A.C.L. Turbulence in the interstellar medium. *Nonlinear Process. Geophys.*, 21 (3): 587–604, May 2014. doi: 10.5194/npg-21-587-2014.

97. Fan, Y. Magnetic fields in the solar convection zone. *Living Rev. Sol. Phys.*, 6 (1): 4, December 2009. doi: 10.12942/lrsp-2009-4.

98. Farag, E., Timmes, F.X., Taylor, M. et al. On stellar evolution in a neutrino Hertzsprung-Russell diagram. *ApJ*, 893 (2): 133, April 2020. doi: 10.3847/1538-4357/ab7f2c.

99. Federrath, C. Inefficient star formation through turbulence, magnetic fields and feedback. *MNRAS*, 450 (4): 4035–4042, July 2015. doi: 10.1093/mnras/stv941.

100. Figer, D.F. An upper limit to the masses of stars. *Nature*, 434 (7030): 192–194, March 2005. doi: 10.1038/nature03293.

101. Firth, R.E., Sullivan, M., Gal-Yam, A. et al. The rising light curves of Type Ia supernovae. *MNRAS*, 446 (4): 3895–3910, February 2015. doi: 10.1093/mnras/stu2314.

102. Fontaine, G. and Brassard, P. The pulsating white dwarf stars. *PASP*, 120 (872): 1043, October 2008. doi: 10.1086/592788.

103. Fossat, E., Regulo, C., Roca Cortes, T. et al. On the acoustic cut-off frequency of the sun. *A&A*, 266 (1): 532–536, December 1992.

104. Fossat, E., Boumier, P., Corbard, T. et al. Asymptotic g modes: Evidence for a rapid rotation of the solar core. *A&A*, 604: A40, August 2017. doi: 10.1051/0004-6361/201730460.

105. Fredslund Andersen, M., Pallé, P., Jessen-Hansen, J. et al. Oscillations in the Sun with SONG: Setting the scale for asteroseismic investigations. *A&A*, 623: L9, March 2019. doi: 10.1051/0004-6361/201935175.

106. Gaia Collaboration, Babusiaux, C., van Leeuwen, F. et al. Gaia data release 2. Observational Hertzsprung-Russell diagrams. *A&A*, 616: A10, August 2018. doi: 10.1051/0004-6361/201832843.

107. Gaia Collaboration, Eyer, L., Rimoldini, L. et al. Gaia data release 2. Variable stars in the colour-absolute magnitude diagram. *A&A*, 623: A110, March 2019. doi: 10.1051/0004-6361/201833304.

108. Galilei, G. *Dialogue Concerning the Two Chief World Systems*. The Modern Library, New York, 1630, as published in 2001 in a Modern Library Paperback Edition. ISBN 0-375-75766-X.

109. García, R. and Ballot, J. Asteroseismology of solar-type stars. *Living Rev. Sol. Phys.*, 16 (1): 4, September 2019. doi: 10.1007/s41116-019-0020-1.

110. García-Berro, E. and Oswalt, T.D. The white dwarf luminosity function. *New A. Rev.*, 72: 1–22, June 2016. doi: 10.1016/j.newar.2016.08.001.

111. Gariazzo, S., Archidiacono, M., de Salas, P.F. et al. Neutrino masses and their ordering: global data, priors and models. *J. Cosmol. Astropart. Phys.*, 2018 (3): 011, March 2018. doi: 10.1088/1475-7516/2018/03/011.

112. Gary, G.A. Plasma beta above a solar active region: rethinking the paradigm. *Sol. Phys.*, 203 (1): 71–86, October 2001. doi: 10.1023/A:1012722021820.

113. Gautschy, A. Helium ignition in the cores of low-mass stars. *arXiv e-prints*, art. arXiv:1208.3870, August 2012.

114. Gautschy, A. The theorem that was none - I. Early history. *arXiv e-prints*, art. arXiv:1504.08188, April 2015.

115. Gautschy, A. The theorem that was none - II. The profound 70s. *arXiv e-prints*, art. arXiv:1812.11864, December 2018.

116. Geldard, R. *Anaxagoras and Universal Mind*. Ralph Waldo Emerson Institute Books, New York, 2007. ISBN 978-1-6058-403-8. doi: 10.1142/5171.

117. Gibson, S.E. Solar prominences: theory and models. Fleshing out the magnetic skeleton. *Living Rev. Sol. Phys.*, 15 (1): 7, October 2018. doi: 10.1007/s41116-018-0016-2.

118. Gies, D.R., Dieterich, S., Richardson, N.D. et al. A spectroscopic orbit for Regulus. *ApJ*, 682 (2): L117, August 2008. doi: 10.1086/591148.

119. Gizon, L., Cameron, R.H., Pourabdian, M. et al. Meridional flow in the Sun's convection zone is a single cell in each hemisphere. *Science*, 368 (6498): 1469–1472, June 2020. doi: 10.1126/science.aaz7119.

120. Glatzmaier, G. *Introduction to Modeling Convection in Planets and Stars: Magnetic Field, Density Stratification, RotationMagnetic Field, Density Stratification, Rotation*. 2014. doi: 10.23943/princeton/9780691141725.001.0001.

121. Gold, T. Rotating neutron stars as the origin of the pulsating radio sources. *Nature*, 218 (5143): 731–732, May 1968. doi: 10.1038/218731a0.

122. González-Avilés, J.J., Guzmán, F.S., Fedun, V. et al. I. Jet formation and evolution due to 3D magnetic reconnection. *ApJ*, 856 (2): 176, April 2018. doi: 10.3847/1538-4357/aab36f.

123. González-Gaitán, S., Tominaga, N., Molina, J. et al. The rise-time of Type II supernovae. *MNRAS*, 451 (2): 2212–2229, August 2015. doi: 10.1093/mnras/stv1097.

124. Gray, D.F. *The Observation and Analysis of Stellar Photospheres*. 2005.

125. Gray, D.F. Spectroscopy of the K0 binary giant α UMa. *ApJ*, 869 (1): 81, December 2018. doi: 10.3847/1538-4357/aae9e6.

126. Greene, J.E., Strader, J. and Ho, L.C. Intermediate-mass black holes. *ARA&A*, 58: 257–312, August 2020. doi: 10.1146/annurev-astro-032620-021835.

127. Grenier, I. and Harding, A.K. Gamma-ray pulsars: A gold mine. *Comptes Rendus Physique*, 16 (6–7): 641–660, August 2015. doi: 10.1016/j.crhy.2015.08.013.

128. Grosdidier, Y., Moffat, A.F.J., Joncas, G. and Acker, A. HST WFPC2/Hα imagery of the nebula M1-67: A clumpy LBV wind imprinting itself on the nebular structure? *ApJ*, 506 (2): L127–L131, October 1998. doi: 10.1086/311647.

129. Großschedl, J.E., Alves, J., Meingast, S. and Hasenberger, B. 3D shape of Orion A with Gaia DR2. An informed view on star formation rates and efficiencies. *arXiv e-prints*, art. arXiv:1812.08024, December 2018.

130. Gudiksen, B.V., Carlsson, M., Hansteen, V.H. et al. The stellar atmosphere simulation code Bifrost. Code description and validation. *A&A*, 531: A154, July 2011. doi: 10.1051/0004-6361/201116520.

131. Gull, T.R., Navarete, F., Corcoran, M.F. et al. Eta Carinae: A tale of two periastron passages. *ApJ*, 923 (1): 102, December 2021. doi: 10.3847/1538-4357/ac22a6.

132. Haberreiter, M., Schmutz, W. and Kosovichev, A.G. Solving the discrepancy between the seismic and photospheric solar radius. *ApJ*, 675 (1): L53, March 2008. doi: 10.1086/529492.

133. Hamacher, D.W. Are supernovae recorded in indigenous astronomical traditions? *J. Astron. Hist. Herit.*, 17 (2): 161–170, July 2014.

134. Hamann, W.R., Gräfener, G., Liermann, A. et al. The galactic WN stars revisited. Impact of Gaia distances on fundamental stellar parameters. *A&A*, 625: A57, May 2019. doi: 10.1051/0004-6361/201834850.

135. Handler, G. Asteroseismology. In Oswalt, T. and Barstow, M.A., editors, *Planets, Stars and Stellar Systems. Volume 4: Stellar Structure and Evolution*, page 207. 2013. doi: 10.1007/978-94-007-5615-1_4.

136. Harper, G.M., Brown, A., Guinan, E.F. et al. An updated 2017 astrometric solution for Betelgeuse. *AJ*, 154 (1): 11, July 2017. doi: 10.3847/1538-3881/aa6ff9.

137. Harper, G.M., Brown, A. and Guinan, E.F. A new VLA-Hipparcos distance to Betelgeuse and its implications. *AJ*, 135 (4): 1430–1440, April 2008. doi: 10.1088/0004-6256/135/4/1430.

138. Hathaway, D. and Upton, L. The solar meridional circulation and sunspot cycle variability. *J. Geophys. Res. Space Phys.*, 119 (5): 3316–3324, May 2014. doi: 10.1002/2013JA019432.

139. Hayashi, C. Stellar evolution in early phases of gravitational contraction. *PASJ*, 13: 450–452, January 1961.

140. Hayashi, C. and Nakano, T. Thermal and dynamical properties of a protostar and its contraction to the stage of quasi-static equilibrium. *Prog. Theor. Phys.*, 34 (5): 754–775, November 1965. doi: 10.1143/PTP.34.754.

141. He, S., Wang, L. and Huang, J.Z. Characterization of type Ia supernova light curves using principal component analysis of sparse functional data. *ApJ*, 857 (2): 110, April 2018. doi: 10.3847/1538-4357/aab0a8.

142. Heger, A., Langer, N. and Woosley, S.E. Presupernova evolution of rotating massive stars. I. Numerical method and evolution of the internal stellar structure. *ApJ*, 528 (1): 368–396, January 2000. doi: 10.1086/308158.

143. Hennebelle, P. and Inutsuka, S. The role of magnetic field in molecular cloud formation and evolution. *FSPAS*, 6: 5, March 2019. doi: 10.3389/fspas.2019. 00005.

144. Hewett, P.C., Warren, S.J., Leggett, S. and Hodgkin, S.T. The UKIRT Infrared Deep Sky Survey ZY JHK photometric system: passbands and synthetic colours. *MNRAS*, 367 (2): 454–468, April 2006. doi: 10.1111/j.1365-2966.2005.09969.x.

145. Hewish, A., Bell, S.J., Pilkington, J.D.H. et al. Observation of a rapidly pulsating radio source. *Nature*, 217 (5130): 709–713, February 1968. doi: 10.1038/ 217709a0.

146. Hopkins, A.M. The Dawes review 8: Measuring the stellar initial mass function. *PASA*, 35: e039, November 2018. doi: 10.1017/pasa.2018.29.

147. Horowitz, C.J., Berry, D.K., Briggs, C.M. et al. Disordered nuclear pasta, magnetic field decay, and crust cooling in neutron stars. *Phys. Rev. Lett.*, 114 (3): 031102, January 2015. doi: 10.1103/PhysRevLett.114.031102.

148. Horowitz, C.J., Piekarewicz, J. and Reed, B. Insights into nuclear saturation density from parity-violating electron scattering. *Phys. Rev. C*, 102 (4): 044321, October 2020. doi: 10.1103/PhysRevC.102.044321.

149. Hosokawa, T. and Omukai, K. Evolution of massive protostars with high accretion rates. *ApJ*, 691 (1): 823–846, January 2009. doi: 10.1088/0004-637X/691/ 1/823.

150. Houdek, G. *Pulsation of solar-type stars*. PhD thesis, June 1996.

151. Howe, R. Solar interior rotation and its variation. *Living Rev. Sol. Phys.*, 6 (1): 1, December 2009. doi: 10.12942/lrsp-2009-1.

152. Iaria, R., D'Aí, A., Lavagetto, G., Di Salvo, T. et al. Chandra observation of Cir X-1 near the periastron passage: evidence for an x-ray jet? *ApJ*, 673 (2): 1033–1043, February 2008. doi: 10.1086/524311.

153. Irwin, J.A. *Astrophysics: Decoding the Cosmos*. Wiley Interscience, 2007.

154. Irwin, Judith Ann. *Astrophysics: Decoding the Cosmos*, 2e. Wiley, 2021.

155. Isern, J., Artigas, A. and García-Berro, E. White dwarf cooling sequences and cosmochronology. In *European Physical Journal Web of Conferences*, volume 43 of *European Physical Journal Web of Conferences*, page 05002, March 2013. doi: 10.1051/epjconf/20134305002.

156. Jackson, J.D. *Classical Electrodynamics*, 3e. John Wiley & Sons, Inc., 1999.

157. Jacoutot, L., Kosovichev, A.G., Wray, A. and Mansour, N.N. Numerical simulation of excitation of solar oscillation modes for different turbulent models. *ApJ*, 682 (2): 1386–1391, August 2008. doi: 10.1086/589226.

158. Jenkins, J.S., Díaz, M., Jones, H.R.A. et al. The observed distribution of spectroscopic binaries from the Anglo-Australian planet search. *MNRAS*, 453 (2): 1439–1457, October 2015. doi: 10.1093/mnras/stv1596.

159. Jetsu, L. and Porceddu, S. Shifting milestones of natural sciences: the ancient Egyptian discovery of Algol's period confirmed. *PLoS ONE*, December 2015. doi: 10.1371/journal.pone.0144140.

160. Jofré, P., Heiter, U., Soubiran, C. et al. Gaia FGK benchmark stars: metallicity. *A&A*, 564: A133, April 2014. doi: 10.1051/0004-6361/201322440.

161. Joyce, S.R.G., Barstow, M.A., Casewell, S.L. et al. Testing the white dwarf mass-radius relation and comparing optical and far-UV spectroscopic results with Gaia DR2, HST, and FUSE. *MNRAS*, 479 (2): 1612–1626, September 2018. doi: 10.1093/mnras/sty1425.

162. Judge, P. Magnetic connections across the chromosphere-corona transition region. *ApJ*, 914 (1): 70, June 2021. doi: 10.3847/1538-4357/abf8ad.

163. Kantor, E. and Gusakov, M.E. The neutrino emission due to plasmon decay and neutrino luminosity of white dwarfs. *MNRAS*, 381 (4): 1702–1710, November 2007. doi: 10.1111/j.1365-2966.2007.12342.x.

164. Kaspi, V.M., Roberts, M.E., Vasisht, G. et al. Chandra x-ray observations of G11.2-0.3: implications for pulsar ages. *ApJ*, 560 (1): 371–377, October 2001. doi: 10.1086/322515.

165. Kaspi, V. and Kramer, M. Radio pulsars: the neutron star population & fundamental physics. *arXiv e-prints*, art. arXiv:1602.07738, February 2016.

166. Kippenhahn, R., Weigert, A. and Weiss, A. *Stellar Structure and Evolution*. Springer, 2012. doi: 10.1007/978-3-642-30304-3.

167. Kirk, B., Conroy, K., Prša, A. et al. Kepler eclipsing binary stars. VII. The catalog of eclipsing binaries found in the entire Kepler data set. *AJ*, 151 (3): 68, March 2016. doi: 10.3847/0004-6256/151/3/68.

168. Kirsch, M.G.F., Becker, W., Benlloch-Garcia, S. et al. Timing accuracy and capabilities of XMM-Newton. In Flanagan, K. and Siegmund, O.H.W., editors, *X-Ray and Gamma-Ray Instrumentation for Astronomy XIII*, volume 5165 of *Society of Photo-Optical Instrumentation Engineers (SPIE) Conference Series*, pages 85–95, February 2004. doi: 10.1117/12.503559.

169. Kissin, Y. and Thompson, C. Rotation and magnetism of massive stellar cores. *ApJ*, 862 (2): 111, August 2018. doi: 10.3847/1538-4357/aab1fb.

170. Klishin, A. and Chilingarian, I. Explaining the stellar initial mass function with the theory of spatial networks. *ApJ*, 824 (1): 17, June 2016. doi: 10.3847/0004-637X/824/1/17.

171. Koester, D. and Chanmugam, G. Review: physics of white dwarf stars. *Rep. Prog. Phys.*, 53 (7): 837–915, July 1990. doi: 10.1088/0034-4885/53/7/001.

172. Komm, R., Howe, R. and Hill, F. Subsurface zonal and meridional flow during cycles 23 and 24. *Sol. Phys.*, 293 (10): 145, October 2018. doi: 10.1007/s11207-018-1365-7.

173. Korenaga, J. Can mantle convection be self-regulated? *Sci. Adv.*, 2 (8): e1601168, 2016. doi: 10.1126/sciadv.1601168.

174. Kosenko, D., Vink, J., Blinnikov, S. and Rasmussen, A. XMM-Newton X-ray spectra of the SNR 0509-67.5: data and models. *A&A*, 490 (1): 223–230, October 2008. doi: 10.1051/0004-6361:200809495.

175. Kozyrev, N. Sources of stellar energy and the theory of the internal constitution of stars. *Progress in Physics*, 1 (3): 61–99, January 2005.

176. Krastev, P. and Li, B. Imprints of the nuclear symmetry energy on the tidal deformability of neutron stars. *J. Phys. G Nucl. Part. Phys.*, 46 (7): 074001, July 2019. doi: 10.1088/1361-6471/ab1a7a.

177. Kroupa, P., Weidner, C., Pflamm-Altenburg, J. et al. The stellar and sub-stellar initial mass function of simple and composite populations. In Oswalt, T. and Gilmore, G. editors, *Planets, Stars and Stellar Systems. Volume 5: Galactic Structure and Stellar Populations*, page 115. 2013. doi: 10.1007/978-94-007-5612-0_4.

178. Kuhn, J.R., Bush, R., Emilio, M. and Scholl, I.F. The precise solar shape and its variability. *Science*, 337 (6102): 1638, September 2012. doi: 10.1126/science.1223231.

179. Kunitomo, M., Guillot, T., Takeuchi, T. and Ida, S. Revisiting the pre-main-sequence evolution of stars. I. Importance of accretion efficiency and deuterium abundance. *A&A*, 599: A49, March 2017. doi: 10.1051/0004-6361/201628260.

180. Kupka, F., Zaussinger, F. and Montgomery, M.H. Mixing and overshooting in surface convection zones of DA white dwarfs: first results from ANTARES. *MNRAS*, 474 (4): 4660–4671, March 2018. doi: 10.1093/mnras/stx3119.

181. Kurtz, D.W. Stellar pulsation: an overview. In Aerts, C. and Sterken, C., editors, *Astrophysics of Variable Stars*, volume 349 of *Astronomical Society of the Pacific Conference Series*, page 101, April 2006.

182. Kurucz, R.L., Furenlid, I., Brault, J. and Testerman, L. *Solar flux atlas from 296 to 1300 nm*. National Solar Observatory, 1984.

183. Kwitter, K. and Henry, R.B.C. Planetary nebulae: sources of enlightenment. *PASP*, 134 (1032): 022001, February 2022. doi: 10.1088/1538-3873/ac32b1.

184. Kwok, S. *Cosmic Butterflies*. Cambridge University Press, 2001.

185. Kwok, S. Planetary nebulae as sources of chemical enrichment of the galaxy. *FSPAS*, 9: 893061, May 2022. doi: 10.3389/fspas.2022.893061.

186. Lamers, H.L.M. and Levesque, E.M. *Understanding Stellar Evolution*. IOP Publishing, 2017. doi: 10.1088/978-0-7503-1278-3.

187. Lang, K.R. *Astrophysical Formulae*. Springer, 1999.

188. Laughlin, G., Bodenheimer, P. and Adams, F.C. The end of the main sequence. *ApJ*, 482 (1): 420–432, June 1997. doi: 10.1086/304125.

189. Leavitt, H. and Pickering, E.C. Periods of 25 variable stars in the small Magellanic cloud. *Harvard College Observatory Circular*, 173: 1–3, March 1912.

190. LeBlanc, F. *An Introduction to Stellar Astrophysics*. Wiley, 2010.

191. Leibundgut, B. and Suntzeff, N.B. Optical light curves of supernovae. In Weiler, K., editor, *Supernovae and Gamma-Ray Bursters*, volume 598, pages 77–90. 2003. doi: 10.1007/3-540-45863-86.

192. Lemen, J.R., Title, A.M., Akin, D.J. et al. The atmospheric imaging assembly (AIA) on the solar dynamics observatory (SDO). *Sol. Phys.*, 275 (1-2): 17–40, January 2012. doi: 10.1007/s11207-011-9776-8.

193. Levesque, E. and Massey, P. Betelgeuse just is not that cool: effective temperature alone cannot explain the recent dimming of Betelgeuse. *ApJ*, 891 (2): L37, March 2020. doi: 10.3847/2041-8213/ab7935.

194. Li, W., Leaman, J., Chornock, R. et al. Nearby supernova rates from the Lick Observatory Supernova Search - II. The observed luminosity functions and fractions of supernovae in a complete sample. *MNRAS*, 412 (3): 1441–1472, April 2011. doi: 10.1111/j.1365-2966.2011.18160.x.

195. Ligi, R., Creevey, O., Mourard, D. et al. Radii, masses, and ages of 18 bright stars using interferometry and new estimations of exoplanetary parameters. *A&A*, 586: A94, February 2016. doi: 10.1051/0004-6361/201527054.

196. Livingstone, M.A., Kaspi, V.M., Gavriil, F.P. et al. New phase-coherent measurements of pulsar braking indices. *Ap&SS*, 308 (1–4): 317–323, April 2007. doi: 10.1007/s10509-007-9320-3.

197. Lopez, L. and Fesen, R.A. The morphologies and kinematics of supernova remnants. *Space Sci. Rev.*, 214 (1): 44, February 2018. doi: 10.1007/s11214-018-0481-x.

198. Lusso, E., Piedipalumbo, E., Risaliti, G. et al. Tension with the flat ΛCDM model from a high-redshift Hubble diagram of supernovae, quasars, and gamma-ray bursts. *A&A*, 628: L4, August 2019. doi: 10.1051/0004-6361/201936223.

199. MacDonald, J. *The Arctic Sky – Inuit Astronomy, Star Lore, and Legend*. Nunavut Research Institute and Royal Ontario Museum, Toronto, Ontario, 1998.

200. Machida, M. and Nakamura, T. Accretion phase of star formation in clouds with different metallicities. *MNRAS*, 448 (2): 1405–1429, April 2015. doi: 10.1093/mnras/stu2633.

201. Maciel, W.J. *Introduction to Stellar Structure*. Springer Nature, 2016. doi: https://doi.org/10.1007/978-3-319-16142-6.

202. Madore, B. and Freedman, W.L. The cepheid distance scale. *PASP*, 103: 933, September 1991. doi: 10.1086/132911.

203. Madore, B.F., Freedman, W. and Moak, S. A method for improving galactic cepheid reddenings and distances. *ApJ*, 842 (1): 42, June 2017. doi: 10.3847/1538-4357/aa6e4d.

204. Mamajek, E.E., Torres, G., Prsa, A. et al. IAU 2015 resolution B2 on recommended zero points for the absolute and apparent bolometric magnitude scales. *arXiv e-prints*, art. arXiv:1510.06262, October 2015.

205. Manchester, R.N. Pulsar timing and its applications. In *Journal of Physics Conference Series*, volume 932 of *Journal of Physics Conference Series*, page 012002, December 2017. doi: 10.1088/1742-6596/932/1/012002.

206. Manchester, R.N., Hobbs, G.B., Teoh, A. and Hobbs, M. The Australia telescope national facility pulsar catalogue. *AJ*, 129 (4): 1993–2006, April 2005. doi: 10.1086/428488.

207. Maoz, D., Mannucci, F. and Nelemans, G. Observational clues to the progenitors of type Ia supernovae. *ARA&A*, 52: 107–170, August 2014. doi: 10.1146/annurev-astro-082812-141031.

208. Marek, A. and Janka, H.T. Delayed neutrino-driven supernova explosions aided by the standing accretion-shock instability. *ApJ*, 694 (1): 664–696, March 2009. doi: 10.1088/0004-637X/694/1/664.

209. Marigo, P., Cummings, J.D., Curtis, J.L. et al. Carbon star formation as seen through the non-monotonic initial-final mass relation. *Nat. Astron.*, 4: 1102–1110, July 2020. doi: 10.1038/s41550-020-1132-1.

210. Martins, L. and Coelho, P. Grids of synthetic stellar spectra. *Canadian Journal of Physics*, 95 (9): 840–842, 2017. doi: 10.1139/cjp-2016-0896. URL https://doi.org/10.1139/cjp-2016-0896.

211. Matsunaga, N. Feast, M. and Menzies, J.W. Period-luminosity relations for type II Cepheids and their application. *MNRAS*, 397 (2): 933–942, August 2009. doi: 10.1111/j.1365-2966.2009.14992.x.

212. Maund, J.R., Crowther, P.A., Janka, H. and Langer, N. Bridging the gap: from massive stars to supernovae. *Philos. Trans. Royal Soc. A*, 375 (2105): 20170025, September 2017. doi: 10.1098/rsta.2017.0025.

213. Maxted, P.F.L., Hutcheon, R.J., Torres, G. et al. Precise mass and radius measurements for the components of the bright solar-type eclipsing binary star V1094 Tauri. *A&A*, 578: A25, June 2015. doi: 10.1051/0004-6361/201525873.

214. Mayor, M. and Queloz, D. A Jupiter-mass companion to a solar-type star. *Nature*, 378 (6555): 355–359, November 1995. doi: 10.1038/378355a0.

215. McAlister, H.A., ten Brummelaar, T.A., Gies, D.R. et al. First results from the CHARA Array. I. An interferometric and spectroscopic study of the fast rotator α Leonis (Regulus). *ApJ*, 628 (1): 439–452, July 2005. doi: 10.1086/430730.

216. McQuillan, A., Mazeh, T. and Aigrain, S. Rotation periods of 34,030 Kepler main-sequence stars: the full autocorrelation sample. *ApJs*, 211 (2): 24, April 2014. doi: 10.1088/0067-0049/211/2/24.

217. Metcalfe, T. and Egeland, R. Understanding the limitations of gyrochronology for old field stars. *ApJ*, 871 (1): 39, January 2019. doi: 10.3847/1538-4357/aaf575.

218. Mihajlov, A.A., Dimitrijevic, M.S., Ignjatovic, L. and Djuric, Z. VizieR online data catalog: ion-atom collisions in sun atmosphere (Mihajlov+ 1994). *VizieR Online Data Catalog*, art. J/A+AS/103/57, November 1993.

219. Miller, J. Compressibility measurements reach white dwarf pressures. *Phys. Today*, 73 (10): 14, 2020. doi: 10.1063/PT.3.4585.

220. Miller Bertolami, M.M., Battich, T., Córsico, A.H. et al. Asteroseismic signatures of the helium core flash. *Nat. Astron.*, 4: 67–71, January 2020. doi: 10.1038/s41550-019-0890-0.

221. Mocák, M., Müller, E., Weiss, A. and Kifonidis, K. The core helium flash revisited. I. One and two-dimensional hydrodynamic simulations. *A&A*, 490 (1): 265–277, October 2008. doi: 10.1051/0004-6361:200810169.

222. Moffat, A.F.J., Lepine, S., Henriksen, R. and Robert, C. First wavelet analysis of emission line variations in Wolf-Rayet stars – turbulence in hot-star outflows. *Ap&SS*, 216 (1-2): 55–65, June 1994. doi: 10.1007/BF00982468.

223. Montargès, M., Cannon, E., Lagadec, E. et al. A dusty veil shading Betelgeuse during its great dimming. *Nature*, 594 (7863): 365–368, June 2021. doi: 10.1038/s41586-021-03546-8.

224. Munoz, M.S., Wade, G.A., Faes, D.M. et al. Untangling magnetic massive star properties with linear polarization variability and the analytic dynamical magnetosphere model. *MNRAS*, 511 (3): 3228–3249, April 2022. doi: 10.1093/mnras/stab3767.

225. Neeley, J.R., Marengo, M., Freedman, W.L. et al. Standard galactic field RR Lyrae II: a Gaia DR2 calibration of the period-Wesenheit-metallicity relation. *MNRAS*, 490 (3): 4254–4270, December 2019. doi: 10.1093/mnras/stz2814.

226. Neuhäuser, R., Torres, G., Mugrauer, M. et al. Colour evolution of Betelgeuse and Antares over two millennia, derived from historical records, as a new constraint on mass and age. *MNRAS*, July 2022. doi: 10.1093/mnras/stac1969.

227. Ng, C.Y., Romani, R.W., Brisken, W.F. et al. The origin and motion of PSR J0538+2817 in S147. *ApJ*, 654 (1): 487–493, January 2007. doi: 10.1086/510576.

228. Nielsen, M.B., Gizon, L., Schunker, H. and Karoff, C. Rotation periods of 12 000 main-sequence Kepler stars: Dependence on stellar spectral type and comparison with v sin i observations. *A&A*, 557: L10, September 2013. doi: 10.1051/0004-6361/201321912.

229. Nielsen, M.B., Gizon, L., Cameron, R. and Miesch, M. Starspot rotation rates versus activity cycle phase: Butterfly diagrams of Kepler stars are unlike that of the Sun. *A&A*, 622: A85, February 2019. doi: 10.1051/0004-6361/201834373.

230. Nordlund, Åke, Stein, R. and Asplund, M. Solar surface convection. *Living Rev. Sol. Phys.*, 6 (1): 2, December 2009. doi: 10.12942/lrsp-2009-2.

231. Norris, R.P., Baron, F.R., Monnier, J.D. et al. Long term evolution of surface features on the red supergiant AZ Cyg. *ApJ*, 919 (2): 124, October 2021. doi: 10.3847/1538-4357/ac0c7e.

232. Offner, S.S.R., Clark, P.C., Hennebelle, P. et al. The origin and universality of the stellar initial mass function. In Beuther, H., Klessen, R.S., Dullemond, C.P., and Henning, T. editors, *Protostars and Planets VI*, page 53, January 2014. doi: 10.2458/azu_uapress_9780816531240-ch003.

233. Olausen, S. and Kaspi, V.M. The McGill magnetar catalog. *ApJs*, 212 (1): 6, May 2014. doi: 10.1088/0067-0049/212/1/6.

234. Orebi Gann, G.D., Zuber, K., Bemmerer, D. and Serenelli, A. The future of solar neutrinos. *arXiv e-prints*, art. arXiv:2107.08613, July 2021.

235. Orlando, S., Ono, M., Nagataki, S. et al. Hydrodynamic simulations unravel the progenitor-supernova-remnant connection in SN 1987A. *A&A*, 636: A22, April 2020. doi: 10.1051/0004-6361/201936718.

236. Ostriker, J. and Gunn, J.E. On the nature of pulsars. I. Theory. *ApJ*, 157: 1395, September 1969. doi: 10.1086/150160.

237. Özel, F. Surface emission from neutron stars and implications for the physics of their interiors. *Rep. Prog. Phys.*, 76 (1): 016901, January 2013. doi: 10.1088/0034-4885/76/1/016901.

238. Ozuyar, D. and Stevens, I.R. The period evolution of V473 Tau. *Information Bulletin on Variable Stars*, 6245: 1, July 2018. doi: 10.22444/IBVS.6245.

239. Paladini, C., Baron, F., Jorissen, A. et al. Large granulation cells on the surface of the giant star π^1 Gruis. *Nature*, 553 (7688): 310–312, January 2018. doi: 10.1038/nature25001.

240. Parker, E.N. Hydromagnetic dynamo models. *ApJ*, 122: 293, September 1955. doi: 10.1086/146087.

241. Pascoli, G. Magnetic fields in circumstellar envelopes of evolved AGB stars. *PASP*, 132 (1009): 034203, March 2020. doi: 10.1088/1538-3873/ab54a2.

242. Passos, D., Charbonneau, P. and Miesch, M. Meridional circulation dynamics from 3D magnetohydrodynamic global simulations of solar convection. *ApJ*, 800 (1): L18, February 2015. doi: 10.1088/2041-8205/800/1/L18.

243. Paxton, B., Bildsten, L., Dotter, A. et al. Modules for experiments in stellar astrophysics (MESA). *ApJs*, 192 (1): 3, January 2011. doi: 10.1088/0067-0049/192/1/3.

244. Paxton, B., Cantiello, M., Arras, P. et al. Modules for experiments in stellar astrophysics (MESA): planets, oscillations, rotation, and massive stars. *ApJs*, 208 (1): 4, September 2013. doi: 10.1088/0067-0049/208/1/4.

245. Paxton, B., Marchant, P., Schwab, J. et al. Modules for experiments in stellar astrophysics (MESA): binaries, pulsations, and explosions. *ApJs*, 220 (1): 15, September 2015. doi: 10.1088/0067-0049/220/1/15.

246. Payne, C.H. On the spectra and temperatures of the B stars. *Nature*, 113 (2848): 783–784, May 1924. doi: 10.1038/113783a0.

247. Percy, J.R. *Understanding Variable Stars*. Cambridge University Press, 2011.

248. Pérez, L.M., Benisty, M., Andrews, S.M. et al. The disk substructures at high angular resolution project (DSHARP). X. Multiple rings, a misaligned inner disk, and a bright arc in the disk around the T Tauri star HD 143006. *ApJ*, 869 (2): L50, December 2018. doi: 10.3847/2041-8213/aaf745.

249. Perlmutter, S. Supernovae, dark energy, and the accelerating universe. *Phys. Today*, 56 (4): 53–62, April 2003. doi: 10.1063/1.1580050.

250. Pethick, C.J., Schaefer, T. and Schwenk, A. Bose-Einstein condensates in neutron stars. *arXiv e-prints*, art. arXiv:1507.05839, July 2015.

251. Phillips, A.C. *The Physics of Stars*, 2e. Wiley, 1999.

252. Phillips, K.J.H. Highly ionized Fe X-ray lines at energies 7.7-8.6 keV. *A&A*, 490 (2): 823–828, November 2008. doi: 10.1051/0004-6361:200810791.

253. Porto de Mello, G. and da Silva, L. HR 6060: the closest ever solar twin? *ApJ*, 482: L89, June 1997. doi: 10.1086/310693.

254. Pradhan, A. and Nahar, S.N. *Atomic Astrophysics and Spectroscopy*. Cambridge University Press, 2011.

255. Prialnik, D. *An Introduction to the Theory of Stellar Structure and Evolution*. Cambridge University Press, 2009.

256. Priest, E.R. Magnetohydrodynamics and solar dynamo action. In Oddbjørn, E., Vial, J.-C. and Skumanich, A., editors, *The Sun as a Guide to Stellar Physics*, pages 239–266. Elsevier, 2019. doi: 10.1016/B978-0-12-814334-6.00009-1.

257. Pudritz, R. and Ray, T.P. The role of magnetic fields in protostellar outflows and star formation. *FSPAS*, 6: 54, July 2019. doi: 10.3389/fspas.2019.00054.

258. Refsdal, S. and Weigert, A. Shell source burning stars with highly condensed cores. *A&A*, 6: 426, July 1970.

259. Remillard, R. and McClintock, J.E. X-ray properties of black-hole binaries. *ARA&A*, 44 (1): 49–92, September 2006. doi: 10.1146/annurev.astro.44.051905.092532.

260. Reynolds, S.P., Borkowski, K.J., Green, D.A. et al. The youngest galactic supernova remnant: G1.9+0.3. *ApJ*, 680 (1): L41, June 2008. doi: 10.1086/589570.

261. Rezzolla, L., Most, E. and Weih, L.R. Using gravitational-wave observations and quasi-universal relations to constrain the maximum mass of neutron stars. *ApJ*, 852 (2): L25, January 2018. doi: 10.3847/2041-8213/aaa401.

262. Ricca, A., Bauschlicher, C.W., Jr., Boersma, C. et al. The infrared spectroscopy of compact polycyclic aromatic hydrocarbons containing up to 384 carbons. *ApJ*, 754 (1): 75, July 2012. doi: 10.1088/0004-637X/754/1/75.

263. Richardson, N.D., Morrison, N.D., Gies, D.R. et al. The Hα variations of the luminous blue variable P Cygni: discrete absorption components and the short S Doradus-phase. *AJ*, 141 (4): 120, April 2011. doi: 10.1088/0004-6256/141/4/120.

264. Richichi, A., Sharma, S., Sinha, T. et al. Further milliarcsecond resolution results on cool giants and binary stars from lunar occultations at Devashtal. *MNRAS*, 498 (2): 2263–2269, October 2020. doi: 10.1093/mnras/staa2403.

265. Riess, A.G., Press, W. and Kirshner, R.P. A precise distance indicator: type IA supernova multicolor light-curve shapes. *ApJ*, 473: 88, December 1996. doi: 10.1086/178129.

266. Rivet, J.P., Siciak, A., de Almeida, E.S.G. et al. Intensity interferometry of P Cygni in the H α emission line: towards distance calibration of LBV supergiant stars. *MNRAS*, 494 (1): 218–227, May 2020. doi: 10.1093/mnras/staa588.

267. Rose, B.M., Garnavich, P. and Berg, M.A. Think global, act local: the influence of environment age and host mass on type Ia supernova light curves. *ApJ*, 874 (1): 32, March 2019. doi: 10.3847/1538-4357/ab0704.

268. Ruiter, A.J. Type Ia supernova sub-classes and progenitor origin. *IAU Symposium*, 357: 1–15, January 2020. doi: 10.1017/S1743921320000587.

269. Russell, H.N. Relations between the spectra and other characteristics of the stars. II. Brightness and spectral class. *Nature*, 93 (2323): 252–258, May 1914.

270. Salaris, M., Althaus, L. and García-Berro, E. Comparison of theoretical white dwarf cooling timescales. *A&A*, 555: A96, July 2013. doi: 10.1051/0004-6361/201220622.

271. Samanta, T., Tian, H., Yurchyshyn, V. et al. Generation of solar spicules and subsequent atmospheric heating. *Science*, 366 (6467): 890–894, November 2019. doi: 10.1126/science.aaw2796.

272. Sandford, S.A., Bernstein, M. and Materese, C.K. The infrared spectra of polycyclic aromatic hydrocarbons with excess peripheral H atoms (H_n-PAHs) and their relation to the 3.4 and 6.9 μm PAH emission features. *ApJs*, 205 (1): 8, March 2013. doi: 10.1088/0067-0049/205/1/8.

273. Schroeder, D.V. *An Introduction to Thermal Physics*. Addison Wesley Longman, 2005.

274. Schwarzschild, K. On the gravitational field of a mass point according to Einstein's theory. *Abh. König. Preuss. Akad. Wissenschaften*, 189–196, January 1916.

275. Schwarzschild, M. *Structure and Evolution of the Stars*. Princeton University Press, 1965.

276. Selvelli, P. and Gilmozzi, R. A UV and optical study of 18 old novae with Gaia DR2 distances: mass accretion rates, physical parameters, and MMRD. *A&A*, 622: A186, February 2019. doi: 10.1051/0004-6361/201834238.

277. Shara, M.M., Mizusawa, T., Zurek, D. et al. The inter-eruption timescale of classical novae from expansion of the Z Camelopardalis shell. *ApJ*, 756 (2): 107, September 2012. doi: 10.1088/0004-637X/756/2/107.

278. Sharma, R., Verth, G. and Erdélyi, R. Dynamic behavior of spicules inferred from perpendicular velocity components. *ApJ*, 840 (2): 96, May 2017. doi: 10.3847/1538-4357/aa6d57.

279. Shivvers, I., Modjaz, M., Zheng, W. et al. Revisiting the Lick Observatory supernova search volume-limited sample: updated classifications and revised stripped-envelope supernova fractions. *PASP*, 129 (975): 054201, May 2017. doi: 10.1088/1538-3873/aa54a6.

280. Shulyak, D., Reiners, A., Engeln, A. et al. Strong dipole magnetic fields in fast rotating fully convective stars. *Nat. Astron.*, 1: 0184, August 2017. doi: 10.1038/s41550-017-0184.

281. Sicilia, A., Lapi, A., Boco, L. et al. The black hole mass function across cosmic times. I. Stellar black holes and light seed distribution. *ApJ*, 924 (2): 56, January 2022. doi: 10.3847/1538-4357/ac34fb.

282. Skowron, D.M., Skowron, J., Mróz, P. et al. A three-dimensional map of the Milky Way using classical Cepheid variable stars. *Science*, 365 (6452): 478–482, August 2019. doi: 10.1126/science.aau3181.

283. Skumanich, A. Time scales for Ca II emission decay, rotational braking, and lithium depletion. *ApJ*, 171: 565, February 1972. doi: 10.1086/151310.

284. Smartt, S.J. Progenitors of core-collapse supernovae. *ARA&A*, 47 (1): 63–106, September 2009. doi: 10.1146/annurev-astro-082708-101737.

285. Soares, E.A. Constraining effective temperature, mass and radius of hot white dwarfs. *arXiv e-prints*, art. arXiv:1701.02295, January 2017.

286. Soker, N. Supernovae Ia in 2019 (review): A rising demand for spherical explosions. *New A. Rev.*, 87: 101535, December 2019. doi: 10.1016/j.newar.2020.101535.

287. Soler, J.D. Using Herschel and Planck observations to delineate the role of magnetic fields in molecular cloud structure. *A&A*, 629: A96, September 2019. doi: 10.1051/0004-6361/201935779.

288. Song, H.F., Maeder, A., Meynet, G. et al. Close-binary evolution. I. Tidally induced shear mixing in rotating binaries. *A&A*, 556: A100, August 2013. doi: 10.1051/0004-6361/201321870.

289. Stevens, A.R.H. and Albrow, M.D. Simulating the role of stellar rotation in the spectroscopic effects of differential limb magnification. *PASA*, 30: e054, November 2013. doi: 10.1017/pasa.2013.33.

290. Strickler, R.R., Cool, A.M., Anderson, J. et al. Helium-core white dwarfs in the globular cluster NGC 6397. *ApJ*, 699 (1): 40–55, July 2009. doi: 10.1088/0004-637X/699/1/40.

291. Sumiyoshi, K., Kojo, T. and Furusawa, S. Equation of state in neutron stars and supernovae. *arXiv e-prints*, art. arXiv:2207.00033, June 2022.

292. Taubenberger, S. The extremes of thermonuclear supernovae. In Alsabti, A. and Murdin, P. editors, *Handbook of Supernovae*, page 317. 2017. doi: 10.1007/978-3-319-21846-537.

293. Taylor, B. and Mohr, P.J. The NIST reference on constants, units and uncertainty, 2002. URL https://physics.nist.gov/cuu/Constants/.

294. Thielemann, F., Isern, J., Perego, A. and von Ballmoos, P. Nucleosynthesis in supernovae. *Space Sci. Rev.*, 214 (3): 62, April 2018. doi: 10.1007/s11214-018-0494-5.

295. Thompson, M.J. Solar interior: helioseismology and the sun's interior. *A&G*, 45 (4): 4.21–4.25, August 2004. doi: 10.1046/j.1468-4004.2003.45421.x.

296. Tian, H. Probing the solar transition region: current status and future perspectives. *Res. Astron. Astrophys.*, 17 (11): 110, October 2017. doi: 10.1088/1674-4527/17/11/110.

297. Tian, H., Harra, L., Baker, D. et al. Upflows in the upper solar atmosphere. *Sol. Phys.*, 296 (3): 47, March 2021. doi: 10.1007/s11207-021-01792-7.

298. Tognelli, E., Prada Moroni, P. and Degl'Innocenti, S. The Pisa pre-main sequence tracks and isochrones. A database covering a wide range of Z, Y, mass, and age values. *A&A*, 533: A109, September 2011. doi: 10.1051/0004-6361/200913913.

299. Tokunaga, A. and Vacca, W.D. The Mauna Kea observatories near-infrared filter set. III. Isophotal wavelengths and absolute calibration. *PASP*, 117 (830): 421–426, April 2005. doi: 10.1086/429382.

300. Torres, S., Rebassa-Mansergas, A., Camisassa, M.E., and Raddi, R. The Gaia DR2 halo white dwarf population: the luminosity function, mass distribution, and its star formation history. *MNRAS*, 502 (2): 1753–1767, April 2021. doi: 10.1093/mnras/stab079.

301. Townes, C.H., Wishnow, E.H., Hale, D. and Walp, B. A systematic change with time in the size of Betelgeuse. *ApJ*, 697 (2): L127–L128, June 2009. doi: 10.1088/0004-637X/697/2/L127.

302. Trampedach, R., Asplund, M., Collet, R. et al. A grid of three-dimensional stellar atmosphere models of solar metallicity. I. General properties, granulation, and atmospheric expansion. *ApJ*, 769 (1): 18, May 2013. doi: 10.1088/0004-637X/769/1/18.

303. Tremblay, P.E., Ludwig, H.G., Freytag, B. et al. Granulation properties of giants, dwarfs, and white dwarfs from the CIFIST 3D model atmosphere grid. *A&A*, 557: A7, September 2013. doi: 10.1051/0004-6361/201321878.

304. Tremblay, P., Fontaine, G., Gentile Fusillo, N.P. et al. Core crystallization and pile-up in the cooling sequence of evolving white dwarfs. *Nature*, 565 (7738): 202–205, January 2019. doi: 10.1038/s41586-018-0791-x.

305. Tzu, Lao. Lao Tzu Quotes; GoodReads, Inc. https://www.goodreads.com/author/quotes/2622245.Lao_Tzu. Accessed: 2022-07-21.

306. Vagnozzi, S. New solar metallicity measurements. *arXiv e-prints*, art. arXiv:1703.10834, March 2017.

307. Vainio, R., Desorgher, L., Heynderickx, D. et al. Dynamics of the Earth's particle radiation environment. *Space Sci. Rev.*, 147 (3–4): 187–231, November 2009. doi: 10.1007/s11214-009-9496-7.

308. van Belle, G. and von Braun, K. Directly determined linear radii and effective temperatures of exoplanet host stars. *ApJ*, 694 (2): 1085–1098, April 2009. doi: 10.1088/0004-637X/694/2/1085.

309. Vibert Douglas, A. *The Life of Arthur Stanley Eddington*. Thomas Nelson and Sons Ltd, 1956.

310. Villante, F. and Serenelli, A. An updated discussion of the solar abundance problem. *arXiv e-prints*, art. arXiv:2004.06365, April 2020.

311. Vincent Van Gogh. The Vincent Van Gogh museum letters #638, 1888. http://vangoghletters.org/vg/letters.html, accessed 2022-01-04.

312. Vink, J., Bleeker, J., van der Heyden, K. et al. The x-ray synchrotron emission of RCW 86 and the implications for its age. *ApJ*, 648 (1): L33–L37, September 2006. doi: 10.1086/507628.

313. Vinyoles, N., Serenelli, A.M., Villante, F.L. et al. A new generation of standard solar models. *ApJ*, 835 (2): 202, February 2017. doi: 10.3847/1538-4357/835/2/202.

314. Vitagliano, E., Tamborra, I. and Raffelt, G. Grand unified neutrino spectrum at Earth: Sources and spectral components. *Rev. Mod. Phys.*, 92 (4): 045006, October 2020. doi: 10.1103/RevModPhys.92.045006.

315. Wade, G. and MiMeS Collaboration. Review: Magnetic fields of O-type stars. In Balega, Yu. Yu., Romanyuk, I. and Kudryavtsev, D.O., editors, *Physics and Evolution of Magnetic and Related Stars*, volume 494 of *Astronomical Society of the Pacific Conference Series*, page 30, April 2015.

316. Wallace, L., Bernath, P., Livingston, W. et al. Water on the sun. *Science*, 268 (5214): 1155–1158, May 1995. doi: 10.1126/science.7761830.

317. Wang, P.F., Wang, C. and Han, J.L. Curvature radiation in rotating pulsar magnetosphere. *MNRAS*, 423 (3): 2464–2475, July 2012. doi: 10.1111/j.1365-2966.2012.21053.x.

318. Wang, W., Liu, R., Wang, Y. et al. Buildup of a highly twisted magnetic flux rope during a solar eruption. *Nat. Comm.*, 8: 1330, November 2017. doi: 10.1038/s41467-017-01207-x.

319. Weiler, M., Jordi, C., Fabricius, C. and Carrasco, J.M. Passband reconstruction from photometry. *A&A*, 615: A24, July 2018. doi: 10.1051/0004-6361/201732489.

320. Werner, K. and Herwig, F. The elemental abundances in bare planetary nebula central stars and the shell burning in AGB stars. *PASP*, 118 (840): 183–204, February 2006. doi: 10.1086/500443.

321. White, C.J., Burrows, A., Coleman, M.S.B., and Vartanyan, D. On the origin of pulsar and magnetar magnetic fields. *ApJ*, 926 (2): 111, February 2022. doi: 10.3847/1538-4357/ac4507.

322. Widrow, L.M., Ryu, D., Schleicher, D.R.G. et al. The first magnetic fields. *Space Sci. Rev.*, 166 (1-4): 37–70, May 2012. doi: 10.1007/s11214-011-9833-5.

323. Wiegelmann, T., Thalmann, J. and Solanki, S.K. The magnetic field in the solar atmosphere. *A&Ar*, 22: 78, November 2014. doi: 10.1007/s00159-014-0078-7.

324. Willmer, C.N.A. The absolute magnitude of the sun in several filters. *ApJs*, 236 (2): 47, June 2018. doi: 10.3847/1538-4365/aabfdf.

325. Woosley, S. and Heger, A. The remarkable deaths of 9–11 solar mass stars. *ApJ*, 810 (1): 34, September 2015. doi: 10.1088/0004-637X/810/1/34.

326. Worley, A., Krastev, P. and Li, B. Nuclear constraints on the moments of inertia of neutron stars. *ApJ*, 685 (1): 390–399, September 2008. doi: 10.1086/589823.

327. Xu, S., Yuan, H., Niu, Z. et al. Stellar loci. V. Photometric metallicities of 27 million FGK stars based on Gaia early data release 3. *ApJs*, 258 (2): 44, February 2022. doi: 10.3847/1538-4365/ac3df6.

328. Yang, F., Long, R.J., Shan, S. et al. A catalog of short period spectroscopic and eclipsing binaries identified from the LAMOST and PTF surveys. *ApJs*, 249 (2): 31, August 2020. doi: 10.3847/1538-4365/ab9b77.

329. Zakhozhay, V.A. Lifetimes of stars in the main sequence and the maximum mass of stars in the galactic disk. *Kinemat. Phys. Celest. Bodies.*, 29 (4): 195–201, July 2013. doi: 10.3103/S0884591313040065.

330. Zhang, J., Zhao, J., Oswalt, T.D. et al. Stellar chromospheric activity and age relation from open clusters in the LAMOST survey. *ApJ*, 887 (1): 84, December 2019. doi: 10.3847/1538-4357/ab4efe.

331. Ziaali, E., Bedding, T.R., Murphy, S.J. et al. The period-luminosity relation for δ Scuti stars using Gaia DR2 parallaxes. *MNRAS*, 486 (3): 4348–4353, July 2019. doi: 10.1093/mnras/stz1110.

332. Ziolkowski, J. Determination of the masses of the components of the HDE 226868/Cyg X-1 binary system. *MNRAS*, 440: L61–L65, May 2014. doi: 10.1093/mnrasl/slu002.

Index

Page numbers in *italics* refer to Figures
Page numbers in **bold** refer to Tables
Page numbers in ***bold italics*** refer to Boxes

Astrophysics: Decoding the Stars, First Edition. Judith Irwin.
© 2023 John Wiley & Sons Ltd. Published 2023 by John Wiley & Sons Ltd.